Computational Geometry – Algorithms and Applications

Springer
Berlin
Heidelberg
New York
Barcelona
Budapest
Hong Kong
London
Milan
Paris
Santa Clara
Singapore
Tokyo

Mark de Berg   Marc van Kreveld
Mark Overmars   Otfried Schwarzkopf

Computational Geometry
Algorithms and Applications

With 370 Illustrations

Springer

Dr. Mark de Berg
Dr. Marc van Kreveld
Prof. Dr. Mark Overmars
Department of Computer Science
Utrecht University
P.O.Box 80.089
3508 TB Utrecht, The Netherlands

Dr. Otfried Schwarzkopf
Department of Computer Science
Pohang University of Science and Technology
Hyoya-Dong, Pohang 790-784, South Korea

Library of Congress Cataloging-in-Publication Data applied for

Die Deutsche Bibliothek – CIP-Einheitsaufnahme

Computational geometry: algorithms and applications/M. de Berg
... - Berlin; Heidelberg; New York; London; Barcelona; Budapest;
Hong Kong; Milan; Paris; Santa Clara; Singapore; Tokyo:
Springer, 1997
ISBN 3-540-61270-X

CR Subject Classification (1991): F.2.2, I.3.5

ISBN 3-540-61270-X Springer-Verlag Berlin Heidelberg New York

Cover Design: Künkel + Lopka, Heidelberg
Typesetting: Camera ready by authors
SPIN 10533916   45/3142 – 5 4 3 2 1 0 – Printed on acid-free paper

Preface

Computational geometry emerged from the field of algorithms design and analysis in the late 1970s. It has grown into a recognized discipline with its own journals, conferences, and a large community of active researchers. The success of the field as a research discipline can on the one hand be explained from the beauty of the problems studied and the solutions obtained, and, on the other hand, by the many application domains—computer graphics, geographic information systems (GIS), robotics, and others—in which geometric algorithms play a fundamental role.

For many geometric problems the early algorithmic solutions were either slow or difficult to understand and implement. In recent years a number of new algorithmic techniques have been developed that improved and simplified many of the previous approaches. In this textbook we have tried to make these modern algorithmic solutions accessible to a large audience. The book has been written as a textbook for a course in computational geometry, but it can also be used for self-study.

Structure of the book. Each of the sixteen chapters (except the introductory chapter) starts with a problem arising in one of the application domains. This problem is then transformed into a purely geometric one, which is solved using techniques from computational geometry. The geometric problem and the concepts and techniques needed to solve it are the real topic of each chapter. The choice of the applications was guided by the topics in computational geometry we wanted to cover; they are not meant to provide a good coverage of the application domains. The purpose of the applications is to motivate the reader; the goal of the chapters is not to provide ready-to-use solutions for them. Having said this, we believe that knowledge of computational geometry is important to solve geometric problems in application areas efficiently. We hope that our book will not only raise the interest of people from the algorithms community, but also from people in the application areas.

For most geometric problems treated we give just one solution, even when a number of different solutions exist. In general we have chosen the solution that is easiest to understand and implement. This is not necessarily the most efficient solution. We also took care that the book contains a good mixture of techniques like divide-and-conquer, plane sweep, and randomized algorithms. We decided not to treat all sorts of variations to the problems; we felt it is more important to introduce all main topics in computational geometry than to give more detailed information about a smaller number of topics.

Several chapters contain one or more sections marked with a star. They contain improvements of the solution, extensions, or explain the relation between various problems. They are not essential for understanding the remainder of the book.

Every chapter concludes with a section that is entitled *Notes and Comments*. These sections indicate where the results described in the chapter originated, mention other solutions, generalizations, and improvements, and provide references. They can be skipped, but do contain useful material for those who want to know more about the topic of the chapter.

At the end of each chapter a number of exercises is provided. These range from easy tests to check whether the reader understands the material to more elaborate questions that extend the material covered. Difficult exercises and exercises about starred sections are indicated with a star.

A course outline.   Even though the chapters in this book are largely independent, they should preferably not be treated in an arbitrary order. For instance, Chapter 2 introduces plane sweep algorithms, and it is best to read this chapter before any of the other chapters that use this technique. Similarly, Chapter 4 should be read before any other chapter that uses randomized algorithms.

For a first course on computational geometry, we advise treating Chapters 1–10 in the given order. They cover the concepts and techniques that, according to us, should be present in any course on computational geometry. When more material can be covered, a selection can be made from the remaining chapters.

Prerequisites.   The book can be used as a textbook for a high-level undergraduate course or a low-level graduate course, depending on the rest of the curriculum. Readers are assumed to have a basic knowledge of the design and analysis of algorithms and data structures: they should be familiar with big-Oh notations and simple algorithmic techniques like sorting, binary search, and balanced search trees. No knowledge of the application domains is required, and hardly any knowledge of geometry. The analysis of the randomized algorithms uses some very elementary probability theory.

Implementations.   The algorithms in this book are presented in a pseudo-code that, although rather high-level, is detailed enough to make it relatively easy to implement them. In particular we have tried to indicate how to handle degenerate cases, which are often a source of frustration when it comes to implementing.

We believe that it is very useful to implement one or more of the algorithms; it will give a feeling for the complexity of the algorithms in practice. Each chapter can be seen as a programming project. Depending on the amount of time available one can either just implement the plain geometric algorithms, or implement the application as well.

To implement a geometric algorithm a number of basic data types—points, lines, polygons, and so on—and basic routines that operate on them are needed.

Implementing these basic routines in a robust manner is not easy, and takes a lot of time. Although it is good to do this at least once, it is useful to have a software library available that contains the basic data types and routines. Pointers to such libraries can be found on our World Wide Web site.

World Wide Web. This book is accompanied by a World Wide Web site, which provides lots of additional material, like an addendum, pointers to geometric software and to an online literature database containing close to 10,000 papers written on computational geometry, and links to other sites that contain information about computational geometry. The address is

```
http://www.cs.ruu.nl/geobook/
```

Even though the book was proofread by a number of people, errors are bound to remain. You can use our WWW page to send us errors you found and any other comments you have about the book. (You can also email directly to `geobook@cs.ruu.nl`.)

Acknowledgements. Writing a textbook is a long process, even with four authors. Over the past three years many people helped us by providing useful advice on what to put in the book and what not, by reading chapters and suggesting changes, and by finding and correcting errors. In particular we would like to thank Pankaj Agarwal, Helmut Alt, Marshall Bern, Jit Bose, Hazel Everett, Gerald Farin, Steve Fortune, Geert-Jan Giezeman, Mordecai Golin, Dan Halperin, Richard Karp, Klara Kedem, René van Oostrum, Sven Skyum, Jack Snoeyink, Gert Vegter, Peter Widmayer, and Chee Yap. Preliminary versions of the book were used in courses in our and other departments. We thank all students who suffered from incomplete versions and errors, and who helped us polish the book into its current shape. We also would like to thank Springer-Verlag for their advice and support during the final stages of the creation of this book.

Finally we would like to acknowledge the support of the Netherlands Organization for Scientific Research (N.W.O.), which supported the project *Computational Geometry and its Application* during which most of this book was written.

Utrecht (the Netherlands), March 1997   *Mark de Berg*
     Marc van Kreveld
     Mark Overmars
Pohang (South Korea), March 1997   *Otfried Schwarzkopf*

Contents

5   Orthogonal Range Searching   93
Querying a Database

6   Point Location   119
Knowing Where You Are

7   Voronoi Diagrams   145
The Post Office Problem

8   Arrangements and Duality   163
Supersampling in Ray Tracing

1 Computational Geometry

Imagine you are walking on the campus of a university and suddenly you realize you have to make an urgent phone call. There are many public phones on campus and of course you want to go to the nearest one. But which one is the nearest? It would be helpful to have a map on which you could look up the nearest public phone, wherever on campus you are. The map should show a subdivision of the campus into regions, and for each region indicate the nearest public phone. What would these regions look like? And how could we compute them?

Even though this is not such a terribly important issue, it describes the basics of a fundamental geometric concept, which plays a role in many applications. The subdivision of the campus is a so-called *Voronoi diagram*, and it will be studied in Chapter 7 in this book. It can be used to model trading areas of different cities, to guide robots, and even to describe and simulate the growth of crystals. Computing a geometric structure like a Voronoi diagram requires geometric algorithms. Such algorithms form the topic of this book.

A second example. Assume you located the closest public phone. With a campus map in hand you will probably have little problem in getting to the phone along a reasonably short path, without hitting walls and other objects. But programming a robot to perform the same task is a lot more difficult. Again, the heart of the problem is geometric: given a collection of geometric obstacles, we have to find a short connection between two points, avoiding collisions with the obstacles. Solving this so-called *motion planning* problem is of crucial importance in robotics. Chapters 13 and 15 deal with geometric algorithms required for motion planning.

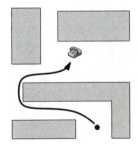

A third example. Assume you don't have one map but two: one with a description of the various buildings, including the public phones, and one indicating the roads on the campus. To plan a motion to the public phone we have to *overlay* these maps, that is, we have to combine the information in the two maps. Overlaying maps is one of the basic operations of geographic information systems. It involves locating the position of objects from one map in the other, computing the intersection of various features, and so on. Chapter 2 deals with this problem.

These are just three examples of geometric problems requiring carefully designed geometric algorithms for their solution. In the 1970s the field of computational geometry emerged, dealing with such geometric problems. It can be defined as the systematic study of algorithms and data structures for geometric objects, with a focus on exact algorithms that are asymptotically fast. Many researchers were attracted by the challenges posed by the geometric problems. The road from problem formulation to efficient and elegant solutions has often been long, with many difficult and sub-optimal intermediate results. Today there is a rich collection of geometric algorithms that are efficient, and relatively easy to understand and implement.

This book describes the most important notions, techniques, algorithms, and data structures from computational geometry in a way that we hope will be attractive to readers who are interested in applying results from computational geometry. Each chapter is motivated with a real computational problem that requires geometric algorithms for its solution. To show the wide applicability of computational geometry, the problems were taken from various application areas: robotics, computer graphics, CAD/CAM, and geographic information systems.

You should not expect ready-to-implement software solutions for major problems in the application areas. Every chapter deals with a single concept in computational geometry; the applications only serve to introduce and motivate the concepts. They also illustrate the process of modeling an engineering problem and finding an exact solution.

1.1 An Example: Convex Hulls

Good solutions to algorithmic problems of a geometric nature are mostly based on two ingredients. One is a thorough understanding of the geometric properties of the problem, the other is a proper application of algorithmic techniques and data structures. If you don't understand the geometry of the problem, all the algorithms of the world won't help you to solve it efficiently. On the other hand, even if you perfectly understand the geometry of the problem, it is hard to solve it effectively if you don't know the right algorithmic techniques. This book will give you a thorough understanding of the most important geometric concepts and algorithmic techniques.

To illustrate the issues that arise in developing a geometric algorithm, this section deals with one of the first problems that was studied in computational geometry: the computation of planar convex hulls. We'll skip the motivation for this problem here; if you are interested you can read the introduction to Chapter 11, where we study convex hulls in 3-dimensional space.

 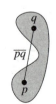

convex   not convex

A subset S of the plane is called *convex* if and only if for any pair of points $p, q \in S$ the line segment $\overline{pq}$ is completely contained in S. The *convex hull* $\mathcal{CH}(S)$ of a set S is the smallest convex set that contains S. To be more precise, it is the intersection of all convex sets that contain S.

We will study the problem of computing the convex hull of a finite set P of n points in the plane. We can visualize what the convex hull looks like by a thought experiment. Imagine that the points are nails sticking out of the plane, take an elastic rubber band, hold it around the nails, and let it go. It will snap around the nails, minimizing its length. The area enclosed by the rubber band is the convex hull of P. This leads to an alternative definition of the convex hull of a finite set P of points in the plane: it is the unique convex polygon whose vertices are points from P and that contains all points of P. Of course we should prove rigorously that this is well defined—that is, that the polygon is unique—and that the definition is equivalent to the one given earlier, but let's skip that in this introductory chapter.

How do we compute the convex hull? Before we can answer this question we must ask another question: what does it mean to compute the convex hull? As we have seen, the convex hull of P is a convex polygon. A natural way to represent a polygon is by listing its vertices in clockwise order, starting with an arbitrary one. So the problem we want to solve is this: given a set $P = \{p_1, p_2, \ldots, p_n\}$ of points in the plane, compute a list that contains those points from P that are the vertices of $\mathcal{CH}(P)$, listed in clockwise order.

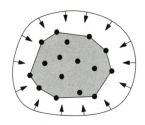

input = set of points:

$p_1, p_2, p_3, p_4, p_5, p_6, p_7, p_8, p_9$

output = representation of the convex hull:

p_4, p_5, p_8, p_2, p_9

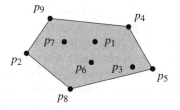

Figure 1.1
Computing a convex hull

The first definition of convex hulls is of little help when we want to design an algorithm to compute the convex hull. It talks about the intersection of all convex sets containing P, of which there are infinitely many. The observation that $\mathcal{CH}(P)$ is a convex polygon is more useful. Let's see what the edges of $\mathcal{CH}(P)$ are. Both endpoints p and q of such an edge are points of P, and if we direct the line through p and q such that $\mathcal{CH}(P)$ lies to the right, then all the points of P must lie to the right of this line. The reverse is also true: if all points of $P \setminus \{p, q\}$ lie to the right of the directed line through p and q, then $\overline{pq}$ is an edge of $\mathcal{CH}(P)$.

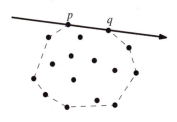

Now that we understand the geometry of the problem a little bit better we can develop an algorithm. We will describe it in a style of pseudocode we will use throughout this book.

Algorithm SLOWCONVEXHULL(P)
Input. A set P of points in the plane.
Output. A list $\mathcal{L}$ containing the vertices of $\mathcal{CH}(P)$ in clockwise order.
1. $E \leftarrow \emptyset$.
2.   **for** all ordered pairs $(p, q) \in P \times P$ with p not equal to q
3.   **do** *valid* $\leftarrow$ **true**

4.   **for** all points $r \in P$ not equal to p or q
5.   **do if** r lies to the left of the directed line from p to q
6.   **then** *valid* $\leftarrow$ **false**.
7.   **if** *valid* **then** Add the directed edge $\vec{pq}$ to E.
8.   From the set E of edges construct a list L of vertices of $\mathcal{CH}(P)$, sorted in clockwise order.

Two steps in the algorithm are perhaps not entirely clear.

The first one is line 5: how do we test whether a point lies to the left or to the right of a directed line? This is one of the primitive operations required in most geometric algorithms. Throughout this book we assume that such operations are available. It is clear that they can be performed in constant time so the actual implementation will not affect the asymptotic running time in order of magnitude. This is not to say that such primitive operations are unimportant or trivial. They are not easy to implement correctly and their implementation will affect the actual running time of the algorithm. Fortunately, software libraries containing such primitive operations are nowadays available. We conclude that we don't have to worry about the test in line 5; we may assume that we have a function available performing the test for us in constant time.

The other step of the algorithm that requires some explanation is the last one. In the loop of lines 2–7 we determine the set E of convex hull edges. From E we can construct the list L as follows. The edges in E are directed, so we can speak about the origin and the destination of an edge. Because the edges are directed such that the other points lie to their right, the destination of an edge comes after its origin when the vertices are listed in clockwise order. Now remove an arbitrary edge $\vec{e_1}$ from E. Put the origin of $\vec{e_1}$ as the first point into L, and the destination as the second point. Find the edge $\vec{e_2}$ in E whose origin is the destination of $\vec{e_1}$, remove it from E, and append its destination to L. Next, find the edge $\vec{e_3}$ whose origin is the destination of $\vec{e_2}$, remove it from E, and append its destination to L. We continue in this manner until there is only one edge left in E. Then we are done; the destination of the remaining edge is necessarily the origin of $\vec{e_1}$, which is already the first point in L. A simple implementation of this procedure takes $O(n^2)$ time. This can easily be improved to $O(n \log n)$, but the time required for the rest of the algorithm dominates the total running time anyway.

Analyzing the time complexity of SLOWCONVEXHULL is easy. We check $n^2 - n$ pairs of points. For each pair we look at the $n - 2$ other points to see whether they all lie on the right side. This will take $O(n^3)$ time in total. The final step takes $O(n^2)$ time, so the total running time is $O(n^3)$. An algorithm with a cubic running time is too slow to be of practical use for anything but the smallest input sets. The problem is that we did not use any clever algorithmic design techniques; we just translated the geometric insight into an algorithm in a brute-force manner. But before we try to do better, it is useful to make several observations about this algorithm.

We have been a bit careless when deriving the criterion of when a pair p, q defines an edge of $\mathcal{CH}(P)$. A point r does not always lie to the right or to the

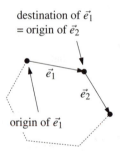

destination of $\vec{e_1}$
= origin of $\vec{e_2}$

$\vec{e_1}$

$\vec{e_2}$

origin of $\vec{e_1}$

left of the line through p and q, it can also happen that it lies *on* this line. What should we do then? This is what we call a *degenerate case*, or a *degeneracy* for short. We prefer to ignore such situations when we first think about a problem, so that we don't get confused when we try to figure out the geometric properties of a problem. However, these situations do arise in practice. For instance, if we create the points by clicking on a screen with a mouse, all points will have small integer coordinates, and it is quite likely that we will create three points on a line.

To make the algorithm correct in the presence of degeneracies we must reformulate the criterion above as follows: a directed edge $\vec{pq}$ is an edge of $\mathcal{CH}(P)$ if and only if all other points $r \in P$ lie either strictly to the right of the directed line through p and q, or they lie on the open line segment $\overline{pq}$. (We assume that there are no coinciding points in P.) So we have to replace line 5 of the algorithm by this more complicated test.

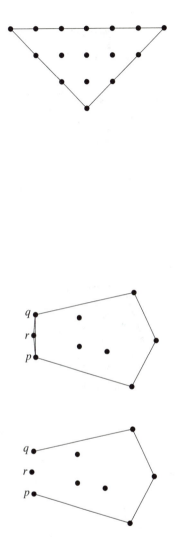

We have been ignoring another important issue that can influence the correctness of the result of our algorithm. We implicitly assumed that we can somehow test exactly whether a point lies to the right or to the left of a given line. This is not necessarily true: if the points are given in floating point coordinates and the computations are done using floating point arithmetic, then there will be rounding errors that may distort the outcome of tests.

Imagine that there are three points p, q, and r, that are nearly collinear, and that all other points lie far to the right of them. Our algorithm tests the pairs (p,q), (r,q), and (p,r). Since these points are nearly collinear, it is possible that the rounding errors lead us to decide that r lies to the right of the line from p to q, that p lies to the right of the line from r to q, and that q lies to the right of the line from p to r. Of course this is geometrically impossible— but the floating point arithmetic doesn't know that! In this case the algorithm will accept all three edges. Even worse, all three tests could give the opposite answer, in which case the algorithm rejects all three edges, leading to a gap in the boundary of the convex hull. And this leads to a serious problem when we try to construct the sorted list of convex hull vertices in the last step of our algorithm. This step assumes that there is exactly one edge starting in every convex hull vertex, and exactly one edge ending there. Due to the rounding errors there can suddenly be two, or no, edges starting in vertex p. This can cause the program implementing our simple algorithm to crash, since the last step has not been designed to deal with such inconsistent data.

Although we have proven the algorithm to be correct and to handle all special cases, it is not *robust*: small errors in the computations can make it fail in completely unexpected ways. The problem is that we have proven the correctness assuming that we can compute exactly with real numbers.

We have designed our first geometric algorithm. It computes the convex hull of a set of points in the plane. However, it is quite slow—its running time is $O(n^3)$—, it deals with degenerate cases in an awkward way, and it is not robust. We should try to do better.

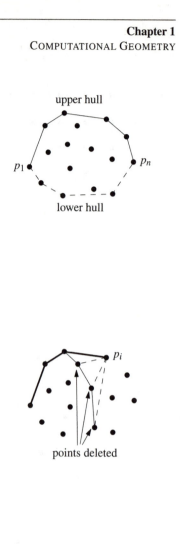

upper hull

p_1 p_n

lower hull

points deleted

p_i

To this end we apply a standard algorithmic design technique: we will develop an *incremental algorithm*. This means that we will add the points in P one by one, updating our solution after each addition. We give this incremental approach a geometric flavor by adding the points from left to right. So we first sort the points by x-coordinate, obtaining a sorted sequence $p_1, \ldots, p_n$, and then we add them in that order. Because we are working from left to right, it would be convenient if the convex hull vertices were also ordered from left to right as they occur along the boundary. But this is not the case. Therefore we first compute only those convex hull vertices that lie on the *upper hull*, which is the part of the convex hull running from the leftmost point p_1 to the rightmost point p_n when the vertices are listed in clockwise order. In other words, the upper hull contains the convex hull edges bounding the convex hull from above. In a second scan, which is performed from right to left, we compute the remaining part of the convex hull, the *lower hull*.

The basic step in the incremental algorithm is the update of the upper hull after adding a point p_i. In other words, given the upper hull of the points $p_1, \ldots, p_{i-1}$, we have to compute the upper hull of $p_1, \ldots, p_i$. This can be done as follows. When we walk around the boundary of a polygon in clockwise order, we make a turn at every vertex. For an arbitrary polygon this can be both a right turn and a left turn, but for a convex polygon every turn must be a right turn. This suggests handling the addition of p_i in the following way. Let $\mathcal{L}_{\text{upper}}$ be a list that stores the upper vertices in left-to-right order. We first append p_i to $\mathcal{L}_{\text{upper}}$. This is correct because p_i is the rightmost point of the ones added so far, so it must be on the upper hull. Next, we check whether the last three points in $\mathcal{L}_{\text{upper}}$ make a right turn. If this is the case there is nothing more to do; $\mathcal{L}_{\text{upper}}$ contains the vertices of the upper hull of $p_1, \ldots, p_i$, and we can proceed to the next point, p_{i+1}. But if the last three points make a left turn, we have to delete the middle one from the upper hull. In this case we are not finished yet: it could be that the new last three points still do not make a right turn, in which case we again have to delete the middle one. We continue in this manner until the last three points make a right turn, or until there are only two points left.

We now give the algorithm in pseudocode. The pseudocode computes both the upper hull and the lower hull. The latter is done by treating the points from right to left, analogous to the computation of the upper hull.

Algorithm CONVEXHULL(P)
Input. A set P of points in the plane.
Output. A list containing the vertices of $\mathcal{CH}(P)$ in clockwise order.
1.   Sort the points by x-coordinate, resulting in a sequence $p_1, \ldots, p_n$.
2.   Put the points p_1 and p_2 in a list $\mathcal{L}_{\text{upper}}$, with p_1 as the first point.
3.   **for** $i \leftarrow 3$ **to** n
4.   **do** Append p_i to $\mathcal{L}_{\text{upper}}$.
5.   **while** $\mathcal{L}_{\text{upper}}$ contains more than two points **and** the last three points in $\mathcal{L}_{\text{upper}}$ do not make a right turn
6.   **do** Delete the middle of the last three points from $\mathcal{L}_{\text{upper}}$.
7.   Put the points p_n and p_{n-1} in a list $\mathcal{L}_{\text{lower}}$, with p_n as the first point.

8.   **for** $i \leftarrow n - 2$ **downto** 1
9.   **do** Append p_i to $\mathcal{L}_{\text{lower}}$.
10.   **while** $\mathcal{L}_{\text{lower}}$ contains more than 2 points **and** the last three points
                  in $\mathcal{L}_{\text{lower}}$ do not make a right turn
11.   **do** Delete the middle of the last three points from $\mathcal{L}_{\text{lower}}$.
12. Remove the first and the last point from $\mathcal{L}_{\text{lower}}$ to avoid duplication of the
      points where the upper and lower hull meet.
13. Append $\mathcal{L}_{\text{lower}}$ to $\mathcal{L}_{\text{upper}}$, and call the resulting list $\mathcal{L}$.
14. **return** $\mathcal{L}$

Once again, when we look closer we realize that the above algorithm is not correct. Without mentioning it, we made the assumption that no two points have the same x-coordinate. If this assumption is not valid the order on x-coordinate is not well defined. Fortunately, this turns out not to be a serious problem. We only have to generalize the ordering in a suitable way: rather than using only the x-coordinate of the points to define the order, we use the lexicographic order. This means that we first sort by x-coordinate, and if points have the same x-coordinate we sort them by y-coordinate.

    Another special case we have ignored is that the three points for which we have to determine whether they make a left or a right turn lie on a straight line. In this case the middle point should not occur on the convex hull, so collinear points must be treated as if they make a left turn. In other words, we should use a test that returns true if the three points make a right turn, and false otherwise. (Note that this is simpler than the test required in the previous algorithm when there were collinear points.)

    With these modifications the algorithm correctly computes the convex hull: the first scan computes the upper hull, which is now defined as the part of the convex hull running from the lexicographically smallest vertex to the lexicographically largest vertex, and the second scan computes the remaining part of the convex hull.

not a right turn

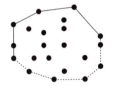

What does our algorithm do in the presence of rounding errors in the floating point arithmetic? When such errors occur, it can happen that a point is removed from the convex hull although it should be there, or that a point inside the real convex hull is not removed. But the structural integrity of the algorithm is unharmed: it will compute a closed polygonal chain. After all, the output is a list of points that we can interpret as the clockwise listing of the vertices of a polygon, and any three consecutive points form a right turn or, because of the rounding errors, they almost form a right turn. Moreover, no point in P can be far outside the computed hull. The only problem that can still occur is that, when three points lie very close together, a turn that is actually a sharp left turn can be interpreted as a right turn. This might result in a dent in the resulting polygon. A way out of this is to make sure that points in the input that are very close together are considered as being the same point, for example by rounding. Hence, although the result need not be exactly correct—but then, we cannot hope for an exact result if we use inexact arithmetic—it does make sense. For many applications this is good enough. Still, it is wise to be careful in the implementation of the basic test to avoid errors as much as possible.

We conclude with the following theorem:

Theorem 1.1 *The convex hull of a set of n points in the plane can be computed in $O(n \log n)$ time.*

Proof. We will prove the correctness of the computation of the upper hull; the lower hull computation can be proved correct using similar arguments. The proof is by induction on the number of point treated. Before the **for**-loop starts, the list $\mathcal{L}_{upper}$ contains the points p_1 and p_2, which trivially form the upper hull of $\{p_1, p_2\}$. Now suppose that $\mathcal{L}_{upper}$ contains the upper hull vertices of $\{p_1, \dots, p_{i-1}\}$ and consider the addition of p_i. After the execution of the **while**-loop and because of the induction hypothesis, we know that the points in $\mathcal{L}_{upper}$ form a chain that only makes right turns. Moreover, the chain starts at the lexicographically smallest point of $\{p_1, \dots, p_i\}$ and ends at the lexicographically largest point, namely p_i. If we can show that all points of $\{p_1, \dots, p_i\}$ that are not in $\mathcal{L}_{upper}$ are below the chain, then $\mathcal{L}_{upper}$ contains the correct points. By induction we know there is no point above the chain that we had before p_i was added. Since the old chain lies below the new chain, the only possibility for a point to lie above the new chain is if it lies in the vertical slab between p_{i-1} and p_i. But this is not possible, since such a point would be in between p_{i-1} and p_i in the lexicographical order. (You should verify that a similar argument holds if p_{i-1} and p_i, or any other points, have the same x-coordinate.)

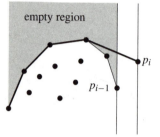

To prove the time bound, we note that sorting the points lexicographically can be done in $O(n \log n)$ time. Now consider the computation of the upper hull. The **for**-loop is executed a linear number of times. The question that remains is how often the **while**-loop inside it is executed. For each execution of the **for**-loop the **while**-loop is executed at least once. For any extra execution a point is deleted from the current hull. As each point can be deleted only once during the construction of the upper hull, the total number of extra executions over all **for**-loops is bounded by n. Similarly, the computation of the lower hull takes $O(n)$ time. Due to the sorting step, the total time required for computing the convex hull is $O(n \log n)$. $\quad\square$

The final convex hull algorithm is simple to describe and easy to implement. It only requires lexicographic sorting and a test whether three consecutive points make a right turn. From the original definition of the problem it was far from obvious that such an easy and efficient solution would exist.

1.2 Degeneracies and Robustness

As we have seen in the previous section, the development of a geometric algorithm often goes through three phases.

In the first phase, we try to ignore everything that will clutter our understanding of the geometric concepts we are dealing with. Sometimes collinear points are a nuisance, sometimes vertical line segments are. When first trying to design or understand an algorithm, it is often helpful to ignore these degenerate cases.

In the second phase, we have to adjust the algorithm designed in the first phase to be correct in the presence of degenerate cases. Beginners tend to do this by adding a huge number of case distinctions to their algorithms. In many situations there is a better way. By considering the geometry of the problem again, one can often integrate special cases with the general case. For example, in the convex hull algorithm we only had to use the lexicographical order instead of the order on *x*-coordinate to deal properly with points with equal *x*-coordinate. For most algorithms in this book we have tried to take this integrated approach to deal with special cases. Still, it is easier not to think about such cases upon first reading. Only after understanding how the algorithm works in the general case should you think about degeneracies.

If you study the computational geometry literature, you will find that many authors ignore special cases, often by formulating specific assumptions on the input. For example, in the convex hull problem we could have ignored special cases by simply stating that we assume that the input is such that no three points are collinear and no two points have the same *x*-coordinate. From a theoretical point of view, such assumptions are usually justified: the goal is then to establish the computational complexity of a problem and, although it is tedious to work out the details, degenerate cases can almost always be handled without increasing the asymptotic complexity of the algorithm. But special cases definitely increase the complexity of the implementations. Most researchers in computational geometry today are aware that their *general position* assumptions are not satisfied in practical applications and that an integrated treatment of the special cases is normally the best way to handle them. Furthermore, there are general techniques—so-called *symbolic perturbation schemes*—that allow one to ignore special cases during the design and implementation, and still have an algorithm that is correct in the presence of degeneracies.

The third phase is the actual implementation. Now one needs to think about the primitive operations, like testing whether a point lies to the left, to the right, or on a directed line. If you are lucky you have a geometric software library available that contains the operations you need, otherwise you must implement them yourself.

Another issue that arises in the implementation phase is that the assumption of doing exact arithmetic with real numbers breaks down, and it is necessary to understand the consequences. Robustness problems are often a cause of frustration when implementing geometric algorithms. Solving robustness problems is not easy. One solution is to use a package providing exact arithmetic (using integers, rationals, or even algebraic numbers, depending on the type of problem) but this will be slow. Alternatively, one can adapt the algorithm to detect inconsistencies and take appropriate actions to avoid crashing the program. In this case it is not guaranteed that the algorithm produces the correct output, and it is important to establish the exact properties that the output has.

This is what we did in the previous section, when we developed the convex hull algorithm: the result might not be a convex polygon but we know that the structure of the output is correct and that the output polygon is very close to the convex hull. Finally, it is possible to predict, based on the input, the precision in the number representation required to solve the problem correctly.

Which approach is best depends on the application. If speed is not an issue, exact arithmetic is preferred. In other cases it is not so important that the result of the algorithm is precise. For example, when displaying the convex hull of a set of points, it is most likely not noticeable when the polygon deviates slightly from the true convex hull. In this case we can use a careful implementation based on floating point arithmetic.

In the rest of this book we will focus on the design phase of geometric algorithm; we won't say much about the problems that arise in the implementation phase.

1.3 Application Domains

As indicated before, we have chosen a motivating example application for every geometric concept, algorithm, or data structure introduced in this book. Most of the applications stem from the areas of computer graphics, robotics, geographic information systems, and CAD/CAM. For those not familiar with these fields, we give a brief description of the areas and indicate some of the geometric problems that arise in them.

Computer graphics. Computer graphics is concerned with creating images of modeled scenes for display on a computer screen, a printer, or other output device. The scenes vary from simple two-dimensional drawings—consisting of lines, polygons, and other primitive objects—to realistic-looking 3-dimensional scenes including light sources, textures, and so on. The latter type of scene can easily contain over a million polygons or curved surface patches.

Because scenes consist of geometric objects, geometric algorithms play an important role in computer graphics.

For 2-dimensional graphics, typical questions involve the intersection of certain primitives, determining the primitive pointed to with the mouse, or determining the subset of primitives that lie within a particular region. Chapters 6, 10, and 16 describe techniques useful for some of these problems.

When dealing with 3-dimensional problems the geometric questions become more complex. A crucial step in displaying a 3-dimensional scene is hidden surface removal: determine the part of a scene visible from a particular viewpoint or, in other words, discard the parts that lie behind other objects. In Chapter 12 we study one approach to this problem.

To create realistic-looking scenes we have to take light into account. This creates many new problems, such as the computation of shadows. Hence, realistic image synthesis requires complicated display techniques, like ray tracing and radiosity. When dealing with moving objects and in virtual reality applications, it is important to detect collisions between objects. All these situations involve geometric problems.

Robotics. The field of robotics studies the design and use of robots. As robots are geometric objects that operate in a 3-dimensional space—the real world—it is obvious that geometric problems arise at many places. At the beginning of this chapter we already introduced the motion planning problem, where a robot has to find a path in an environment with obstacles. In Chapters 13 and 15 we study some simple cases of motion planning. Motion planning is one aspect of the more general problem of task planning. One would like to give a robot high-level tasks—"vacuum the room"—and let the robot figure out the best way to execute the task. This involves planning motions, planning the order in which to perform subtasks, and so on.

Other geometric problems occur in the design of robots and work cells in which the robot has to operate. Most industrial robots are robot arms with a fixed base. The parts operated on by the robot arm have to be supplied in such a way that the robot can easily grasp them. Some of the parts may have to be immobilized so that the robot can work on them. They may also have to be turned to a known orientation before the robot can work on them. These are all geometric problems, sometimes with a kinematic component. Some of the algorithms described in this book are applicable in such problems. For example, the smallest enclosing disc problem, treated in Section 4.7, can be used for optimal placement of robot arms.

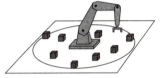

Geographic information systems. A geographic information system, or GIS for short, stores geographical data like the shape of countries, the height of mountains, the course of rivers, the type of vegetation at different locations, population density, or rainfall. They can also store human-made structures such as cities, roads, railways, electricity lines, or gas pipes. A GIS can be used to extract information about certain regions and, in particular, to obtain information about the relation between different types of data. For example, a biologist may want to relate the average rainfall to the existence of certain plants, and a civil engineer may need to query a GIS to determine whether there are any gas pipes underneath a lot where excavation works are to be performed.

As most geographic information concerns properties of points and regions on the earth's surface, geometric problems occur in abundance here. Moreover, the amount of data is so large that efficient algorithms are a must. Below we mention the GIS-related problems treated in this book.

A first question is how to store geographic data. Suppose for instance that we want to develop a car guidance system, which shows the driver at any moment where she is. This requires storing a huge map of roads and other data. At every moment we have to be able to determine the position of the car on the map and to quickly select a small portion of the map for display on the on-board computer. Efficient data structures are needed for these operations. Chapters 6, 10, and 16 describe computational geometry solutions to these problems.

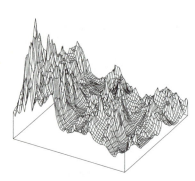

The information about the height in some mountainous terrain is usually only available at certain sample points. For other positions we have to obtain the heights by interpolating between nearby sample points. But which sample points should we choose? Chapter 9 deals with this problem.

The combination of different types of data is one of the most important operations in a GIS. For example, we may want to check which houses lie in a forest, locate all bridges by checking where roads cross rivers, or determine a good location for a new golf course by finding a slightly hilly, rather cheap area not too far from a particular town. A GIS usually stores different types of data in separate maps. To combine the data we have to overlay different maps. Chapter 2 deals with a problem arising when we want to compute the overlay.

Finally, we mention the same example we gave at the beginning of this chapter: the location of the nearest public phone (or hospital, or any other facility). This requires the computation of a Voronoi diagram, a structure studied in detail in Chapter 7.

CAD/CAM. Computer aided design (CAD) concerns itself with the design of products with a computer. The products can vary from printed circuit boards, machine parts, or furniture, to complete buildings. In all cases the resulting product is a geometric entity and, hence, it is to be expected that all sorts of geometric problems appear. Indeed, CAD packages have to deal with intersections and unions of objects, with decomposing objects and object boundaries into simpler shapes, and with visualizing the designed products.

To decide whether a design meets the specifications certain tests are needed. Often one does not need to build a prototype for these tests, and a simulation suffices. Chapter 14 deals with a problem arising in the simulation of heat emission by a printed circuit board.

Once an object has been designed and tested, it has to be manufactured. Computer aided manufacturing (CAM) packages can be of assistance here. CAM involves many geometric problems. Chapter 4 studies one of them.

A recent trend is *design for assembly*, where assembly decisions are already taken into account during the design stage. A CAD system supporting this would allow designers to test their design for feasibility, answering questions like: can the product be built easily using a certain manufacturing process? Many of these questions require geometric algorithms to be answered.

Other applications domains. There are many more application domains where geometric problems occur and geometric algorithms and data structures can be used to solve them.

For example, in molecular modeling, molecules are often represented by collections of intersecting balls in space, one ball for each atom. Typical questions are to compute the union of the atom balls to obtain the molecule surface, or to compute where two molecules can touch each other.

Another area is pattern recognition. Consider for example an optical character recognition system. Such a system scans a paper with text on it with the

caffeine

goal of recognizing the text characters. A basic step is to match the image of a character against a collection of stored characters to find the one that best fits it. This leads to a geometric problem: given two geometric objects, determine how well they resemble each other.

Even certain areas that at first sight do not seem to be geometric can benefit from geometric algorithms, because it is often possible to formulate non-geometric problem in geometric terms. In Chapter 5, for instance, we will see how records in a database can be interpreted as points in a higher-dimensional space, and we will present a geometric data structure such that certain queries on the records can be answered efficiently.

We hope that the above collection of geometric problems makes it clear that computational geometry plays a role in many different areas of computer science. The algorithms, data structures, and techniques described in this book will provide you with the tools needed to attack such geometric problems successfully.

1.4 Notes and Comments

Every chapter of this book ends with a section entitled *Notes and Comments*. These sections indicate where the results described in the chapter came from, indicate generalizations and improvements, and provide references. They can be skipped but do contain useful material for those who want to know more about the topic of the chapter. More information can also be found in the *Handbook of Computational Geometry* [293] and the *Handbook of Discrete and Computational Geometry* [159].

In this chapter the geometric problem treated in detail was the computation of the convex hull of a set of points in the plane. This is a classic topic in computational geometry and the amount of literature about it is huge. The algorithm described in this chapter is commonly known as *Graham's scan*, and is based on a modification by Andrew [8] of one of the earliest algorithms by Graham [160]. This is only one of the many $O(n \log n)$ algorithms available for solving the problem. A divide-and-conquer approach was given by Preparata and Hong [285]. Also an incremental method exists that inserts the points one by one in $O(\log n)$ time per insertion [284]. Overmars and van Leeuwen generalized this to a method in which points could be both inserted and deleted in $O(\log^2 n)$ time [269]. Other results on dynamic convex hulls were for example obtained by Hersberger and Suri [177].

Even though an $\Omega(n \log n)$ lower bound is known for the problem [340] many authors have tried to improve the result. This makes sense because in many applications the number of points that appear on the convex hull is relatively small, while the lower bound result assumes that (almost) all points show up on the convex hull. Hence, it is useful to look at algorithms whose running time depends on the complexity of the convex hull. Jarvis [187] introduced a wrapping technique, often referred to as *Jarvis's march*, that computes the convex hull in $O(h \cdot n)$ time where h is the complexity of the convex hull. The same worst-case performance is achieved by the algorithm of Overmars and van Leeuwen [267], based on earlier work by Bykat [59], Eddy [127], and Green and Silverman [161]. This algorithm has the advantage that its expected running time is linear for many distributions of points. Finally, Kirkpatrick and Seidel [202] improved the result to $O(n \log h)$, and recently Chan [62] discovered a much simpler algorithm to achieve the same result.

The convex hull can be defined in any dimension. Convex hulls in 3-dimensional space can still be computed in $O(n \log n)$ time, as we will see in Chapter 11. For dimensions higher than 3, however, the complexity of the convex hull is no longer linear in the number of points. See the notes and comments of Chapter 11 for more details.

In the past years a number of general methods for handling special cases have been suggested. These *symbolic perturbation schemes* perturb the input in such a way that all degeneracies disappear. However, the perturbation is only done symbolically. This technique was introduced by Edelsbrunner and Mücke [135] and later refined by Yap [342] and Emiris and Canny [142, 143]. Symbolic perturbation relieves the programmer of the burden of degeneracies, but it has some drawbacks: the use of a symbolic perturbation library slows down the algorithm, and sometimes one needs to recover the 'real result' from the 'perturbed result', which is not always easy. These drawbacks led Burnikel et al. [57] to claim that it is both simpler (in terms of programming effort) and more efficient (in terms of running time) to deal directly with degenerate inputs.

Robustness in geometric algorithms is a topic that has recently received a lot of interest. Most geometric comparisons can be formulated as computing the sign of some determinant. A possible way to deal with the inexactness in floating point arithmetic when evaluating this sign is to choose a small threshold value ε and to say that the determinant is zero when the outcome of the floating point computation is less than ε. When implemented naively, this can lead to inconsistencies (for instance, for three points a, b, c we may decide that $a = b$ and $b = c$ but $a \neq c$) that cause the program to fail. Guibas et al. [166] showed that combining such an approach with interval arithmetic and backwards error analysis can give robust algorithms. Another option is to use *exact arithmetic*. Here one computes as many bits of the determinant as are needed to determine its sign. This will slow down the computation, but techniques have been developed to keep the performance penalty relatively small [151, 343]. Besides these general approaches, there have been a number papers dealing with robust computation in specific problems [22, 25, 61, 119, 149, 150, 186, 245].

We gave a brief overview of the application domains from which we took our examples to motivate the different geometric notions and algorithms studied in this book. Below are some references to textbooks you can consult if you want to know more about the application domains. Of course there are many more good books about these domains than the few we mention.

There is a large number of books on computer graphics. The book by Foley et al. [148] is very extensive and generally considered one of the best books on the topic. Another good book, focusing on 3D computer graphics, is the one by Watt [330].

An extensive overview the motion planning problem can be found in the book of Latombe [207] or the somewhat older book of Hopcroft, Schwartz, and Sharir [184]. More general information about robotics is provided by the books of Schilling [300] and McKerrow [237]. Papers on many algorithmic issues of robotics and manipulation can be found in the two volumes of *Algorithmic Foundations of Robotics* [158, 208].

There is a large collection of books about geographic information systems, but most of them do not consider algorithmic issues in much detail. Some general textbooks are the ones by Burrough [58], Clarke [107], and Maguire, Goodchild, and Rhind [225]. Many data structure aspects are described in the two books of Samet [296, 297].

The books by Faux and Pratt [144], Mortenson [251], and Hoffmann [183] are good introductory texts on CAD/CAM and geometric modeling.

1.5 Exercises

1.1 The convex hull of a set S is defined to be the intersection of all convex sets that contain S. For the convex hull of a set of points it was indicated that the convex hull is the convex set with smallest perimeter. We want to show that these are equivalent definitions.

    a.   Prove that the intersection of two convex sets is again convex. This implies that the intersection of a finite family of convex sets is convex as well.

    b.   Prove that the smallest perimeter polygon $\mathcal{P}$ containing a set of points P is convex.

    c.   Prove that any convex set containing the set of points P contains the smallest perimeter polygon $\mathcal{P}$.

1.2 Let P be a set of points in the plane. Let $\mathcal{P}$ be the convex polygon whose vertices are points from P and that contains all points in P. Prove that this polygon $\mathcal{P}$ is uniquely defined, and that it is the intersection of all convex sets containing P.

1.3 Let E be an unsorted set of n segments that are the edges of a convex polygon. Describe an $O(n \log n)$ algorithm that computes from E a list containing all vertices of the polygon, sorted in clockwise order.

1.4 For the convex hull algorithm we have to be able to test whether a point r lies left or right of the directed line through two points p and q. Let $p = (p_x, p_y)$, $q = (q_x, q_y)$, and $r = (r_x, r_y)$.

a.   Show that the sign of the determinant

$$D = \begin{vmatrix} 1 & p_x & p_y \\ 1 & q_x & q_y \\ 1 & r_x & r_y \end{vmatrix}$$

determines whether r lies left or right of the line.

b.   Show that $|D|$ in fact is twice the surface of the triangle determined by p, q, and r.

c.   Why is this an attractive way to implement the basic test in algorithm CONVEXHULL? Give an argument for both integer and floating point coordinates.

1.5 Verify that the algorithm CONVEXHULL with the indicated modifications correctly computes the convex hull, also of degenerate sets of points. Consider for example such nasty cases as a set of points that all lie on one (vertical) line.

1.6 In many situations we need to compute convex hulls of objects other than points.

a.   Let S be a set of n line segments in the plane. Prove that the convex hull of S is exactly the same as the convex hull of the $2n$ endpoints of the segments.

b.*   Let $\mathcal{P}$ be a non-convex polygon. Describe an algorithm that computes the convex hull of $\mathcal{P}$ in $O(n)$ time. *Hint:* Use a variant of algorithm CONVEXHULL where the vertices are not treated in lexicographical order, but in some other order.

1.7 Consider the following alternative approach to computing the convex hull of a set of points in the plane: We start with the rightmost point. This is the first point p_1 of the convex hull. Now imagine that we start with a vertical line and rotate it clockwise until it hits another point p_2. This is the second point on the convex hull. We continue rotating the line but this time around p_2 until we hit a point p_3. In this way we continue until we reach p_1 again.

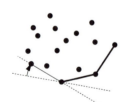

a.   Give pseudocode for this algorithm.

b.   What degenerate cases can occur and how can we deal with them?

c.   Prove that the algorithm correctly computes the convex hull.

d.   Prove that the algorithm can be implemented to run in time $O(n \cdot h)$, where h is the complexity of the convex hull.

e.   What problems might occur when we deal with inexact floating point arithmetic?

1.8 The $O(n \log n)$ algorithm to compute the convex hull of a set of n points in the plane that was described in this chapter is based on the paradigm of incremental construction: add the points one by one, and update the convex hull after each addition. In this exercise we shall develop an algorithm based on another paradigm, namely divide-and-conquer.

a. Let $\mathcal{P}_1$ and $\mathcal{P}_2$ be two disjoint convex polygons with n vertices in total. Give an $O(n)$ time algorithm that computes the convex hull of $\mathcal{P}_1 \cup \mathcal{P}_2$.

b. Use the algorithm from part a to develop an $O(n \log n)$ time divide-and-conquer algorithm to compute the convex hull of a set of n points in the plane.

1.9 Suppose that we have a subroutine CONVEXHULL available for computing the convex hull of a set of points in the plane. Its output is a list of convex hull vertices, sorted in clockwise order. Now let $\{x_1, x_2, \ldots, x_n\}$ be a set of n numbers. Show that S can be sorted in $O(n)$ time plus the time needed for one call to CONVEXHULL. Since the sorting problem has an $\Omega(n \log n)$ lower bound, this implies that the convex hull problem has an $\Omega(n \log n)$ lower bound as well. Hence, the algorithm presented in this chapter is asymptotically optimal.

2   Line Segment Intersection

Thematic Map Overlay

When you are visiting a country, maps are an invaluable source of information. They tell you where tourist attractions are located, they indicate the roads and railway lines to get there, they show small lakes, and so on. Unfortunately, they can also be a source of frustration, as it is often difficult to find the right information: even when you know the approximate position of a small town,

Figure 2.1
Cities, rivers, railroads, and their overlay in western Canada

it can still be difficult to spot it on the map. To make maps more readable, geographic information systems split them into several *layers*. Each layer is a thematic map, that is, it stores only one type of information. Thus there will be a layer storing the roads, a layer storing the cities, a layer storing the rivers,

19

grizzly bear

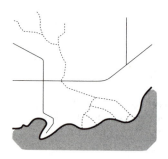

and so on. The theme of a layer can also be more abstract. For instance, there could be a layer for the population density, for average precipitation, habitat of the grizzly bear, or for vegetation. The type of geometric information stored in a layer can be very different: the layer for a road map could store the roads as collections of line segments (or curves, perhaps), the layer for cities could contain points labeled with city names, and the layer for vegetation could store a subdivision of the map into regions labeled with the type of vegetation.

Users of a geographic information system can select one of the thematic maps for display. To find a small town you would select the layer storing cities, and you would not be distracted by information such as the names of rivers and lakes. After you have spotted the town, you probably want to know how to get there. To this end geographic information systems allow users to view an *overlay* of several maps—see Figure 2.1. Using an overlay of the road map and the map storing cities you can now figure out how to get to the town. When two or more thematic map layers are shown together, intersections in the overlay are positions of special interest. For example, when viewing the overlay of the layer for the roads and the layer for the rivers, it would be useful if the intersections were clearly marked. In this example the two maps are basically networks, and the intersections are points. In other cases one is interested in the intersection of complete regions. For instance, geographers studying the climate could be interested in finding regions where there is pine forest and the annual precipitation is between 1000 mm and 1500 mm. These regions are the intersections of the regions labeled "pine forest" in the vegetation map and the regions labeled "1000–1500" in the precipitation map.

2.1 Line Segment Intersection

We first study the simplest form of the map overlay problem, where the two map layers are networks represented as collections of line segments. For example, a map layer storing roads, railroads, or rivers at a small scale. Note that curves can be approximated by a number of small segments. At first we won't be interested in the regions induced by these line segments. Later we shall look at the more complex situation where the maps are not just networks, but subdivisions of the plane into regions that have an explicit meaning. To solve the network overlay problem we first have to state it in a geometric setting. For the overlay of two networks the geometric situation is the following: given two sets of line segments, compute all intersections between a segment from one set and a segment from the other. This problem specification is not quite precise enough yet, as we didn't define when two segments intersect. In particular, do two segments intersect when an endpoint of one of them lies on the other? In other words, we have to specify whether the input segments are open or closed. To make this decision we should go back to the application, the network overlay problem. Roads in a road map and rivers in a river map are represented by chains of segments, so a crossing of a road and a river corresponds to the interior of one chain intersecting the interior of another chain.

This does not mean that there is an intersection between the interior of two segments: the intersection point could happen to coincide with an endpoint of a segment of a chain. In fact, this situation is not uncommon because windy rivers are represented by many small segments and coordinates of endpoints may have been rounded when maps are digitized. We conclude that we should define the segments to be closed, so that an endpoint of one segment lying on another segment counts as an intersection.

To simplify the description somewhat we shall put the segments from the two sets into one set, and compute all intersections among the segments in that set. This way we certainly find all the intersections we want. We may also find intersections between segments from the same set. Actually, we certainly will, because in our application the segments from one set form a number of chains, and we count coinciding endpoints as intersections. These other intersections can be filtered out afterwards by simply checking for each reported intersection whether the two segments involved belong to the same set. So our problem specification is as follows: given a set S of n closed segments in the plane, report all intersection points among the segments in S.

This doesn't seem like a challenging problem: we can simply take each pair of segments, compute whether they intersect, and, if so, report their intersection point. This brute-force algorithm clearly requires $O(n^2)$ time. In a sense this is optimal: when each pair of segments intersects any algorithm must take $\Omega(n^2)$ time, because it has to report all intersections. A similar example can be given when the overlay of two networks is considered. In practical situations, however, most segments intersect no or only a few other segments, so the total number of intersection points is much smaller than quadratic. It would be nice to have an algorithm that is faster in such situations. In other words, we want an algorithm whose running time depends not only on the number of segments in the input, but also on the number of intersection points. Such an algorithm is called an *output-sensitive algorithm*: the running time of the algorithm is sensitive to the size of the output. We could also call such an algorithm *intersection-sensitive*, since the number of intersections is what determines the size of the output.

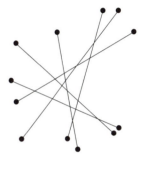

How can we avoid testing all pairs of segments for intersection? Here we must make use of the geometry of the situation: segments that are close together are candidates for intersection, unlike segments that are far apart. Below we shall see how we can use this observation to obtain an output-sensitive algorithm for the line segment intersection problem.

Let $S := \{s_1, s_2, \ldots, s_n\}$ be the set of segments for which we want to compute all intersections. We want to avoid testing pairs of segments that are far apart. But how can we do this? Let's first try to rule out an easy case. Define the y-interval of a segment to be its orthogonal projection onto the y-axis. When the y-intervals of a pair of segments do not overlap—we could say that they are far apart in the y-direction—then they cannot intersect. Hence, we only need to test pairs of segments whose y-intervals overlap, that is, pairs for which there exists a horizontal line that intersects both segments. To find these pairs we

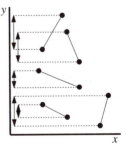

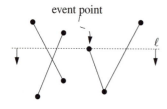

event point

imagine sweeping a line ℓ downwards over the plane, starting from a position above all segments. While we sweep the imaginary line, we keep track of all segments intersecting it—the details of this will be explained later—so that we can find the pairs we need.

This type of algorithm is called a *plane sweep algorithm* and the line ℓ is called the *sweep line*. The *status* of the sweep line is the set of segments intersecting it. The status changes while the sweep line moves downwards, but not continuously. Only at particular points is an update of the status required. We call these points the *event points* of the plane sweep algorithm. In this algorithm the event points are the endpoints of the segments.

The moments at which the sweep line reaches an event point are the only moments when the algorithm actually does something: it updates the status of the sweep line and performs some intersection tests. In particular, if the event point is the upper endpoint of a segment, then a new segment starts intersecting the sweep line and must be added to the status. This segment is tested for intersection against the ones already intersecting the sweep line. If the event point is a lower endpoint, a segment stops intersecting the sweep line and must be deleted from the status. This way we only test pairs of segments for which there is a horizontal line that intersects both segments. Unfortunately, this is not enough: there are still situations where we test a quadratic number of pairs, whereas there is only a small number of intersection points. A simple example is a set of vertical segments that all intersect the x-axis. So the algorithm is not output-sensitive. The problem is that two segments that intersect the sweep line can still be far apart in the horizontal direction.

Let's order the segments from left to right as they intersect the sweep line, to include the idea of being close in the horizontal direction. We shall only test segments when they are adjacent in the horizontal ordering. This means that we only test any new segment against two segments, namely, the ones immediately left and right of the upper endpoint. Later, when the sweep line has moved downwards to another position, a segment can become adjacent to other segments against which it will be tested. Our new strategy should be reflected in the status of our algorithm: the status now corresponds to the *ordered* sequence of segments intersecting the sweep line. The new status not only changes at endpoints of segments; it also changes at intersection points, where the order of the intersected segments changes. When this happens we must test the two segments that change position against their new neighbors. This is a new type of event point.

Before trying to turn these ideas into an efficient algorithm, we should convince ourselves that the approach is correct. We have reduced the number of pairs to be tested, but do we still find all intersections? In other words, if two segments s_i and s_j intersect, is there always a position of the sweep line ℓ where s_i and s_j are adjacent along ℓ? Let's first ignore some nasty cases: assume that no segment is horizontal, that any two segments intersect in at most one point—they do not overlap—, and that no three segments meet in a common point. Later we shall see that these cases are easy to handle, but for now it is convenient to forget about them. The intersections where an endpoint of

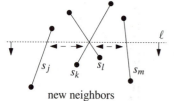

s_j s_k s_l s_m

new neighbors

a segment lies on another segment can easily be detected when the sweep line reaches the endpoint. So the only question is whether intersections between the interiors of segments are always detected.

Lemma 2.1 *Let s_i and s_j be two non-horizontal segments whose interiors intersect in a single point p, and assume there is no third segment passing through p. Then there is an event point above p where s_i and s_j become adjacent and are tested for intersection.*

Proof. Let ℓ be a horizontal line slightly above p. If ℓ is close enough to p then s_i and s_j must be adjacent along ℓ. (To be precise, we should take ℓ such that there is no event point on ℓ, nor in between ℓ and the horizontal line through p.) In other words, there is a position of the sweep line where s_i and s_j are adjacent. On the other hand, s_i and s_j are not yet adjacent when the algorithm starts, because the sweep line starts above all line segments and the status is empty. Hence, there must be an event point q where s_i and s_j become adjacent and are tested for intersection. □

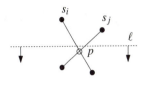

So our approach is correct, at least when we forget about the nasty cases mentioned earlier. Now we can proceed with the development of the plane sweep algorithm. Let's briefly recap the overall approach. We imagine moving a horizontal sweep line ℓ downwards over the plane. The sweep line halts at certain event points; in our case these are the endpoints of the segments, which we know beforehand, and the intersection points, which are computed on the fly. While the sweep line moves we maintain the ordered sequence of segments intersected by it. When the sweep line halts at an event point the sequence of segments changes and, depending on the type of event point, we have to take several actions to update the status and detect intersections.

When the event point is the upper endpoint of a segment, there is a new segment intersecting the sweep line. This segment must be tested for intersection against its two neighbors along the sweep line. Only intersection points below the sweep line are important; the ones above the sweep line have been detected already. For example, if segments s_i and s_k are adjacent on the sweep line, and a new upper endpoint of a segment s_j appears in between, then we have to test s_j for intersection with s_i and s_k. If we find an intersection below the sweep line, we have found a new event point. After the upper endpoint is handled we continue to the next event point.

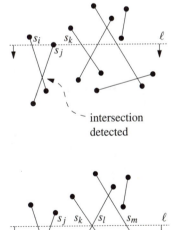

intersection
detected

When the event point is an intersection, the two segments that intersect change their order. Each of them gets (at most) one new neighbor against which it is tested for intersection. Again, only intersections below the sweep line are still interesting. Suppose that four segments s_j, s_k, s_l, and s_m appear in this order on the sweep line when the intersection point of s_k and s_l is reached. Then s_k and s_l switch position and we must test s_l and s_j for intersection below the sweep line, and also s_k and s_m. The new intersections that we find are, of course, also new event points for the algorithm.

When the event point is the lower endpoint of a segment, its two neighbors now become adjacent and must be tested for intersection. If they intersect

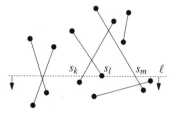

below the sweep line, then their intersection point is a new event point. Assume three segments s_k, s_l, and s_m appear in this order on the sweep line when the lower endpoint of s_l is encountered. Then s_k and s_m will become adjacent and we test them for intersection.

After we have swept the whole plane—more precisely, after we have treated the last event point—we have computed all intersection points. This is guaranteed by the following invariant, which holds at any time during the plane sweep: all intersection points above the sweep line have been computed correctly.

After this sketch of the algorithm, it's time to go into more detail. It's also time to look at the degenerate cases that can arise, like three or more segments meeting in a point. We should first specify what we expect from the algorithm in these cases. We could require the algorithm to simply report each intersection point once, but it seems more useful if it reports for each intersection point a list of segments that pass through it or have it as an endpoint. There is another special case for which we should define the required output more carefully, namely that of two partially overlapping segments, but for simplicity we shall ignore this case in the rest of this section.

We start by describing the data structures the algorithm uses.

First of all we need a data structure—called the *event queue*—that stores the events. We denote the event queue by Q. We need an operation that removes the next event that will occur from Q, and returns it so that it can be treated. This event is the highest event below the sweep line. If two event points have the same y-coordinate, then the one with smaller x-coordinate will be returned. In other words, event points on the same horizontal line are treated from left to right. This implies that we should consider the left endpoint of a horizontal segment to be its upper endpoint, and its right endpoint to be its lower endpoint. You can also think about our convention as follows: instead of having a horizontal sweep line, imagine it is sloping just a tiny bit upward. As a result the sweep line reaches the left endpoint of a horizontal segment just before reaching the right endpoint. The event queue should allow for insertions, because new events will be computed on the fly. Notice that two event points can coincide. For example, the upper endpoints of two distinct segments may coincide. It is convenient to treat this as one event point. Hence, an insertion must be able to check whether an event is already present in Q.

We implement the event queue as follows. Define an order $\prec$ on the event points that represents the order in which they will be handled. Hence, if p and q are two event points then we have $p \prec q$ if and only if $p_y > q_y$ holds or $p_y = q_y$ and $p_x < q_x$ holds. We store the event points in a balanced binary search tree, ordered according to $\prec$. With each event point p in Q we will store the segments starting at p, that is, the segments whose upper endpoint is p. This information will be needed to handle the event. Both operations—fetching the next event and inserting an event—take $O(\log m)$ time, where m is the number of events in Q. (We do not use a heap to implement the event queue, because we have to be able to test whether a given event is already present in Q.)

Second, we need to maintain the status of the algorithm. This is the ordered sequence of segments intersecting the sweep line. The status structure, denoted by $\mathcal{T}$, is used to access the neighbors of a given segment s, so that they can be tested for intersection with s. The status structure must be dynamic: as segments start or stop to intersect the sweep line, they must be inserted into or deleted from the structure. Because there is a well-defined order on the segments in the status structure we can use a balanced binary search tree as status structure. When you are only used to binary search trees that store numbers, this may be surprising. But binary search trees can store any set of elements, as long as there is an order on the elements.

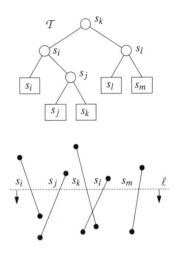

In more detail, we store the segments intersecting the sweep line ordered in the leaves of a balanced binary search tree $\mathcal{T}$. The left-to-right order of the segments along the sweep line corresponds to the left-to-right order of the leaves in $\mathcal{T}$. We must also store information in the internal nodes to guide the search down the tree to the leaves. At each internal node, we store the segment from the rightmost leaf in its left subtree. (Alternatively, we could store the segments only in interior nodes. This will save some storage. However, it is conceptually simpler to think about the segments in interior nodes as values to guide the search, not as data items. Storing the segments in the leaves also makes some algorithms simpler to describe.) Suppose we search in $\mathcal{T}$ for the segment immediately to the left of some point p which lies on the sweep line. At each internal node v we simply test whether p lies left or right of the segment stored at v. Depending on the outcome we descend to the left or right subtree of v, eventually ending up in a leaf. Either this leaf, or the leaf immediately to the left of it, stores the segment we are searching for. In a similar way we can find the segment immediately to the right of p, or the segments containing p. It follows that each update and neighbor search operation takes $O(\log n)$ time.

The event queue Q and the status structure $\mathcal{T}$ are the only two data structures we need. The global algorithm can now be described as follows.

Algorithm FINDINTERSECTIONS(S)
Input. A set S of line segments in the plane.
Output. The set of intersection points among the segments in S, with for each
   intersection point the segments that contain it.
1.   Initialize an empty event queue Q. Next, insert the segment endpoints
      into Q; when an upper endpoint is inserted, the corresponding segment
      should be stored with it.
2.   Initialize an empty status structure $\mathcal{T}$.
3.   **while** Q is not empty
4.   **do** Determine the next event point p in Q and delete it.
5.   HANDLEEVENTPOINT(p)

We have already seen how events are handled: at endpoints of segments we have to insert or delete segments from the status structure $\mathcal{T}$, and at intersection points we have to change the order of two segments. In both cases we also have to do intersection tests between segments that become neighbors after the event. In degenerate cases—where several segments are involved in one event point—the details are a little bit more tricky. The next procedure describes how to handle event points correctly; it is illustrated in Figure 2.2.

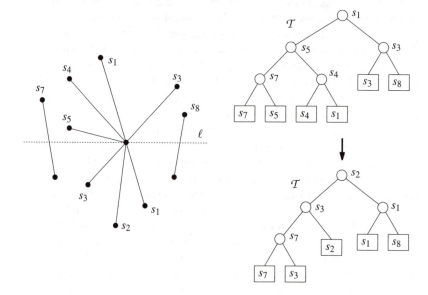

Figure 2.2
An event point and the changes in the status structure

HANDLEEVENTPOINT(p)
1. Let $U(p)$ be the set of segments whose upper endpoint is p; these segments are stored with the event point p. (For horizontal segments, the upper endpoint is by definition the left endpoint.)
2. Search in T for the set $S(p)$ of all segments that contain p; they are adjacent in T. Let $L(p) \subset S(p)$ be the set of segments whose lower endpoint is p, and let $C(p) \subset S(p)$ be the set of segments that contain p in their interior.
3. **if** $L(p) \cup U(p) \cup C(p)$ contains more than one segment
4.   **then** Report p as an intersection, together with $L(p)$, $U(p)$, and $C(p)$.
5. Delete the segments in $L(p) \cup C(p)$ from T.
6. Insert the segments in $U(p) \cup C(p)$ into T. The order of the segments in T should correspond to the order in which they are intersected by a sweep line just below p. If there is a horizontal segment, it comes last among all segments containing p.
7. (∗ Deleting and re-inserting the segments of $C(p)$ reverses their order. ∗)
8. **if** $U(p) \cup C(p) = \emptyset$
9.   **then** Let s_l and s_r be the left and right neighbors of p in T.
10.   FINDNEWEVENT(s_l, s_r, p)
11.   **else** Let s' be the leftmost segment of $U(p) \cup C(p)$ in T.
12.   Let s_l be the left neighbor of s' in T.
13.   FINDNEWEVENT(s_l, s', p)
14.   Let s'' be the rightmost segment of $U(p) \cup C(p)$ in T.
15.   Let s_r be the right neighbor of s'' in T.
16.   FINDNEWEVENT(s'', s_r, p)

Note that in lines 8–16 we assume that s_l and s_r actually exist. If they do not exist the corresponding steps should obviously not be performed.

The procedures for finding the new intersections are easy: they simply test two segments for intersection. The only thing we need to be careful about is, when we find an intersection, whether this intersection has already been handled earlier or not. When there are no horizontal segments, then the intersection has not been handled yet when the intersection point lies below the sweep line. But how should we deal with horizontal segments? Recall our convention that events with the same y-coordinate are treated from left to right. This implies that we are still interested in intersection points lying to the right of the current event point. Hence, the procedure FINDNEWEVENT is defined as follows.

FINDNEWEVENT(s_l, s_r, p)
1. **if** s_l and s_r intersect below the sweep line, or on it and to the right of the current event point p, and the intersection is not yet present as an event in Q
2. **then** Insert the intersection point as an event into Q.

What about the correctness of our algorithm? It is clear that FINDINTERSECTIONS only reports true intersection points, but does it find all of them? The next lemma states that this is indeed the case.

Lemma 2.2 *Algorithm* FINDINTERSECTIONS *computes all intersection points and the segments that contain it correctly.*

Proof. Recall that the priority of an event is given by its y-coordinate, and that when two events have the same y-coordinate the one with smaller x-coordinate is given higher priority. We shall prove the lemma by induction on the priority of the event points.

Let p be an intersection point and assume that all intersection points q with a higher priority have been computed correctly. We shall prove that p and the segments that contain p are computed correctly. Let $U(p)$ be the set of segments that have p as their upper endpoint (or, for horizontal segments, their left endpoint), let $L(p)$ be the set of segments having p as their lower endpoint (or, for horizontal segments, their right endpoint), and let $C(p)$ be the set of segments having p in their interior.

First, assume that p is an endpoint of one or more of the segments. In that case p is stored in the event queue Q at the start of the algorithm. The segments from $U(p)$ are stored with p, so they will be found. The segments from $L(p)$ and $C(p)$ are stored in $\mathcal{T}$ when p is handled, so they will be found in line 2 of HANDLEEVENTPOINT. Hence, p and all the segments involved are determined correctly when p is an endpoint of one or more of the segments.

Now assume that p is not an endpoint of a segment. All we need to show is that p will be inserted into Q at some moment. Note that all segments that are involved have p in their interior. Order these segments by angle around p, and let s_i and s_j be two neighboring segments. Following the proof of Lemma 2.1

we see that there is an event point with a higher priority than p such that s_i and s_j become adjacent when q is passed. In Lemma 2.1 we assumed for simplicity that s_i and s_j are non-horizontal, but it is straightforward to adapt the proof for horizontal segments. By induction, the event point q was handled correctly, which means that p is detected and stored into Q. ⬚

So we have a correct algorithm. But did we succeed in developing an output-sensitive algorithm? The answer is yes: the running time of the algorithm is $O((n+k)\log n)$, where k is the size of the output. The following lemma states an even stronger result: the running time is $O((n+I)\log n)$, where I is the number of intersections. This is stronger, because for one intersection point the output can consist of a large number of segments, namely in the case where many segments intersect in a common point.

Lemma 2.3 *The running time of Algorithm* FINDINTERSECTIONS *for a set S of n line segments in the plane is $O(n\log n + I\log n)$, where I is the number of intersection points of segments in S.*

Proof. The algorithm starts by constructing the event queue on the segment endpoints. Because we implemented the event queue as a balanced binary search tree, this takes $O(n\log n)$ time. Initializing the status structure takes constant time. Then the plane sweep starts and all the events are handled. To handle an event we perform three operations on the event queue Q: the event itself is deleted from Q in line 4 of FINDINTERSECTIONS, and there can be one or two calls to FINDNEWEVENT, which may cause at most two new events to be inserted into Q. Deletions and insertions on Q take $O(\log n)$ time each. We also perform operations—insertions, deletions, and neighbor finding—on the status structure $\mathcal{T}$, which take $O(\log n)$ time each. The number of operations is linear in the number $m(p) := \mathrm{card}(L(p) \cup U(p) \cup C(p))$ of segments that are involved in the event. If we denote the sum of all $m(p)$, over all event points p, by m, the running time of the algorithm is $O(m\log n)$.

It is clear that $m = O(n+k)$, where k is the size of the output; after all, whenever $m(p) > 1$ we report all segments involved in the event, and the only events involving one segment are the endpoints of segments. But we want to prove that $m = O(n+I)$, where I is the number of intersection points. To show this, we will interpret the set of segments as a planar graph embedded in the plane. (If you are not familiar with planar graph terminology, you should read the first paragraphs of Section 2.2 first.) Its vertices are the endpoints of segments and intersection points of segments, and its edges are the pieces of the segments connecting vertices. Consider an event point p. It is a vertex of the graph, and $m(p)$ is bounded by the degree of the vertex. Consequently, m is bounded by the sum of the degrees of all vertices of our graph. Every edge of the graph contributes one to the degree of exactly two vertices (its endpoints), so m is bounded by $2n_e$, where n_e is the number of edges of the graph. Let's bound n_e in terms of n and I. By definition, n_v, the number of vertices, is at most $2n+I$. It is well known that in planar graphs $n_e = O(n_v)$, which proves our claim. But, for completeness, let us give the argument here. Every face

of the planar graph is bounded by at least three edges—provided that there are at least three segments—and an edge can bound at most two different faces. Therefore n_f, the number of faces, is at most $2n_e/3$. We now use *Euler's formula*, which states that for any planar graph with n_v vertices, n_e edges, and n_f faces, the following relation holds:

$$n_v - n_e + n_f \geqslant 2.$$

Equality holds if and only if the graph is connected. Plugging the bounds on n_v and n_f into this formula, we get

$$2 \leqslant (2n + I) - n_e + \frac{2n_e}{3} = (2n + I) - n_e/3.$$

So $n_e \leqslant 6n + 3I - 6$, and $m \leqslant 12n + 6I - 12$, and the bound on the running time follows. ◻

We still have to analyze the other complexity aspect, the amount of storage used by the algorithm. The tree $\mathcal{T}$ stores a segment at most once, so it uses $O(n)$ storage. The size of Q can be larger, however. The algorithm inserts intersection points in Q when they are detected and it removes them when they are handled. When it takes a long time before intersections are handled, it could happen that Q gets very large. Of course its size is always bounded by $O(n + I)$, but it would be better if the working storage were always linear.

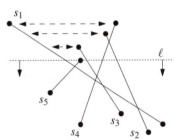

There is a relatively simple way to achieve this: only store intersection points of pairs of segments that are currently adjacent on the sweep line. The algorithm given above also stores intersection points of segments that have been horizontally adjacent, but aren't anymore. By storing only intersections among adjacent segments, the number of event points in Q is never more than linear. The modification required in the algorithm is that the intersection point of two segments must be deleted when they stop being adjacent. These segments must become adjacent again before the intersection point is reached, so the intersection point will still be reported correctly. The total time taken by the algorithm remains $O(n \log n + I \log n)$. We obtain the following theorem:

Theorem 2.4 *Let S be a set of n line segments in the plane. All intersection points in S, with for each intersection point the segments involved in it, can be reported in $O(n \log n + I \log n)$ time and $O(n)$ space, where I is the number of intersection points.*

2.2 The Doubly-Connected Edge List

We have solved the easiest case of the map overlay problem, where the two maps are networks represented as collections of line segments. In general, maps have a more complicated structure: they are subdivisions of the plane into labeled regions. A thematic map of forests in Canada, for instance, would be

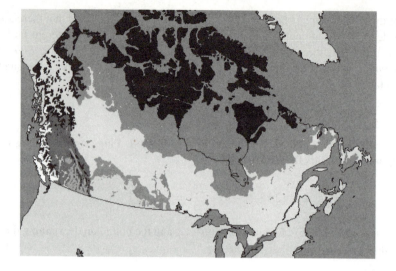

Figure 2.3
Types of forest in Canada

a subdivision of Canada into regions with labels such as "pine", "deciduous", "birch", and "mixed".

Before we can give an algorithm for computing the overlay of two subdivisions, we must develop a suitable representation for a subdivision. Storing a subdivision as a collection of line segments is not such a good idea. Operations like reporting the boundary of a region would be rather complicated. It is better to incorporate structural, topological information: which segments bound a given region, which regions are adjacent, and so on.

The maps we consider are *planar subdivisions* induced by planar embeddings of graphs. Such a subdivision is *connected* if the underlying graph is connected. The embedding of a node of the graph is called a *vertex*, and the embedding of an arc is called an *edge*. We only consider embeddings where every edge is a straight line segment. In principle, edges in a subdivision need not be straight. A subdivision need not even be a planar embedding of a graph, as it may have unbounded edges. In this section, however, we don't consider such more general subdivisions. We consider an edge to be open, that is, its endpoints—which are vertices of the subdivision—are not part of it. A *face* of the subdivision is a maximal connected subset of the plane that doesn't contain a point on an edge or a vertex. Thus a face is an open polygonal region whose boundary is formed by edges and vertices from the subdivision. The *complexity* of a subdivision is defined as the sum of the number of vertices, the number of edges, and the number of faces it consists of. If a vertex is the endpoint of an edge, then we say that the vertex and the edge are *incident*. Similarly, a face and an edge on its boundary are incident, and a face and a vertex of its boundary are incident.

What should we require from a representation of a subdivision? An operation one could ask for is to determine the face containing a given point. This is definitely useful in some applications—indeed, in a later chapter we shall design a data structure for this—but it is a bit too much to ask from a basic

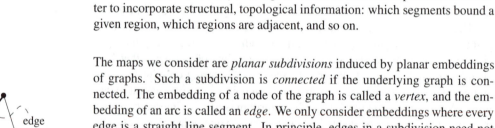

edge

vertex

face

disconnected
subdivision

representation. The things we can ask for should be more local. For example, it is reasonable to require that we can walk around the boundary of a given face, or that we can access one face from an adjacent one if we are given a common edge. Another operation that could be useful is to visit all the edges around a given vertex. The representation that we shall discuss supports these operations. It is called the doubly-connected edge list.

A *doubly-connected edge list* contains a record for each face, edge, and vertex of the subdivision. Besides the geometric and topological information—to be described shortly—each record may also store additional information. For instance, if the subdivision represents a thematic map for vegetation, the doubly-connected edge list would store in each face record the type of vegetation of the corresponding region. The additional information is also called *attribute information*. The geometric and topological information stored in the doubly-connected edge list should enable us to perform the basic operations mentioned earlier. To be able to walk around a face in counterclockwise order we store a pointer from each edge to the next. It can also come in handy to walk around a face the other way, so we also store a pointer to the previous edge. An edge usually bounds two faces, so we need two pairs of pointers for it. It is convenient to view the different sides of an edge as two distinct *half-edges*, so that we have a unique next half-edge and previous half-edge for every half-edge. This also means that a half-edge bounds only one face. The two half-edges we get for a given edge are called *twins*. Defining the next half-edge of a given half-edge with respect to a counterclockwise traversal of a face induces an orientation on each half-edge: it is oriented such that the face that it bounds lies to its left for an observer walking along the edge. Because half-edges are oriented we can speak of the *origin* and the *destination* of a half-edge. If a half-edge $\vec{e}$ has v as its origin and w as its destination, then its twin $Twin(\vec{e})$ has w as its origin and v as its destination. To reach the boundary of a face we just need to store one pointer in the face record to an arbitrary half-edge bounding the face. Starting from that half-edge, we can step from each half-edge to the next and walk around the face.

What we just said does not quite hold for the boundaries of holes in a face: if they are traversed in counterclockwise order then the face lies to the right. It will be convenient to orient half-edges such that their face always lies to the same side, so we change the direction of traversal for the boundary of a hole to clockwise. Now a face always lies to the left of any half-edge on its boundary. Another consequence is that twin half-edges always have opposite orientations. The presence of holes in a face also means that one pointer from the face to an arbitrary half-edge on its boundary is not enough to visit the whole boundary: we need a pointer to a half-edge in every boundary component. If a face has isolated vertices that don't have any incident edge, we can store pointers to them as well. For simplicity we'll ignore this case.

Let's summarize. The doubly-connected edge list consists of three collections of records: one for the vertices, one for the faces, and one for the half-edges. These records store the following geometric and topological information:

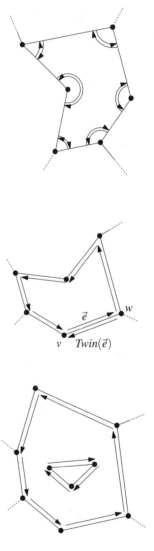

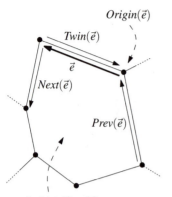

Origin($\vec{e}$)

Twin($\vec{e}$)

$\vec{e}$

Next($\vec{e}$)

Prev($\vec{e}$)

IncidentFace($\vec{e}$)

- The vertex record of a vertex v stores the coordinates of v in a field called *Coordinates(v)*. It also stores a pointer *IncidentEdge(v)* to an arbitrary half-edge that has v as its origin.

- The face record of a face f stores a pointer *OuterComponent(f)* to some half-edge on its outer boundary. For the unbounded face this pointer is **nil**. It also stores a list *InnerComponents(f)*, which contains for each hole in the face a pointer to some half-edge on the boundary of the hole.

- The half-edge record of a half-edge $\vec{e}$ stores a pointer *Origin($\vec{e}$)* to its origin, a pointer *Twin($\vec{e}$)* to its twin half-edge, and a pointer *IncidentFace($\vec{e}$)* to the face that it bounds. We don't need to store the destination of an edge, because it is equal to *Origin(Twin($\vec{e}$))*. The origin is chosen such that *IncidentFace($\vec{e}$)* lies to the left of $\vec{e}$ when it is traversed from origin to destination. The half-edge record also stores pointers *Next($\vec{e}$)* and *Prev($\vec{e}$)* to the next and previous edge on the boundary of *IncidentFace($\vec{e}$)*. Thus *Next($\vec{e}$)* is the unique half-edge on the boundary of *IncidentFace($\vec{e}$)* that has the destination of $\vec{e}$ as its origin, and *Prev($\vec{e}$)* is the unique half-edge on the boundary of *IncidentFace($\vec{e}$)* that has *Origin($\vec{e}$)* as its destination.

A constant amount of information is used for each vertex and edge. A face may require more storage, since the list *InnerComponents(f)* has as many elements as there are holes in the face. Because any half-edge is pointed to at most once from all *InnerComponents(f)* lists together, we conclude that the amount of storage is linear in the complexity of the subdivision. An example of a doubly-connected edge list for a simple subdivision is given below. The two half-edges corresponding to an edge e_i are labeled $\vec{e}_{i,1}$ and $\vec{e}_{i,2}$.

Vertex	Coordinates	IncidentEdge
v_1	$(0,4)$	$\vec{e}_{1,1}$
v_2	$(2,4)$	$\vec{e}_{4,2}$
v_3	$(2,2)$	$\vec{e}_{2,1}$
v_4	$(1,1)$	$\vec{e}_{2,2}$

Face	OuterComponent	InnerComponents
f_1	**nil**	$\vec{e}_{1,1}$
f_2	$\vec{e}_{4,1}$	**nil**

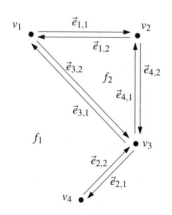

v_1 $\vec{e}_{1,1}$ v_2

$\vec{e}_{1,2}$

$\vec{e}_{3,2}$ f_2 $\vec{e}_{4,2}$

$\vec{e}_{4,1}$

$\vec{e}_{3,1}$

f_1 v_3

$\vec{e}_{2,2}$

$\vec{e}_{2,1}$

v_4

Half-edge	Origin	Twin	IncidentFace	Next	Prev
$\vec{e}_{1,1}$	v_1	$\vec{e}_{1,2}$	f_1	$\vec{e}_{4,2}$	$\vec{e}_{3,1}$
$\vec{e}_{1,2}$	v_2	$\vec{e}_{1,1}$	f_2	$\vec{e}_{3,2}$	$\vec{e}_{4,1}$
$\vec{e}_{2,1}$	v_3	$\vec{e}_{2,2}$	f_1	$\vec{e}_{2,2}$	$\vec{e}_{4,2}$
$\vec{e}_{2,2}$	v_4	$\vec{e}_{2,1}$	f_1	$\vec{e}_{3,1}$	$\vec{e}_{2,1}$
$\vec{e}_{3,1}$	v_3	$\vec{e}_{3,2}$	f_1	$\vec{e}_{1,1}$	$\vec{e}_{2,2}$
$\vec{e}_{3,2}$	v_1	$\vec{e}_{3,1}$	f_2	$\vec{e}_{4,1}$	$\vec{e}_{1,2}$
$\vec{e}_{4,1}$	v_3	$\vec{e}_{4,2}$	f_2	$\vec{e}_{1,2}$	$\vec{e}_{3,2}$
$\vec{e}_{4,2}$	v_2	$\vec{e}_{4,1}$	f_1	$\vec{e}_{2,1}$	$\vec{e}_{1,1}$

The information stored in the doubly-connected edge list is enough to perform the basic operations. For example, we can walk around the outer boundary of a given face f by following $Next(\vec{e})$ pointers, starting from the half-edge $OuterComponent(f)$. We can also visit all edges incident to a vertex v. It is a good exercise to figure out for yourself how to do this.

We described a fairly general version of the doubly-connected edge list. In applications where the vertices carry no attribute information we could store their coordinates directly in the $Origin()$ field of the edge; there is no strict need for a separate type of vertex record. Even more important is to realize that in many applications the faces of the subdivision carry no interesting meaning (think of the network of rivers or roads that we looked at before). If that is the case, we can completely forget about the face records, and the $IncidentFace()$ field of half-edges. As we will see, the algorithm of the next section doesn't need these fields (and is actually simpler to implement if we don't need to update them). Some implementations of doubly-connected edge lists may also insist that the graph formed by the vertices and edges of the subdivision be connected. This can always be achieved by introducing dummy edges, and has two advantages. Firstly, a simple graph transversal can be used to visit all half-edges, and secondly, the $InnerComponents()$ list for faces is not necessary.

2.3 Computing the Overlay of Two Subdivisions

Now that we have designed a good representation of a subdivision, we can tackle the general map overlay problem. We define the overlay of two subdivisions S_1 and S_2 to be the subdivision $O(S_1, S_2)$ such that there is a face f in $O(S_1, S_2)$ if and only if there are faces f_1 in S_1 and f_2 in S_2 such that f is a maximal connected subset of $f_1 \cap f_2$. This sounds more complicated than it is: what it means is that the overlay is the subdivision of the plane induced by the edges from S_1 and S_2. Figure 2.4 illustrates this. The general map over-

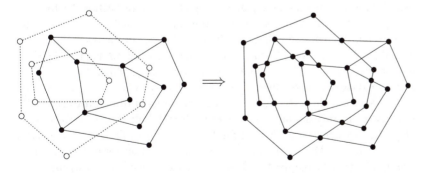

Figure 2.4
Overlaying two subdivisions

lay problem is to compute a doubly-connected edge list for $O(S_1, S_2)$, given the doubly-connected edge lists of S_1 and S_2. We require that each face in $O(S_1, S_2)$ be labeled with the labels of the faces in S_1 and S_2 that contain it. This way we have access to the attribute information stored for these faces. In an overlay of a vegetation map and a precipitation map this would mean that we know for each region in the overlay the type of vegetation and the amount of precipitation.

Let's first see how much information from the doubly-connected edge lists for S_1 and S_2 we can re-use in the doubly-connected edge list for $O(S_1, S_2)$. Consider the network of edges and vertices of S_1. This network is cut into pieces by the edges of S_2. These pieces are for a large part re-usable; only the edges that have been cut by the edges of S_2 should be renewed. But does this also hold for the half-edge records in the doubly-connected edge list that correspond to the pieces? If the orientation of a half-edge would change, we would still have to change the information in these records. Fortunately, this is not the case. The half-edges are oriented such that the face that they bound lies to the left; the shape of the face may change in the overlay, but it will remain to the same side of the half-edge. Hence, we can re-use half-edge records corresponding to edges that are not intersected by edges from the other map. Stated differently, the only half-edge records in the doubly-connected edge list for $O(S_1, S_2)$ that we cannot borrow from S_1 or S_2 are the ones that are incident to an intersection between edges from different maps.

This suggest the following approach. First, copy the doubly-connected edge lists of S_1 and S_2 into one new doubly-connected edge list. The new doubly-connected edge list is not a valid doubly-connected edge list, of course, in the sense that it does not yet represent a planar subdivision. This is the task of the overlay algorithm: it must transform the doubly-connected edge list into a valid doubly-connected edge list for $O(S_1, S_2)$ by computing the intersections between the two networks of edges, and linking together the appropriate parts of the two doubly-connected edge lists.

We did not talk about the new face records yet. The information for these records is more difficult to compute, so we leave this for later. We first describe in a little more detail how the vertex and half-edge records of the doubly-connected edge list for $O(S_1, S_2)$ are computed.

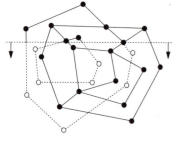

Our algorithm is based on the plane sweep algorithm of Section 2.1 for computing the intersections in a set of line segments. We run this algorithm on the set of segments that is the union of the sets of edges of the two subdivisions S_1 and S_2. Here we consider the edges to be closed. Recall that the algorithm is supported by two data structures: an event queue Q, which stores the event points, and the status structure T, which is a balanced binary search tree storing the segments intersecting the sweep line, ordered from left to right. We now also maintain a doubly-connected edge list D. Initially, D contains a copy of the doubly-connected edge list for S_1 and a copy of the doubly-connected edge list for S_2. During the plane sweep we shall transform D to a correct doubly-connected edge list for $O(S_1, S_2)$. That is to say, as far as the vertex and half-edge records are concerned; the face information will be computed later. We keep cross pointers between the edges in the status structure T and the half-edge records in D that correspond to them. This way we can access the part of D that needs to be changed when we encounter an intersection point.

The invariant that we maintain is that at any time during the sweep, the part of the overlay above the sweep line has been computed correctly.

Now, let's consider what we must do when we reach an event point. First of all, we update $\mathcal{T}$ and $\mathcal{Q}$ as in the line segment intersection algorithm. If the event involves only edges from one of the two subdivisions, this is all; the event point is a vertex that can be re-used. If the event involves edges from both subdivisions, we must make local changes to $\mathcal{D}$ to link the doubly-connected edge lists of the two original subdivisions at the intersection point. This is tedious but not difficult.

the geometric situation and the
two doubly-connected edge lists
before handling the intersection

the doubly-connected edge list
after handling the intersection

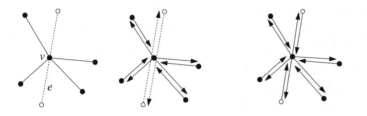

Figure 2.5
An edge of one subdivision passing through a vertex of the other

We describe the details for one of the possible cases, namely when an edge e of S_1 passes through a vertex v of S_2, see Figure 2.5. The edge e must be replaced by two edges denoted e' and e''. In the doubly-connected edge list, the two half-edges for e must become four. We create two new half-edge records, both with v as the origin. The two existing half-edges for e keep the endpoints of e as their origin, as shown in Figure 2.5. Then we pair up the existing half-edges with the new half-edges by setting their $Twin()$ pointers. So e' is represented by one new and one existing half-edge, and the same holds for e''. Now we must set a number of $Prev()$ and $Next()$ pointers. We first deal with the situation around the endpoints of e; later we'll worry about the situation around v. The $Next()$ pointers of the two new half-edges each copy the $Next()$ pointer of the old half-edge that is not its twin. The half-edges to which these pointers point must also update their $Prev()$ pointer and set it to the new half-edges. The correctness of this step can be verified best by looking at a figure.

It remains to correct the situation around vertex v. We must set the $Next()$ and $Prev()$ pointers of the four half-edges representing e' and e'', and of the four half-edges incident from S_2 to v. We locate these four half-edges from S_2 by testing where e' and e'' should be in the cyclic order of the edges around vertex v. There are four pairs of half-edges that become linked by a $Next()$ pointer from the one and a $Prev()$ pointer from the other. Consider the half-edge for e' that has v as its destination. It must be linked to the first half-edge, seen clockwise from e', with v as its origin. The half-edge for e' with v as its origin must be linked to the first counterclockwise half-edge with v as its destination. The same statements hold for e''.

Most of the steps in the description above take only constant time. Only locating where e' and e'' appear in the cyclic order around v may take longer:

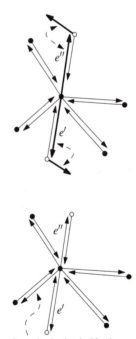

first clockwise half-edge
from e' with v as its origin

it will take time linear in the degree of v. The other cases that can arise—crossings of two edges from different maps, and coinciding vertices—are not more difficult than the case we just discussed. These cases also take time $O(m)$, where m is the number of edges incident to the event point. This means that updating $\mathcal{D}$ does not increase the running time of the line segment intersection algorithm asymptotically. Notice that every intersection that we find is a vertex of the overlay. It follows that the vertex records and the half-edge records of the doubly-connected edge list for $O(S_1, S_2)$ can be computed in $O(n \log n + k \log n)$ time, where n denotes the sum of the complexities of S_1 and S_2, and k is the complexity of the overlay.

After the fields involving vertex and half-edge records have been set, it remains to compute the information about the faces of $O(S_1, S_2)$. More precisely, we have to create a face record for each face f in $O(S_1, S_2)$, we have to make *OuterComponent(f)* point to a half-edge on the outer boundary of f, and we have to make a list *InnerComponents(f)* of pointers to half-edges on the boundaries of the holes inside f. Furthermore, we must set the *IncidentFace()* fields of the half-edges on the boundary of f so that they point to the face record of f. Finally, each of the new faces must be labeled with the names of the faces in the old subdivisions that contain it.

How many face records will there be? Well, except for the unbounded face, every face has a unique outer boundary, so the number of face records we have to create is equal to the number of outer boundaries plus one. From the part of the doubly-connected edge list we have constructed so far we can easily extract all boundary cycles. But how do we know whether a cycle is an outer boundary or the boundary of a hole in a face? This can be decided by looking at the leftmost vertex v of the cycle, or, in case of ties, at the lowest of the leftmost ones. Recall that half-edges are directed in such a way that their incident face locally lies to the left. Consider the two half-edges of the cycle that are incident to v. Because we know that the incident face lies to the left, we can compute the angle these two half-edges make inside the incident face. If this angle is smaller than 180° then the cycle is an outer boundary, and otherwise it is the boundary of a hole. This property holds for the leftmost vertex of a cycle, but not necessarily for other vertices of that cycle.

To decide which boundary cycles bound the same face we construct a graph $\mathcal{G}$. For every boundary cycle—inner and outer—there is a node in $\mathcal{G}$. There is also one node for the imaginary outer boundary of the unbounded face. There is an arc between two cycles if and only if one of the cycles is the boundary of a hole and the other cycle has a half-edge immediately to the left of the leftmost vertex of that hole cycle. If there is no half-edge to the left of the leftmost vertex of a cycle, then the node representing the cycle is linked to the node of the unbounded face. Figure 2.6 gives an example. The dotted segments in the figure indicate the linking of the hole cycles to other cycles. The graph corresponding to the subdivision is also shown in the figure. The hole cycles are shown as single circles, and the outer boundary cycles are shown as double circles. Observe that C_3 and C_6 are in the same connected component as C_2. This indicates that C_3 and C_6 are hole cycles in the face whose outer

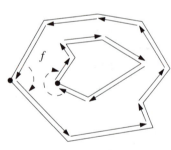

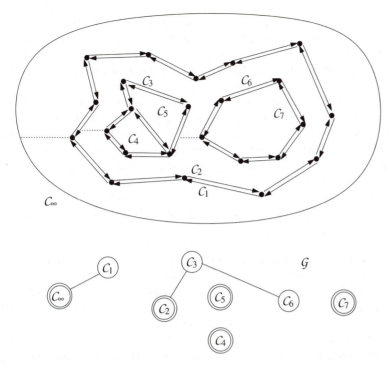

Figure 2.6
A subdivision and the corresponding
graph G

boundary is C_2. If there is only one hole in a face f, then the graph G links
the boundary cycle of the hole to the outer boundary of f. In general this need
not be the case: a hole can also be linked to another hole, as you can see in
Figure 2.6. This hole, which lies in the same face f, may be linked to the outer
boundary of f, or it may be linked to yet another hole. But eventually we must
end up linking a hole to the outer boundary, as the next lemma shows.

Lemma 2.5 *Each connected component of the graph G corresponds exactly
to the set of cycles incident to one face.*

Proof. Consider a cycle C bounding a hole in a face f. Because f lies locally
to the left of the leftmost vertex of C, C must be linked to another cycle that
also bounds f. It follows that cycles in the same connected component of G
bound the same face.

   To finish the proof, we show that every cycle bounding a hole in f is in
the same connected component as the outer boundary of f. Suppose there is
a cycle for which this is not the case. Let C be the leftmost such cycle, that
is, the one whose the leftmost vertex is leftmost. By definition there is an arc
between the C and another cycle C' that lies partly to the left of the leftmost
vertex of C. Hence, C' is in the same connected component as C, which is
not the component of the outer boundary of f. This contradicts the definition
of C.   ◻

Lemma 2.5 shows that once we have the graph G, we can create a face record
for every component. Then we can set the *IncidentFace*() pointers of the half-

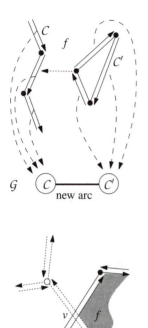

edges that bound each face f, and we can construct the list *InnerComponents*(f) and the set *OuterComponent*(f). How can we construct G? Recall that in the plane sweep algorithm for line segment intersection we always looked for the segments immediately to the left of an event point. (They had to be tested for intersection against the leftmost edge through the event point.) Hence, the information we need to construct G is determined during the plane sweep. So, to construct G, we first make a node for every cycle. To find the arcs of G, we consider the leftmost vertex v of every cycle bounding a hole. If $\vec{e}$ is the half-edge immediately left of v, then we add an arc between the two nodes in G representing the cycle containing $\vec{e}$ and the hole cycle of which v is the leftmost vertex. To find these nodes in G efficiently we need pointers from every half-edge record to the node in G representing the cycle it is in. So the face information of the doubly-connected edge list can be set in $O(n+k)$ additional time, after the plane sweep.

One thing remains: each face f in the overlay must be labeled with the names of the faces in the old subdivisions that contained it. To find these faces, consider an arbitrary vertex v of f. If v is the intersection of an edge e_1 from S_1 and an edge e_2 from S_2 then we can decide which faces of S_1 and S_2 contain f by looking at the *IncidentFace*() pointer of the appropriate half-edges corresponding to e_1 and e_2. If v is not an intersection but a vertex of, say, S_1, then we only know the face of S_1 containing f. To find the face of S_2 containing f, we have to do some more work: we have to determine the face of S_2 that contains v. In other words, if we knew for each vertex of S_1 in which face of S_2 it lay, and vice versa, then we could label the faces of $O(S_1, S_2)$ correctly. How can we compute this information? The solution is to apply the paradigm that has been introduced in this chapter, plane sweep, once more. However, we won't explain this final step here. It is a good exercise to test your understanding of the plane sweep approach to design the algorithm yourself. (In fact, it is not necessary to compute this information in a separate plane sweep. It can also be done in the sweep that computes the intersections.)

Putting everything together we get the following algorithm.

Algorithm MAPOVERLAY(S_1, S_2)
Input. Two planar subdivisions S_1 and S_2 stored in doubly-connected edge lists.
Output. The overlay of S_1 and S_2 stored in a doubly-connected edge list $\mathcal{D}$.
1. Copy the doubly-connected edge lists for S_1 and S_2 to a new doubly-connected edge list $\mathcal{D}$.
2. Compute all intersections between edges from S_1 and S_2 with the plane sweep algorithm of Section 2.1. In addition to the actions on $\mathcal{T}$ and Q required at the event points, do the following:

   - Update $\mathcal{D}$ as explained above if the event involves edges of both S_1 and S_2. (This was explained for the case where an edge of S_1 passes through a vertex of S_2.)
   - Store the half-edge immediately to the left of the event point at the vertex in $\mathcal{D}$ representing it.

3. (∗ Now $\mathcal{D}$ is the doubly-connected edge list for $O(S_1, S_2)$, except that the information about the faces has not been computed yet. ∗)
4. Determine the boundary cycles in $O(S_1, S_2)$ by traversing $\mathcal{D}$.
5. Construct the graph G whose nodes correspond to boundary cycles and whose arcs connect each hole cycle to the cycle to the left of its leftmost vertex, and compute its connected components. (The information to determine the arcs of G has been computed in line 2, second item.)
6. **for** each connected component in G
7. **do** Let C be the unique outer boundary cycle in the component and let f denote the face bounded by the cycle. Create a face record for f, set *OuterComponent*(f) to some half-edge of C, and construct the list *InnerComponents*(f) consisting of pointers to one half-edge in each hole cycle in the component. Let the *IncidentFace*$()$ pointers of all half-edges in the cycles point to the face record of f.
8. Label each face of $O(S_1, S_2)$ with the names of the faces of S_1 and S_2 containing it, as explained above.

Theorem 2.6 *Let S_1 be a planar subdivision of complexity n_1, let S_2 be a subdivision of complexity n_2, and let $n := n_1 + n_2$. The overlay of S_1 and S_2 can be constructed in $O(n \log n + k \log n)$ time, where k is the complexity of the overlay.*

Proof. Copying the doubly-connected edge lists in line 1 takes $O(n)$ time, and the plane sweep of line 2 takes $O(n \log n + k \log n)$ time by Lemma 2.3. Steps 4–7, where we fill in the face records, takes time linear in the complexity of $O(S_1, S_2)$. (The connected components of a graph can be determined in linear time by a simple depth first search.) Finally, labeling each face in the resulting subdivision with the faces of the original subdivisions that contain it can be done in $O(n \log n + k \log n)$ time. ∎

2.4 Boolean Operations

The map overlay algorithm is a powerful instrument that can be used for various other applications. One particular useful one is performing the Boolean operations union, intersection, and difference on two polygons $\mathcal{P}_1$ and $\mathcal{P}_2$. See Figure 2.7 for an example. Note that the output of the operations might no longer be a polygon. It can consist of a number of polygonal regions, some with holes.

To perform the Boolean operation we regard the polygons as planar maps whose bounded faces are labeled $\mathcal{P}_1$ and $\mathcal{P}_2$, respectively. We compute the overlay of these maps, and we extract the faces in the overlay whose labels correspond to the particular Boolean operation we want to perform. If we want to compute the intersection $\mathcal{P}_1 \cap \mathcal{P}_2$, we extract the faces in the overlay that are labeled with $\mathcal{P}_1$ and $\mathcal{P}_2$. If we want to compute the union $\mathcal{P}_1 \cup \mathcal{P}_2$, we extract the faces in the overlay that are labeled with $\mathcal{P}_1$ or $\mathcal{P}_2$. And if we want to compute

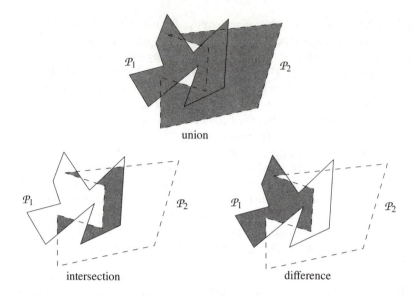

Figure 2.7
The Boolean operations union,
intersection and difference on two
polygons $\mathcal{P}_1$ and $\mathcal{P}_2$

the difference $\mathcal{P}_1 \setminus \mathcal{P}_2$, we extract the faces in the overlay that are labeled with $\mathcal{P}_1$ and not with $\mathcal{P}_2$.

Because every intersection point of an edge of $\mathcal{P}_1$ and an edge of $\mathcal{P}_2$ is a vertex of $\mathcal{P}_1 \cap \mathcal{P}_2$, the running time of the algorithm is $O(n \log n + k \log n)$, where n is the total number of vertices in $\mathcal{P}_1$ and $\mathcal{P}_2$, and k is the complexity of $\mathcal{P}_1 \cap \mathcal{P}_2$. The same holds for the other Boolean operations: every intersection of two edges is a vertex of the final result, no matter which operation we want to perform. We immediately get the following result.

Corollary 2.7 *Let $\mathcal{P}_1$ be a polygon with n_1 vertices and $\mathcal{P}_2$ a polygon with n_2 vertices, and let $n := n_1 + n_2$. Then $\mathcal{P}_1 \cap \mathcal{P}_2$, $\mathcal{P}_1 \cup \mathcal{P}_2$, and $\mathcal{P}_1 \setminus \mathcal{P}_2$ can each be computed in $O(n \log n + k \log n)$ time, where k is the complexity of the output.*

2.5 Notes and Comments

The line segment intersection problem is one of the fundamental problems in computational geometry. The $O(n \log n + k \log n)$ solution presented in this chapter was given by Bentley and Ottmann [35] in 1979. (A few years earlier, Shamos and Hoey [312] had solved the *detection* problem, where one is only interested in deciding whether there is at least one intersection, in $O(n \log n)$ time.) The method for reducing the working storage from $O(n + k)$ to $O(n)$ described in this chapter is taken from Pach and Sharir [276]. Brown [56] describes an alternative method to achieve the reduction.

The lower bound for the problem of reporting all line segment intersections is $\Omega(n \log n + k)$, so the plane sweep algorithm described in this chapter is not optimal when k is large. A first step towards an optimal algorithm was taken by Chazelle [67], who gave an algorithm with $O(n \log^2 n / \log \log n + k)$ running time. In 1988 Chazelle and Edelsbrunner [78, 79] presented the first

$O(n \log n + k)$ time algorithm. Unfortunately, it requires $O(n+k)$ storage. Later Clarkson and Shor [111] and Mulmuley [254] gave randomized incremental algorithms whose expected running time is also $O(n \log n + k)$. (See Chapter 4 for an explanation of randomized algorithms.) The working storage of these algorithms is $O(n)$ and $O(n+k)$, respectively. Unlike the algorithm of Chazelle and Edelsbrunner, these randomized algorithms also work for computing intersections in a set of curves. Recently Balaban [23] gave a new deterministic algorithm for the segment intersection problem. His algorithm works in $O(n \log n + k)$ time and it uses $O(n)$ space. Hence, it is the first deterministic algorithm that is optimal both with respect to time and with respect to storage. It also works for curves.

There are cases of the line segment intersection problem that are easier than the general case. One such case is where we have two sets of segments, say red segments and blue segments, such that no two segments from the same set intersect each other. (This is, in fact, exactly the network overlay problem. In the solution described in this chapter, however, the fact that the segments came from two sets of non-intersecting segments was not used.) This so-called red-blue line segment intersection problem was solved in $O(n \log n + k)$ time and $O(n)$ storage by Mairson and Stolfi [226] before the general problem was solved optimally. Other optimal red-blue intersection algorithms were given by Chazelle et al. [80] and by Palazzi and Snoeyink [278]. If the two sets of segments form connected subdivisions then the situation is even better: in this case the overlay can be computed in $O(n+k)$ time, as has been shown by Finke and Hinrichs [145]. Their result generalizes and improves previous results on map overlay by Nievergelt and Preparata [258], Guibas and Seidel [168], and Mairson and Stolfi [226].

Plane sweep is one of the most important paradigms for designing geometric algorithms. The first algorithms in computational geometry based on this paradigm are by Shamos and Hoey [312], Lee and Preparata [215], and Bentley and Ottmann [35]. Plane sweep algorithms are especially suited for finding intersections in sets of objects, but they can also be used for solving many other problems. In Chapter 3 plane sweep solves part of the polygon triangulation problem, and in Chapter 7 we will see a plane sweep algorithm to compute the so-called Voronoi diagram of a set of points. The algorithm presented in the current chapter sweeps a horizontal line downwards over the plane. For some problems it is more convenient to sweep the plane in another way. For instance, we can sweep the plane with a rotating line—see Chapter 15 for an example—or with a pseudo-line (a line that need not be straight, but otherwise behaves more or less as a line) [130]. The plane sweep technique can also be used in higher dimensions: here we sweep the space with a hyperplane [180, 275, 287]. Such algorithms are called space sweep algorithms.

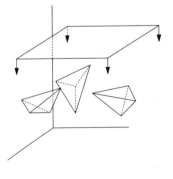

In this chapter we described a data structure for storing subdivisions: the doubly-connected edge list. This structure, or in fact a variant of it, was described by Muller and Preparata [252]. There are also other data structures for storing subdivisions, such as the winged edge structure by Baumgart [28] and the quad edge structure by Guibas and Stolfi [171]. The difference between all these structures is small. They all have more or less the same functionality, but some save a few bytes of storage per edge.

2.6 Exercises

2.1 Let S be a set of n disjoint line segments whose upper endpoints lie on the line $y = 1$ and whose lower endpoints lie on the line $y = 0$. These segments partition the horizontal strip $[-\infty : \infty] \times [0 : 1]$ into $n + 1$ regions. Give an $O(n \log n)$ time algorithm to build a binary search tree on the segments in S such that the region containing a query point can be determined in $O(\log n)$ time. Also describe the query algorithm in detail.

2.2 The intersection detection problem for a set S of n line segments is to determine whether there exists a pair of segments that intersect. Give a plane sweep algorithm that solves the intersection detection problem in $O(n \log n)$ time.

2.3 Change the code of Algorithm FINDINTERSECTIONS (and of the procedures that it calls) such that the working storage is $O(n)$ instead of $O(n + k)$.

2.4 Let S be a set of n line segments in the plane that may (partly) overlap each other. For example, S could contain the segments $\overline{(0,0)(1,0)}$ and $\overline{(-1,0)(2,0)}$. We want to compute all intersections in S. More precisely, we want to compute each proper intersection of two segments in S (that is, each intersection of two non-parallel segments) and for each endpoint of a segment all segments containing the point. Adapt algorithm FIND-INTERSECTIONS to this end.

2.5 Which of the following equalities are always true?

$$
\begin{aligned}
\mathit{Twin}(\mathit{Twin}(\vec{e})) &= \vec{e} \\
\mathit{Next}(\mathit{Prev}(\vec{e})) &= \vec{e} \\
\mathit{Twin}(\mathit{Prev}(\mathit{Twin}(\vec{e}))) &= \mathit{Next}(\vec{e}) \\
\mathit{IncidentFace}(\vec{e}) &= \mathit{IncidentFace}(\mathit{Next}(\vec{e}))
\end{aligned}
$$

2.6 Give an example of a doubly-connected edge list where for an edge e the faces $\mathit{IncidentFace}(\vec{e})$ and $\mathit{IncidentFace}(\mathit{Twin}(\vec{e}))$ are the same.

2.7 Given a subdivision in doubly-connected edge list representation where $\mathit{Twin}(\vec{e}) = \mathit{Next}(\vec{e})$ holds for every half-edge $\vec{e}$, how many faces can the subdivision have?

2.8 Give pseudocode to list all vertices incident to a given vertex v in a doubly-connected edge list. Also, give pseudocode to list all edges that bound a face in a not necessarily connected subdivision.

2.9 Suppose that a doubly-connected edge list of a connected subdivision is given. Give pseudocode to list all faces with vertices that appear on the outer boundary.

2.10 Let S be a subdivision of complexity n, and let P be a set of m points. Give a plane sweep algorithm that computes for every point in P in which face of S it is contained. Show that your algorithm runs in $O((n+m)\log(n+m))$ time.

2.11 Let S be a set of n circles in the plane. Describe a plane sweep algorithm to compute all intersection points between the circles. (Because we deal with circles, not discs, two circles do not intersect if one lies entirely inside the other.) Your algorithm should run in $O((n+k)\log n)$ time, where k is the number of intersection points.

2.12* Let S be a set of n disjoint triangles in the plane. We want to find a set of $n-1$ segments with the following properties:

- Each segment connects a point on the boundary of one triangle to a point on the boundary of another triangle.
- The interiors of the segments are pairwise disjoint and they are disjoint from the triangles.
- Together they connect all triangles to each other, that is, by walking along the segments and the triangle boundaries it must be is possible to walk from a triangle to any other triangle.

Develop a plane sweep algorithm for this problem that runs in $O(n\log n)$ time. State the events and the data structures that you use explicitly, and describe the cases that arise and the actions required for each of them. Also state the sweep invariant.

2.13 Let S_1 be a set of n disjoint horizontal line segments and let S_2 be a set of m disjoint vertical line segments. Give an $O((n+m)\log(n+m))$ time algorithm to count how many intersections there are in $S_1 \cup S_2$.

2.14 Let S be a set of n disjoint line segments in the plane, and let p be a point not on any of the line segments of S. We wish to determine all line segments of S that p can see, that is, all line segments of S that contain some point q so that the open segment $\overline{pq}$ doesn't intersect any line segment of S. Give an $O(n\log n)$ time algorithm that makes use of a rotating half-line with its endpoint at p.

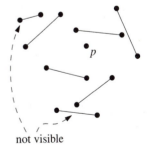

not visible

3 Polygon Triangulation

Guarding an Art Gallery

Works of famous painters are not only popular among art lovers, but also among criminals. They are very valuable, easy to transport, and apparently not so difficult to sell. Art galleries therefore have to guard their collections

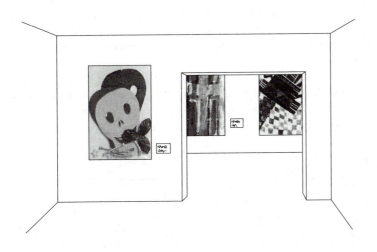

Figure 3.1
An art gallery

carefully. During the day the attendants can keep a look-out, but at night this has to be done by video cameras. These cameras are usually hung from the ceiling and they rotate about a vertical axis. The images from the cameras are sent to TV screens in the office of the night watch. Because it is easier to keep an eye on few TV screens rather than on many, the number of cameras should be as small as possible. An additional advantage of a small number of cameras is that the cost of the security system will be lower. On the other hand we cannot have too few cameras, because every part of the gallery must be visible to at least one of them. So we should place the cameras at strategic positions, such that each of them guards a large part of the gallery. This gives rise to what is usually referred to as the *Art Gallery Problem*: how many cameras do we need to guard a given gallery and how do we decide where to place them?

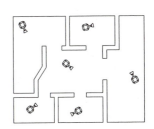

45

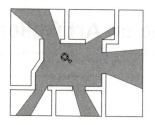

3.1 Guarding and Triangulations

If we want to define the art gallery problem more precisely, we should first formalize the notion of gallery. A gallery is, of course, a 3-dimensional space, but a floor plan gives us enough information to place the cameras. Therefore we model a gallery as a polygonal region in the plane. We further restrict ourselves to regions that are *simple polygons*, that is, regions enclosed by a single closed polygonal chain that does not intersect itself. Thus we do not allow regions with holes. A camera position in the gallery corresponds to a point in the polygon. A camera sees those points in the polygon to which it can be connected with an open segment that lies in the interior of the polygon.

How many cameras do we need to guard a simple polygon? This clearly depends on the polygon at hand: the more complex the polygon, the more cameras are required. We shall therefore express the bound on the number of cameras needed in terms of n, the number of vertices of the polygon. But even when two polygons have the same number of vertices, one can be easier to guard than the other. A convex polygon, for example, can always be guarded with one camera. To be on the safe side we shall look at the worst-case scenario, that is, we shall give a bound that is good for any simple polygon with n vertices. (It would be nice if we could find the minimum number of cameras for the specific polygon we are given, not just a worst-case bound. Unfortunately, the problem of finding the minimum number of cameras for a given polygon is NP-hard.)

Let $\mathcal{P}$ be a simple polygon with n vertices. Because $\mathcal{P}$ may be a complicated shape, it seems difficult to say anything about the number of cameras we need to guard $\mathcal{P}$. Hence, we first decompose $\mathcal{P}$ into pieces that are easy to guard, namely triangles. We do this by drawing *diagonals* between pairs

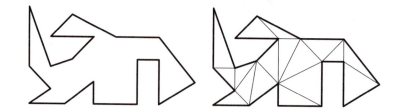

Figure 3.2
A simple polygon and a possible
triangulation of it

of vertices. A diagonal is an open line segment that connects two vertices of $\mathcal{P}$ and lies in the interior of $\mathcal{P}$. A decomposition of a polygon into triangles by a maximal set of non-intersecting diagonals is called a *triangulation* of the polygon—see Figure 3.2. (We require that the set of non-intersecting diagonals be maximal to ensure that no triangle has a polygon vertex in the interior of one of its edges. This could happen if the polygon has three consecutive collinear vertices.) Triangulations are usually not unique; the polygon in Figure 3.2, for example, can be triangulated in many different ways. We can guard $\mathcal{P}$ by placing a camera in every triangle of a triangulation $\mathcal{T}_{\mathcal{P}}$ of $\mathcal{P}$. But does a triangulation always exist? And how many triangles can there be in a triangulation? The following theorem answers these questions.

Theorem 3.1 *Every simple polygon admits a triangulation, and any triangulation of a simple polygon with n vertices consists of exactly $n-2$ triangles.*

Proof. We prove this theorem by induction on n. When $n = 3$ the polygon itself is a triangle and the theorem is trivially true. Let $n > 3$, and assume that the theorem is true for all $m < n$. Let $\mathcal{P}$ be a polygon with n vertices. We first prove the existence of a diagonal in $\mathcal{P}$. Let v be the leftmost vertex of $\mathcal{P}$. (In case of ties, we take the lowest leftmost vertex.) Let u and w be the two neighboring vertices of v on the boundary of $\mathcal{P}$. If the open segment $\overline{uw}$ lies in the interior of $\mathcal{P}$, we have found a diagonal. Otherwise, there are one or more vertices inside the triangle defined by u, v, and w, or on the diagonal $\overline{uw}$. Of those vertices, let v' be the one farthest from $\overline{uw}$. The segment connecting v' to v cannot intersect an edge of $\mathcal{P}$, because such an edge would have an endpoint inside the triangle that is farther from $\overline{uw}$, contradicting the definition of v'. Hence, $\overline{vv'}$ is a diagonal.

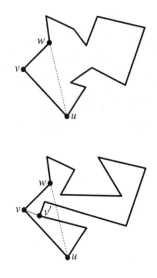

So a diagonal exists. Any diagonal cuts $\mathcal{P}$ into two simple subpolygons $\mathcal{P}_1$ and $\mathcal{P}_2$. Let m_1 be the number of vertices of $\mathcal{P}_1$ and m_2 the number of vertices of $\mathcal{P}_2$. Both m_1 and m_2 must be smaller than n, so by induction $\mathcal{P}_1$ and $\mathcal{P}_2$ can be triangulated. Hence, $\mathcal{P}$ can be triangulated as well.

It remains to prove that any triangulation of $\mathcal{P}$ consists of $n-2$ triangles. To this end, consider an arbitrary diagonal in some triangulation $\mathcal{T}_{\mathcal{P}}$. This diagonal cuts $\mathcal{P}$ into two subpolygons with m_1 and m_2 vertices, respectively. Every vertex of $\mathcal{P}$ occurs in exactly one of the two subpolygons, except for the vertices defining the diagonal, which occur in both subpolygons. Hence, $m_1 + m_2 = n + 2$. By induction, any triangulation of $\mathcal{P}_i$ consists of $m_i - 2$ triangles, which implies that $\mathcal{T}_{\mathcal{P}}$ consists of $(m_1 - 2) + (m_2 - 2) = n - 2$ triangles. $\square$

Theorem 3.1 implies that any simple polygon with n vertices can be guarded with $n-2$ cameras. But placing a camera inside every triangle seems overkill. A camera placed on a diagonal, for example, will guard two triangles, so by placing the cameras on well-chosen diagonals we might be able to reduce the number of cameras to roughly $n/2$. Placing cameras at vertices seems even better, because a vertex can be incident to many triangles, and a camera at that vertex guards all of them. This suggests the following approach.

Let $\mathcal{T}_{\mathcal{P}}$ be a triangulation of $\mathcal{P}$. Select a subset of the vertices of $\mathcal{P}$, such that any triangle in $\mathcal{T}_{\mathcal{P}}$ has at least one selected vertex, and place the cameras at the selected vertices. To find such a subset we assign each vertex of $\mathcal{P}$ a color: white, gray, or black. The coloring will be such that any two vertices connected by an edge or a diagonal have different colors. This is called a *3-coloring* of a triangulated polygon. In a 3-coloring of a triangulated polygon, every triangle has a white, a gray, and a black vertex. Hence, if we place cameras at all gray vertices, say, we have guarded the whole polygon. By choosing the smallest color class to place the cameras, we can guard $\mathcal{P}$ using at most $\lfloor n/3 \rfloor$ cameras.

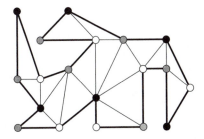

But does a 3-coloring always exist? The answer is yes. To see this, we look at what is called the *dual graph* of $\mathcal{T}_{\mathcal{P}}$. This graph $\mathcal{G}(\mathcal{T}_{\mathcal{P}})$ has a node for every triangle in $\mathcal{T}_{\mathcal{P}}$. We denote the triangle corresponding to a node ν by $t(\nu)$. There is an arc between two nodes ν and μ if $t(\nu)$ and $t(\mu)$ share

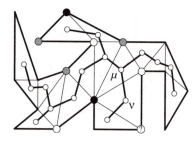

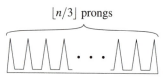

$\lfloor n/3 \rfloor$ prongs

a diagonal. The arcs in $G(\mathcal{T}_P)$ correspond to diagonals in $\mathcal{T}_P$. Because any diagonal cuts $\mathcal{P}$ into two, the removal of an edge from $G(\mathcal{T}_P)$ splits the graph into two. Hence, $G(\mathcal{T}_P)$ is a tree. (Notice that this is not true for a polygon with holes.) This means that we can find a 3-coloring using a simple graph traversal, such as depth first search. Next we describe how to do this. While we do the depth first search, we maintain the following invariant: all vertices of the already encountered triangles have been colored white, gray, or black, and no two connected vertices have received the same color. The invariant implies that we have computed a valid 3-coloring when all triangles have been encountered. The depth first search can be started from any node of $G(\mathcal{T}_P)$; the three vertices of the corresponding triangle are colored white, gray, and black. Now suppose that we reach a node v in G, coming from node μ. Hence, $t(v)$ and $t(\mu)$ share a diagonal. Since the vertices of $t(\mu)$ have already been colored, only one vertex of $t(v)$ remains to be colored. There is one color left for this vertex, namely the color that is not used for the vertices of the diagonal between $t(v)$ and $t(\mu)$. Because $G(\mathcal{T}_P)$ is a tree, the other nodes adjacent to v have not been visited yet, and we still have the freedom to give the vertex the remaining color.

We conclude that a triangulated simple polygon can always be 3-colored. As a result, any simple polygon can be guarded with $\lfloor n/3 \rfloor$ cameras. But perhaps we can do even better. After all, a camera placed at a vertex may guard more than just the incident triangles. Unfortunately, for any n there are simple polygons that require $\lfloor n/3 \rfloor$ cameras. An example is a comb-shaped polygon with a long horizontal base edge and $\lfloor n/3 \rfloor$ prongs made of two edges each. The prongs are connected by horizontal edges. The construction can be made such that there is no position in the polygon from which a camera can look into two prongs of the comb simultaneously. So we cannot hope for a strategy that always produces less than $\lfloor n/3 \rfloor$ cameras. In other words, the 3-coloring approach is optimal in the worst case.

We just proved the Art Gallery Theorem, a classical result from combinatorial geometry.

Theorem 3.2 (Art Gallery Theorem) *For a simple polygon with n vertices, $\lfloor n/3 \rfloor$ cameras are occasionally necessary and always sufficient to have every point in the polygon visible from at least one of the cameras.*

Now we know that $\lfloor n/3 \rfloor$ cameras are always sufficient. But we don't have an efficient algorithm to compute the camera positions yet. What we need is a fast algorithm for triangulating a simple polygon. The algorithm should deliver a suitable representation of the triangulation—a doubly-connected edge list, for instance—so that we can step in constant time from a triangle to its neighbors. Given such a representation, we can compute a set of at most $\lfloor n/3 \rfloor$ camera positions in linear time with the method described above: use depth first search on the dual graph to compute a 3-coloring and take the smallest color class to place the cameras. In the coming sections we describe how to compute a triangulation in $O(n \log n)$ time. Anticipating this, we already state the final result about guarding a polygon.

Theorem 3.3 *Let $\mathcal{P}$ be a simple polygon with n vertices. A set of $\lfloor n/3 \rfloor$ camera positions in $\mathcal{P}$ such that any point inside $\mathcal{P}$ is visible from at least one of the cameras can be computed in $O(n \log n)$ time.*

3.2 Partitioning a Polygon into Monotone Pieces

Let $\mathcal{P}$ be a simple polygon with n vertices. We saw in Theorem 3.1 that a triangulation of $\mathcal{P}$ always exists. The proof of that theorem is constructive and leads to a recursive triangulation algorithm: find a diagonal and triangulate the two resulting subpolygons recursively. To find the diagonal we take the leftmost vertex of $\mathcal{P}$ and try to connect its two neighbors u and w; if this fails we connect v to the vertex farthest from $\overline{uw}$ inside the triangle defined by u, v, and w. This way it takes linear time to find a diagonal. This diagonal may split $\mathcal{P}$ into a triangle and a polygon with $n-1$ vertices. Indeed, if we succeed to connect u and w this will always be the case. As a consequence, the triangulation algorithm will take quadratic time in the worst case. Can we do better? For some classes of polygons we surely can. Convex polygons, for instance, are easy: Pick one vertex of the polygon and draw diagonals from this vertex to all other vertices except its neighbors. This takes only linear time. So a possible approach to triangulate a non-convex polygon would be to first decompose $\mathcal{P}$ into convex pieces, and then triangulate the pieces. Unfortunately, it is as difficult to partition a polygon into convex pieces as it is to triangulate it. Therefore we shall decompose $\mathcal{P}$ into so-called monotone pieces, which turns out to be a lot easier.

A simple polygon is called *monotone with respect to a line ℓ* if for any line ℓ' perpendicular to ℓ the intersection of the polygon with ℓ' is connected. In other words, the intersection should be a line segment, a point, or empty. A polygon that is monotone with respect to the y-axis is called *y-monotone*. The following property is characteristic for y-monotone polygons: if we walk from a topmost to a bottommost vertex along the left (or the right) boundary chain, then we always move downwards or horizontally, never upwards.

Our strategy to triangulate the polygon $\mathcal{P}$ is to first partition $\mathcal{P}$ into y-monotone pieces, and then triangulate the pieces. We can partition a polygon into monotone pieces as follows. Imagine walking from the topmost vertex of $\mathcal{P}$ to the bottommost vertex on the left or right boundary chain. A vertex where the direction in which we walk switches from downward to upward or from upward to downward is called a *turn vertex*. To partition $\mathcal{P}$ into y-monotone pieces we should get rid of these turn vertices. This can be done by adding diagonals. If at a turn vertex v both incident edges go down and the interior of the polygon locally lies above v, then we must choose a diagonal that goes up from v. The diagonal splits the polygon into two. The vertex v will appear in both pieces. Moreover, in both pieces v has an edge going down (namely on original edge of $\mathcal{P}$) and an edge going up (the diagonal). Hence, v cannot be a turn vertex anymore in either of them. If both incident edges of a turn vertex go up

and the interior locally lies below it, we have to choose a diagonal that goes down. Apparently there are different types of turn vertices. Let's make this more precise.

If we want to define the different types of turn vertices carefully, we should pay special attention to vertices with equal y-coordinate. We do this by defining the notions of "below" and "above" as follows: a point p is below another point q if $p_y < q_y$ or $p_y = q_y$ and $p_x > q_x$, and p is above q if $p_y > q_y$ or $p_y = q_y$ and $p_x < q_x$. (You can imagine rotating the plane slightly in clockwise direction with respect to the coordinate system, such that no two points have the same y-coordinate; the above/below relation we just defined is the same as the above/below relation in this slightly rotated plane.)

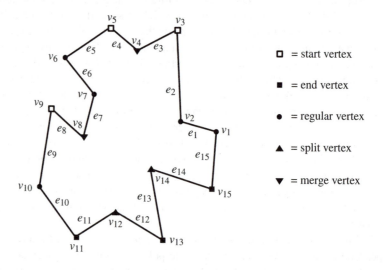

Figure 3.3
Five types of vertices

We distinguish five types of vertices in $\mathcal{P}$—see Figure 3.3. Four of these types are turn vertices: start vertices, split vertices, end vertices, and merge vertices. They are defined as follows. A vertex v is a *start vertex* if its two neighbors lie below it and the interior angle at v is less than π; if the interior angle is greater than π then v is a *split vertex*. (If both neighbors lie below v, then the interior angle cannot be exactly π.) A vertex is an *end vertex* if its two neighbors lie above it and the interior angle at v is less than π; if the interior angle is greater than π then v is a *merge vertex*. The vertices that are not turn vertices are *regular vertices*. Thus a regular vertex has one of its neighbors above it, and the other neighbor below it. These names have been chosen because the algorithm will use a downward plane sweep, maintaining the intersection of the sweep line with the polygon. When the sweep line reaches a split vertex, a component of the intersection splits, when it reaches a merge vertex, two components merge, and so on.

The split and merge vertices are sources of local non-monotonicity. The following, stronger statement is even true.

Lemma 3.4 *A polygon is y-monotone if it has no split vertices or merge vertices.*

Proof. Suppose $\mathcal{P}$ is not y-monotone. We have to prove that $\mathcal{P}$ contains a split or a merge vertex.

Since $\mathcal{P}$ is not monotone, there is a horizontal line ℓ that intersects $\mathcal{P}$ in more than one connected component. We can choose ℓ such that the leftmost component is a segment, not a single point. Let p be the left endpoint of this segment, and let q be the right endpoint. Starting at q, we follow the boundary of $\mathcal{P}$ such that $\mathcal{P}$ lies to the left of the boundary. (This means that we go up from q.) At some point, let's call it r, the boundary will intersect ℓ again. If $r \neq p$, as in Figure 3.4(a), then the highest vertex we encountered while going from q to r must be a split vertex, and we are done.

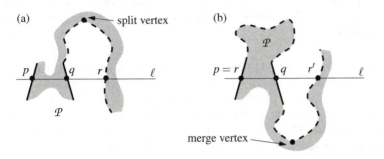

Figure 3.4
Two cases in the proof of Lemma 3.4

If $r = p$, as in Figure 3.4(b), we again follow the boundary of $\mathcal{P}$ starting at q, but this time in the other direction. As before, the boundary will intersect ℓ. Let r' be the point where this happens. We cannot have $r' = p$, because that would mean that the boundary of $\mathcal{P}$ intersects ℓ only twice, contradicting that ℓ intersects $\mathcal{P}$ in more than one component. So we have $r' \neq p$, implying that the lowest vertex we have encountered while going from q to r' must be a merge vertex. $\quad\square$

Lemma 3.4 implies that $\mathcal{P}$ has been partitioned into y-monotone pieces once we get rid of its split and merge vertices. We do this by adding a diagonal going upward from each split vertex and a diagonal going downward from each merge vertex. These diagonals should not intersect each other, of course. Once we have done this, $\mathcal{P}$ has been partitioned into y-monotone pieces.

Let's first see how we can add the diagonals for the split vertices. We use a plane sweep method for this. Let $v_1, v_2, \ldots, v_n$ be a counterclockwise enumeration of the vertices of $\mathcal{P}$. Let $e_1, \ldots, e_n$ be the set of edges of $\mathcal{P}$, where $e_i = \overline{v_i v_{i+1}}$ for $1 \leqslant i < n$ and $e_n = \overline{v_n v_1}$. The plane sweep algorithm moves an imaginary sweep line ℓ downward over the plane. The sweep line halts at certain event points. In our case these will be the vertices of $\mathcal{P}$; no new event points will be created during the sweep. The event points are stored in a event queue Q. The event queue is a priority queue, where the priority of a vertex is its y-coordinate. If two vertices have the same y-coordinate then the leftmost one has higher priority. This way the next event to be handled can be found in $O(\log n)$ time.

The goal of the sweep is to add diagonals from each split vertex to a vertex lying above it. Suppose that the sweep line reaches a split vertex v_i. To which vertex should we connect v_i? A good candidate is a vertex close to v_i, because we can probably connect v_i to this vertex without intersecting any edge of $\mathcal{P}$. Let's make this more precise. Let e_j be the edge immediately to the left of v_i on the sweep line, and let e_k be the edge immediately to the right of v_i on the sweep line. Then we can always connect v_i to the lowest vertex in between e_j and e_k, and above v_i. If there is no such vertex then we can connect v_i to the upper endpoint of e_j or to the upper endpoint of e_k. We call this vertex the *helper* of e_j and denote it by *helper*(e_j). Formally, *helper*(e_j) is defined as the lowest vertex above the sweep line such that the horizontal segment connecting the vertex to e_j lies inside $\mathcal{P}$. Note that *helper*(e_j) can be the upper endpoint of e_j itself.

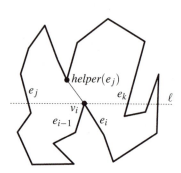

Now we know how to get rid of split vertices: connect them to the helper of the edge to their left. What about merge vertices? They seem more difficult to get rid of, because they need a diagonal to a vertex that is lower than they are. Since the part of $\mathcal{P}$ below the sweep line has not been explored yet, we cannot add such a diagonal when we encounter a merge vertex. Fortunately, this problem is easier than it seems at first sight. Suppose the sweep line reaches a merge vertex v_i. Let e_j and e_k be the edges immediately to the right and to the left of v_i on the sweep line, respectively. Observe that v_i becomes the new helper of e_j when we reach it. We would like to connect v_i to the highest vertex below the sweep line in between e_j and e_k. This is exactly the opposite of what we did for split vertices, which we connected to the lowest vertex above the sweep line in between e_j and e_k. This is not surprising: merge vertices are split vertices upside down. Of course we don't know the highest vertex below the sweep line when we reach v_i. But it is easy to find later on: when we reach a vertex v_m that replaces v_i as the helper of e_j, then this is the vertex we are looking for. So whenever we replace the helper of some edge, we check whether the old helper is a merge vertex and, if so, we add the diagonal between the old helper and the new one. This diagonal is always added when the new helper is a split vertex, to get rid of the split vertex. If the old helper was a merge vertex, we thus get rid of a split vertex and a merge vertex with the same diagonal. It can also happen that the helper of e_j is not replaced anymore below v_i. In this case we can connect v_i to the lower endpoint of e_j.

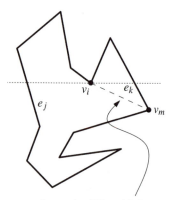

diagonal will be added
when the sweep line
reaches v_m

In the approach above, we need to find the edge to the left of each vertex. Therefore we store the edges of $\mathcal{P}$ intersecting the sweep line in the leaves of a dynamic binary search tree $\mathcal{T}$. The left-to-right order of the leaves of $\mathcal{T}$ corresponds to the left-to-right order of the edges. Because we are only interested in edges to the left of split and merge vertices we only need to store edges in $\mathcal{T}$ that have the interior of $\mathcal{P}$ to their right. With each edge in $\mathcal{T}$ we store its helper. The tree $\mathcal{T}$ and the helpers stored with the edges form the status of the sweep line algorithm. The status changes as the sweep line moves: edges start or stop intersecting the sweep line, and the helper of an edge may be replaced.

The algorithm partitions $\mathcal{P}$ into subpolygons that have to be processed further in a later stage. To have easy access to these subpolygons we shall store the subdivision induced by $\mathcal{P}$ and the added diagonals in a doubly-connected edge list $\mathcal{D}$. We assume that $\mathcal{P}$ is initially specified as a doubly-connected edge list; if $\mathcal{P}$ is given in another form—by a counterclockwise list of its vertices, for example—we first construct a doubly-connected edge list for $\mathcal{P}$. The diagonals computed for the split and merge vertices are added to the doubly-connected edge list. To access the doubly-connected edge list we use cross-pointers between the edges in the status structure and the corresponding edges in the doubly-connected edge list. Adding a diagonal can then be done in constant time with some simple pointer manipulations. The global algorithm is now as follows.

Algorithm MAKEMONOTONE($\mathcal{P}$)
Input. A simple polygon $\mathcal{P}$ stored in a doubly-connected edge list $\mathcal{D}$.
Output. A partitioning of $\mathcal{P}$ into monotone subpolygons, stored in $\mathcal{D}$.
1. Construct a priority queue Q on the vertices of $\mathcal{P}$, using their y-coordinates as priority. If two points have the same y-coordinate, the one with smaller x-coordinate has higher priority.
2. Initialize an empty binary search tree $\mathcal{T}$.
3. **while** Q is not empty
4.   **do** Remove the vertex v_i with the highest priority from Q.
5.   Call the appropriate procedure to handle the vertex, depending on its type.

We next describe more precisely how to handle the event points. You should first read these algorithms without thinking about degenerate cases, and check only later that they are also correct in degenerate cases. (To this end you should give an appropriate meaning to "directly left of" in line 1 of HANDLESPLIT-VERTEX and line 2 of HANDLEMERGEVERTEX.) There are always two things we must do when we handle a vertex. First, we must check whether we have to add a diagonal. This is always the case for a split vertex, and also when we replace the helper of an edge and the previous helper was a merge vertex. Second, we must update the information in the status structure $\mathcal{T}$. The precise algorithms for each type of event are given below. You can use the example figure in the margin on the next page to see what happens in each of the different cases.

HANDLESTARTVERTEX(v_i)
1. Insert e_i in $\mathcal{T}$ and set *helper*(e_i) to v_i.

At the start vertex v_5 in the example figure, for instance, we insert e_5 into the tree $\mathcal{T}$.

HANDLEENDVERTEX(v_i)
1. **if** *helper*(e_{i-1}) is a merge vertex
2.   **then** Insert the diagonal connecting v_i to *helper*(e_{i-1}) in $\mathcal{D}$.
3. Delete e_{i-1} from $\mathcal{T}$.

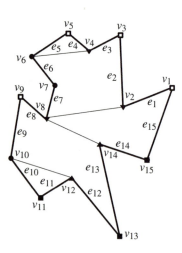

In the running example, when we reach end vertex v_{15}, the helper of the edge e_{14} is v_{14}. v_{14} is not a merge vertex, so we don't need to insert a diagonal.

HANDLESPLITVERTEX(v_i)
1.   Search in $\mathcal{T}$ to find the edge e_j directly left of v_i.
2.   Insert the diagonal connecting v_i to $helper(e_j)$ in $\mathcal{D}$.
3. $helper(e_j) \leftarrow v_i$
4.   Insert e_i in $\mathcal{T}$ and set $helper(e_i)$ to v_i.

For split vertex v_{14} in our example, e_9 is the edge to the left. Its helper is v_8, so we add a diagonal from v_{14} to v_8.

HANDLEMERGEVERTEX(v_i)
1.   **if** $helper(e_{i-1})$ is a merge vertex
2.   **then** Insert the diagonal connecting v_i to $helper(e_{i-1})$ in $\mathcal{D}$.
3.   Delete e_{i-1} from $\mathcal{T}$.
4.   Search in $\mathcal{T}$ to find the edge e_j directly left of v_i.
5.   **if** $helper(e_j)$ is a merge vertex
6.   **then** Insert the diagonal connecting v_i to $helper(e_j)$ in $\mathcal{D}$.
7. $helper(e_j) \leftarrow v_i$

For the merge vertex v_8 in our example, the helper v_2 of edge e_7 is a merge vertex, so we add a diagonal from v_8 to v_2.

The only routine that remains to be described is the one to handle a regular vertex. The actions we must take at a regular vertex depend on whether $\mathcal{P}$ lies locally to its left or to its right.

HANDLEREGULARVERTEX(v_i)
1.   **if** the interior of $\mathcal{P}$ lies to the right of v_i
2.   **then if** $helper(e_{i-1})$ is a merge vertex
3.   **then** Insert the diagonal connecting v_i to $helper(e_{i-1})$ in $\mathcal{D}$.
4.   Delete e_{i-1} from $\mathcal{T}$.
5.   Insert e_i in $\mathcal{T}$ and set $helper(e_i)$ to v_i.
6.   **else** Search in $\mathcal{T}$ to find the edge e_j directly left of v_i.
7.   **if** $helper(e_j)$ is a merge vertex
8.   **then** Insert the diagonal connecting v_i to $helper(e_j)$ in $\mathcal{D}$.
9. $helper(e_j) \leftarrow v_i$

For instance, at the regular vertex v_6 in our example, we add a diagonal from v_6 to v_4.

It remains to prove that MAKEMONOTONE correctly partitions $\mathcal{P}$ into monotone pieces.

Lemma 3.5 *Algorithm* MAKEMONOTONE *adds a set of non-intersecting diagonals that partitions $\mathcal{P}$ into monotone subpolygons.*

Proof. It is easy to see that the pieces into which $\mathcal{P}$ is partitioned contain no split or merge vertices. Hence, they are monotone by Lemma 3.4. It remains to prove that the added segments are valid diagonals (that is, that they don't intersect the edges of $\mathcal{P}$) and that they don't intersect each other. To this end we will show that when a segment is added, it intersects neither an edge of $\mathcal{P}$ nor any of the previously added segments. We shall prove this for the segment added in HANDLESPLITVERTEX; the proof for the segments added in HANDLEEND-VERTEX, HANDLEREGULARVERTEX, and HANDLEMERGEVERTEX is similar. We assume that no two vertices have the same y-coordinate; the extension to the general case is fairly straightforward.

Consider a segment $\overline{v_m v_i}$ that is added by HANDLESPLITVERTEX when v_i is reached. Let e_j be the edge to the left of v_i, and let e_k be the edge to the right of v_i. Thus $helper(e_j) = v_m$ when we reach v_i.

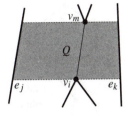

We first argue that $\overline{v_m v_i}$ does not intersect an edge of $\mathcal{P}$. To see this, consider the quadrilateral Q bounded by the horizontal lines through v_m and v_i, and by e_j and e_k. There are no vertices of $\mathcal{P}$ inside Q, otherwise v_m would not be the helper of e_j. Now suppose there would be an edge of $\mathcal{P}$ intersecting $\overline{v_m v_i}$. Since the edge cannot have an endpoint inside Q and polygon edges do not intersect each other, it would have to intersect the horizontal segment connecting v_m to e_j or the horizontal segment connecting v_i to e_j. Both are impossible, since for both v_m and v_i, the edge e_j lies immediately to the left. Hence, no edge of $\mathcal{P}$ can intersect $\overline{v_m v_i}$.

Now consider a previously added diagonal. Since there are no vertices of $\mathcal{P}$ inside Q, and any previously added diagonal must have both of it endpoints above v_i, it cannot intersect $\overline{v_m v_i}$. $\square$

We now analyze the running time of the algorithm. Constructing the priority queue Q takes linear time and initializing $\mathcal{T}$ takes constant time. To handle an event during the sweep, we perform one operation on Q, at most one query, one insertion, and one deletion on $\mathcal{T}$, and we insert at most two diagonals into $\mathcal{D}$. Priority queues and balanced search trees allow for queries and updates in $O(\log n)$ time, and the insertion of a diagonal into $\mathcal{D}$ takes $O(1)$ time. Hence, handling an event takes $O(\log n)$ time, and the total algorithm runs in $O(n \log n)$ time. The amount of storage used by the algorithm is clearly linear: every vertex is stored at most once in Q, and every edge is stored at most once in $\mathcal{T}$. Together with Lemma 3.5 this implies the following theorem.

Theorem 3.6 *A simple polygon with n vertices can be partitioned into y-monotone polygons in $O(n \log n)$ time with an algorithm that uses $O(n)$ storage.*

3.3 Triangulating a Monotone Polygon

We have just seen how to partition a simple polygon into y-monotone pieces in $O(n \log n)$ time. In itself this is not very interesting. But in this section we show that monotone polygons can be triangulated in linear time. Together these results imply that any simple polygon can be triangulated in $O(n \log n)$ time, a nice improvement over the quadratic time algorithm that we sketched at the beginning of the previous section.

Let $\mathcal{P}$ be a y-monotone polygon with n vertices. For the moment we assume that $\mathcal{P}$ is *strictly y-monotone*, that is, we assume that $\mathcal{P}$ is y-monotone and does not contain horizontal edges. Thus we always go down when we walk on the left or right boundary chain of $\mathcal{P}$ from the highest vertex of $\mathcal{P}$ to the lowest one. This is the property that makes triangulating a monotone polygon easy: we can work our way through $\mathcal{P}$ from top to bottom on both chains, adding diagonals whenever this is possible. Next we describe the details of this greedy triangulation algorithm.

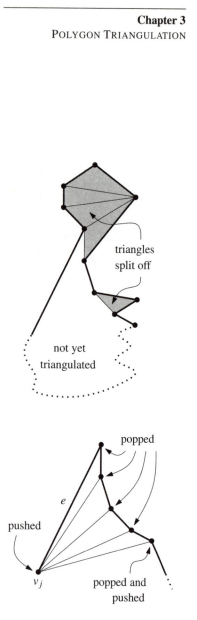

triangles split off

not yet triangulated

The algorithm handles the vertices in order of decreasing y-coordinate. If two vertices have the same y-coordinate, then the leftmost one is handled first. The algorithm requires a stack S as auxiliary data structure. Initially the stack is empty; later it contains the vertices of $\mathcal{P}$ that have been encountered, but may still need more diagonals. When we handle a vertex we add as many diagonals from this vertex to vertices on the stack as possible. These diagonals split off triangles from $\mathcal{P}$. The vertices that have been handled but not split off—the vertices on the stack—are on the boundary of the part of $\mathcal{P}$ that still needs to be triangulated. The lowest of these vertices, which is the one encountered last, is on top of the stack, the second lowest is second on the stack, and so on. The part of $\mathcal{P}$ that still needs to be triangulated, and lies above the last vertex that has been encountered so far, has a particular shape: it looks like a funnel turned upside down. One boundary of the funnel consists of a part of a single edge of $\mathcal{P}$, and the other boundary is a chain consisting of reflex vertices, that is, the interior angle at these vertices is at least 180°. Only the highest vertex, which is at the bottom of the stack, is convex. This property remains true after we have handled the next vertex. Hence, it is an invariant of the algorithm.

    Now, let's see which diagonals we can add when we handle the next vertex. We distinguish two cases: v_j, the next vertex to be handled, lies on the same chain as the reflex vertices on the stack, or it lies on the opposite chain. If v_j lies on the opposite chain, it must be the lower endpoint of the single edge e bounding the funnel. Due to the shape of the funnel, we can add diagonals from v_j to all vertices currently on the stack, except for the last one; the last vertex on the stack is the upper vertex of e, so it is already connected to v_j. All these vertices are popped from the stack. The untriangulated part of the polygon above v_j is bounded by the diagonal that connects v_j to the vertex previously on top of the stack and the edge of $\mathcal{P}$ extending downward from this vertex, so it looks like a funnel and the invariant is preserved. This vertex and v_j remain part of the not yet triangulated polygon, so they are pushed onto the stack.

    The other case is when v_j is on the same chain as the reflex vertices on the stack. This time we may not be able to draw diagonals from v_j to all vertices on the stack. Nevertheless, the ones to which we can connect v_j are all consecutive and they are on top of the stack, so we can proceed as follows. First, pop one vertex from the stack; this vertex is already connected to v_j by an edge of $\mathcal{P}$. Next, pop vertices from the stack and connect them to v_j until we encounter one where this is not possible. Checking whether a diagonal can be drawn from v_j to a vertex v_k on the stack can be done by looking at v_j, v_k, and the previous

popped

e

pushed

v_j

popped and pushed

vertex that was popped. When we find a vertex to which we cannot connect v_j, we push the last vertex that has been popped back onto the stack. This is either the last vertex to which a diagonal was added or, if no diagonals have been added, it is the neighbor of v_j on the boundary of $\mathcal{P}$—see Figure 3.5. After this

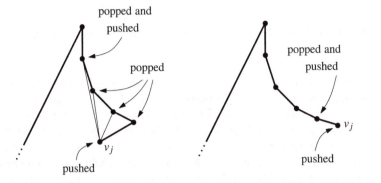

has been done we push v_j onto the stack. In both cases the invariant is restored: one side of the funnel is bounded by a part of a single edge, and the other side is bounded by a chain of reflex vertices. We get the following algorithm. (The algorithm is actually similar to the convex hull algorithm of Chapter 1.)

Algorithm TRIANGULATEMONOTONEPOLYGON($\mathcal{P}$)

Input. A strictly y-monotone polygon $\mathcal{P}$ stored in a doubly-connected edge list $\mathcal{D}$.

Output. A triangulation of $\mathcal{P}$ stored in the doubly-connected edge list $\mathcal{D}$.

1. Merge the vertices on the left chain and the vertices on the right chain of $\mathcal{P}$ into one sequence, sorted on decreasing y-coordinate. If two vertices have the same y-coordinate, then the leftmost one comes first. Let $u_1, \ldots, u_n$ denote the sorted sequence.
2. Push u_1 and u_2 onto the stack S.
3. **for** $j \leftarrow 3$ **to** $n - 1$
4.   **do if** u_j and the vertex on top of S are on different chains
5.   **then** Pop all vertices from S.
6.   Insert into $\mathcal{D}$ a diagonal from u_j to each popped vertex, except the last one.
7.   Push u_{j-1} and u_j onto S.
8.   **else** Pop one vertex from S.
9.   Pop the other vertices from S as long as the diagonals from u_j to them are inside $\mathcal{P}$. Insert these diagonals into $\mathcal{D}$. Push the last vertex that has been popped back onto S.
10.   Push u_j onto S.
11. Add diagonals from u_n to all stack vertices except the first and the last one.

How much time does the algorithm take? Step 1 takes linear time and Step 2 takes constant time. The **for**-loop is executed $n - 3$ times, and one execution may take linear time. But at every execution of the **for**-loop at most two vertices are pushed. Hence, the total number of pushes, including the two in Step 2, is bounded by $2n - 4$. Because the number of pops cannot exceed the number of pushes, the total time for all executions of the **for**-loop is $O(n)$. The last step of the algorithm also takes at most linear time, so the total algorithm runs in $O(n)$ time.

Theorem 3.7 *A strictly y-monotone polygon with n vertices can be triangulated in linear time.*

We wanted a triangulation algorithm for monotone polygons as a subroutine for triangulating arbitrary simple polygons. The idea was to first decompose a polygon into monotone pieces and then to triangulate these pieces. It seems that we have all the ingredients we need. There is one problem, however: in this section we have assumed that the input is a *strictly* y-monotone polygon, whereas the algorithm of the previous section may produce monotone pieces with horizontal edges. Recall that in the previous section we treated vertices with the same y-coordinates from left to right. This had the same effect as a slight rotation of the plane in clockwise direction such that no two vertices are on a horizontal line. It follows that the monotone subpolygons produced by the algorithm of the previous section are strictly monotone in this slightly rotated plane. Hence, the triangulation algorithm of the current section operates correctly if we treat vertices with the same y-coordinate from left to right (which corresponds to working in the rotated plane). So we can combine the two algorithms to obtain a triangulation algorithm that works for any simple polygon. We get the following result.

Theorem 3.8 *A simple polygon with n vertices can be triangulated in $O(n \log n)$ time with an algorithm that uses $O(n)$ storage.*

We have seen how to triangulate simple polygons. But what about polygons with holes, can they also be triangulated easily? The answer is yes. In fact, the algorithm we have seen also works for polygons with holes: nowhere in the algorithm for splitting a polygon into monotone pieces did we use the fact that the polygon was simple. It even works in a more general setting: Suppose we have a planar subdivision S and we want to triangulate that subdivision. More precisely, if B is a bounding box containing all edges of S in its interior, we want to find a maximal set of non-intersecting diagonals—line segments connecting vertices of S or B that do not intersect the edges of S—that partitions B into triangles. Figure 3.6 shows a triangulated subdivision. The edges of the subdivisions and of the bounding box are shown bold. To compute such a triangulation we can use the algorithm of this chapter: first split the subdivision into monotone pieces, and then triangulate the pieces. This leads to the following theorem.

Theorem 3.9 *A planar subdivision with n vertices in total can be triangulated in $O(n \log n)$ time with an algorithm that uses $O(n)$ storage.*

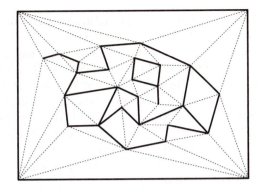

Figure 3.6
A triangulated subdivision

3.4 Notes and Comments

The Art Gallery Problem was posed in 1973 by Victor Klee in a conversation with Vasek Chvátal. In 1975 Chvátal [105] gave the first proof that $\lfloor n/3 \rfloor$ cameras are always sufficient and sometimes necessary; a result that became known as the *Art Gallery Theorem* or the *Watchman Theorem*. Chvátal's proof is quite complicated. The much simpler proof presented in this chapter was discovered by Fisk [147]. His proof is based on the *Two Ears Theorem* by Meisters [243], from which the 3-colorability of the graph that is a triangulation of a simple polygon follows easily. The algorithmic problem of finding the minimum number of guards for a given simple polygon was shown to be NP-hard by Aggarwal [4] and Lee and Lin [211]. The book by O'Rourke [262] and the overview by Shermer [315] contain an extensive treatment of the Art Gallery Problem and numerous variations.

A decomposition of a polygon, or any other region, into simple pieces is useful in many problems. Often the simple pieces are triangles, in which case we call the decomposition a triangulation, but sometimes other shapes such as quadrilaterals or trapezoids are used—see also Chapters 6, 9, and 14. We only discuss the results on triangulating polygons here. The linear time algorithm to triangulate a monotone polygon described in this chapter was given by Garey et al. [156], and the plane sweep algorithm to partition a polygon into monotone pieces is due to Lee and Preparata [215]. Avis and Toussaint [20] and Chazelle [64] described other algorithms for triangulating a simple polygon in $O(n \log n)$ time.

For a long time one of the main open problems in computational geometry was whether simple polygons can be triangulated in $o(n \log n)$ time. (For triangulating subdivisions with holes there is an $\Omega(n \log n)$ lower bound.) In this chapter we have seen that this is indeed the case for monotone polygons. Linear-time triangulation algorithms were also found for other special classes of polygons [87, 88, 141, 153, 181] but the problem for general simple polygons remained open for a number of years. In 1988 Tarjan and Van Wyk [321] broke the $O(n \log n)$ barrier by presenting an $O(n \log \log n)$ algorithm. Their algorithm was later simplified by Kirkpatrick et al. [201]. Randomization—an approach used in Chapters 4, 6, and 9—proved to be a good tool in developing even faster algorithms: Clarkson et al. [112], Devillers [118], and Seidel [308] presented algorithms with $O(n \log^* n)$ running time, where $\log^* n$ is the iterated logarithm of n (being the number of times you can take the logarithm before the result is smaller than 1). These algorithms are not only slightly faster than the $O(n \log \log n)$ algorithm, but also simpler. Seidel's algorithm is closely related to the algorithm for constructing a trapezoidal decomposition of a planar subdivision described in Chapter 6. However, the question whether a simple polygon can be triangulated in linear time was still open. In 1990 this problem was finally settled by Chazelle [71, 73], who gave a (quite complicated) deterministic linear time algorithm.

The 3-dimensional equivalent to the polygon triangulation problem is this: decompose a given polytope into non-overlapping tetrahedra, where the vertices of the tetrahedra must be vertices of the original polytope. Such a decomposition is called a *tetrahedralization* of the polytope. This problem is much more difficult than the two-dimensional version. In fact, it is not always possible to decompose a polytope into tetrahedra without using additional vertices. Chazelle [65] has shown that for a simple polytope with n vertices, $\Theta(n^2)$ additional vertices may be needed and are always sufficient to obtain a decomposition into tetrahedra. This bound was refined by Chazelle and Palios [89] to $\Theta(n + r^2)$, where r is the number of reflex edges of the polytope. The algorithm to compute the decomposition runs in $O(nr + r^2 \log r)$ time. Deciding whether a given simple polytope can be tetrahedralized without additional vertices is NP-complete [292].

3.5 Exercises

3.1 Prove that any polygon admits a triangulation, even if it has holes. Can you say anything about the number of triangles in the triangulation?

3.2 A *rectilinear polygon* is a simple polygon of which all edges are horizontal or vertical. Let P be a rectilinear polygon with n vertices. Give an example to show that $\lfloor n/4 \rfloor$ cameras are sometimes necessary to guard it.

3.3 Prove or disprove: The dual graph of the triangulation of a monotone polygon is always a chain, that is, any node in this graph has degree at most two.

3.4 Suppose that a simple polygon P with n vertices is given, together with a set of diagonals that partitions P into convex quadrilaterals. How many cameras are sufficient to guard P? Why doesn't this contradict the Art Gallery Theorem?

3.5 Give the pseudo-code of the algorithm to compute a 3-coloring of a triangulated simple polygon. The algorithm should run in linear time.

3.6 Give an algorithm that computes in $O(n \log n)$ time a diagonal that splits a simple polygon with n vertices into two simple polygons each with at most $\lfloor 3n/4 \rfloor$ vertices. *Hint:* Use the dual graph of a triangulation.

3.7 Can the algorithm of this chapter also be used to triangulate a set of n points? If so, explain how to do this efficiently.

3.8 Give an efficient algorithm to determine whether a polygon P with n vertices is monotone with respect to some line, not necessarily a horizontal or vertical one.

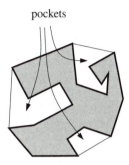
pockets

3.9 The *pockets* of a simple polygon are the areas outside the polygon, but inside its convex hull. Let P_1 be a simple polygon with m vertices, and assume that a triangulation of P_1 as well as its pockets is given. Let P_2 be a convex polygon with n vertices. Show that the intersection $P_1 \cap P_2$ can be computed in $O(m + n)$ time.

3.10 The *stabbing number* of a triangulated simple polygon P is the maximum number of diagonals intersected by any line segment interior to P. Give an algorithm that computes a triangulation of a convex polygon that has stabbing number $O(\log n)$.

3.11 Given a simple polygon P with n vertices and a point p inside it, show how to compute the region inside P that is visible from p.

4 Linear Programming

Manufacturing with Molds

Most objects we see around us today—from car bodies to plastic cups and cutlery—are made using some form of automated manufacturing. Computers play an important role in this process, both in the design phase and in the construction phase; CAD/CAM facilities are a vital part of any modern factory. The construction process used to manufacture a specific object depends on factors such as the material the object should be made of, the shape of the object, and whether the object will be mass produced. In this chapter we study some geometric aspects of manufacturing with molds, a commonly used process for plastic or metal objects. For metal objects this process is often referred to as *casting*.

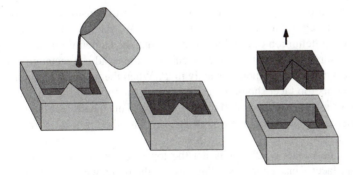

Figure 4.1
The casting process

Figure 4.1 illustrates the casting process: liquid metal is poured into a mold, it solidifies, and then the object is removed from the mold. The last step is not always as easy as it seems; the object could be stuck in the mold, so that it cannot be removed without breaking the mold. Sometimes we can get around this problem by using a different mold. There are also objects, however, for which no good mold exists; a sphere is an example. This is the problem we shall study in this chapter: given an object, is there a mold for it from which it can be removed?

We shall confine ourselves to the following situation. First of all, we assume that the object to be constructed is polyhedral. Secondly, we only consider molds of one piece, not molds consisting of two or more pieces. (Using molds consisting of two pieces, it is possible to manufacture objects such as spheres, which cannot be manufactured using a mold of a single piece.) Finally, we only allow the object to be removed from the mold by a single translation. This means that we will not be able to remove a screw from its mold. Fortunately, translational motions suffice for many objects.

4.1 The Geometry of Casting

If we want to determine whether an object can be manufactured by casting, we have to find a suitable mold for it. The shape of the cavity in the mold is determined by the shape of the object, but different orientations of the object give rise to different molds. Choosing the orientation can be crucial: some orientations may give rise to molds from which the object cannot be removed, while other orientations allow removal of the object. One obvious restriction on the orientation is that the object must have a horizontal *top facet*. This facet will be the only one not in contact with the mold. Hence, there are as many potential orientations—or, equivalently, possible molds—as the object has facets. We call an object *castable* if it can be removed from its mold for at least one of these orientations. In the following we shall concentrate on determining whether an object is removable by a translation from a specific given mold. To decide on the castability of the object we then simply try every potential orientation.

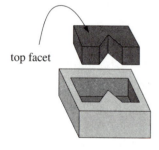

top facet

Let $\mathcal{P}$ be a 3-dimensional polyhedron—that is, a 3-dimensional solid bounded by planar facets—with a designated top facet. (We shall not try to give a precise, formal definition of a polyhedron. Giving such a definition is tricky and not necessary in this context.) We assume that the mold is a rectangular block with a cavity that corresponds exactly to $\mathcal{P}$. When the polyhedron is placed in the mold, its top facet should be coplanar with the topmost facet of the mold, which we assume to be parallel to the xy-plane. This means that the mold has no unnecessary parts sticking out on the top that might prevent $\mathcal{P}$ from being removed.

We call a facet of $\mathcal{P}$ that is not the top facet an *ordinary facet*. Every ordinary facet f has a corresponding facet in the mold, which we denote by $\hat{f}$.

We want to decide whether $\mathcal{P}$ can be removed from its mold by a single translation. In other words, we want to decide whether a direction $\vec{d}$ exists such that $\mathcal{P}$ can be translated to infinity in direction $\vec{d}$ without intersecting the interior of the mold during the translation. Note that we allow $\mathcal{P}$ to slide along the mold. Because the facet of $\mathcal{P}$ not touching the mold is its top facet, the removal direction has to be upward, that is, it must have a positive z-component. This is only a necessary condition on the removal direction; we need more constraints to be sure that a direction is valid.

Let f be an ordinary facet of $\mathcal{P}$. This facet must move away from, or slide along, its corresponding facet $\hat{f}$ of the mold. To make this constraint precise, we need to define the angle of two vectors in 3-space. We do this as follows. Take the plane spanned by the vectors (we assume both vectors are rooted at the origin); the angle of the vectors is the smaller of the two angles measured in this plane. Now $\hat{f}$ blocks any translation in a direction making an angle of less than 90° with $\vec{\eta}(f)$, the outward normal of f. So a necessary condition on $\vec{d}$ is that it makes an angle of at least 90° with the outward normal of every ordinary facet of $\mathcal{P}$. The next lemma shows that this condition is also sufficient.

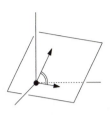

Lemma 4.1 *The polyhedron $\mathcal{P}$ can be removed from its mold by a translation in direction $\vec{d}$ if and only if $\vec{d}$ makes an angle of at least 90° with the outward normal of all ordinary facets of $\mathcal{P}$.*

Proof. The "only if" part is easy: if $\vec{d}$ would make an angle less than 90° with some outward normal $\vec{\eta}(f)$, then any point q in the interior of f collides with the mold when translated in direction $\vec{d}$.

To prove the "if" part, suppose that at some moment $\mathcal{P}$ collides with the mold when translated in direction $\vec{d}$. We have to show that there must be an outward normal making an angle of less than 90° with $\vec{d}$. Let p be a point of $\mathcal{P}$ that collides with a facet $\hat{f}$ of the mold. This means that p is about to move into the interior of the mold, so $\vec{\eta}(\hat{f})$, the outward normal of $\hat{f}$, must make an angle greater than 90° with $\vec{d}$. But then $\vec{d}$ makes an angle less than 90° with the outward normal of the ordinary facet f of $\mathcal{P}$ that corresponds to $\hat{f}$.   ▱

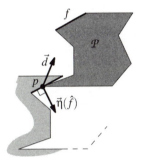

Lemma 4.1 has an interesting consequence: if $\mathcal{P}$ can be removed by a sequence of small translations, then it can be removed by a single translation. So allowing for more than one translation does not help in removing the object from its mold.

We are left with the task of finding a direction $\vec{d}$ that makes an angle of at least 90° with the outward normal of each ordinary facet of $\mathcal{P}$. A direction in 3-dimensional space can be represented by a vector rooted at the origin. We already know that we can restrict our attention to directions with a positive z-component. We can represent all such directions as points in the plane $z = 1$, where the point $(x, y, 1)$ represents the direction of the vector $(x, y, 1)$. This way every point in the plane $z = 1$ represents a unique direction, and every direction with a positive z-value is represented by a unique point in that plane.

Lemma 4.1 gives necessary and sufficient conditions on the removal direction $\vec{d}$. How do these conditions translate into our plane of directions? Let $\vec{\eta} = (\vec{\eta}_x, \vec{\eta}_y, \vec{\eta}_z)$ be the outward normal of an ordinary facet. The direction $\vec{d} = (d_x, d_y, 1)$ makes an angle at least 90° with $\vec{\eta}$ if and only if the dot product of $\vec{d}$ and $\vec{\eta}$ is non-positive. Hence, an ordinary facet induces a constraint of the form

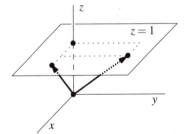

$$\vec{\eta}_x d_x + \vec{\eta}_y d_y + \vec{\eta}_z \leqslant 0.$$

This inequality describes a half-plane on the plane $z = 1$, that is, the area left or the area right of a line on the plane. (This last statement is not true for horizontal facets, which have $\vec{\eta}_x = \vec{\eta}_y = 0$. In this case the constraint is either impossible to satisfy or always satisfied, which is easy to test.) Hence, every non-horizontal facet of $\mathcal{P}$ defines a closed half-plane on the plane $z = 1$, and any point in the common intersection of these half-planes corresponds to a direction in which $\mathcal{P}$ can be removed. The common intersection of these half-planes may be empty; in this case $\mathcal{P}$ cannot be removed from the given mold.

We have transformed our manufacturing problem to a purely geometric problem in the plane: given a set of half-planes, find a point in their common intersection or decide that the common intersection is empty. If the polyhedron to be manufactured has n facets, then the planar problem has at most $n - 1$ half-planes (the top facet does not induce a half-plane). In the next sections we will see that the planar problem just stated can be solved in expected linear time—see Section 4.4, where also the meaning of "expected" is explained.

Recall that the geometric problem corresponds to testing whether $\mathcal{P}$ can be removed from a given mold. If this is impossible, there can still be other molds, corresponding to different choices of the top facet, from which $\mathcal{P}$ is removable. In order to test whether $\mathcal{P}$ is castable, we try all its facets as top facets. This leads to the following result.

Theorem 4.2 *Let $\mathcal{P}$ be a polyhedron with n facets. In $O(n^2)$ expected time and using $O(n)$ storage it can be decided whether $\mathcal{P}$ is castable. Moreover, if $\mathcal{P}$ is castable, a mold and a valid direction for removing $\mathcal{P}$ from it can be computed in the same amount of time.*

4.2 Half-Plane Intersection

Let $H = \{h_1, h_2, \ldots, h_n\}$ be a set of linear constraints in two variables, that is, constraints of the form

$$a_i x + b_i y \leqslant c_i,$$

where a_i, b_i, and c_i are constants such that at least one of a_i and b_i is non-zero. Geometrically, we can interpret such a constraint as a closed half-plane in $\mathbb{R}^2$, bounded by the line $a_i x + b_i y = c_i$. The problem we consider in this section is to find the set of all points $(x, y) \in \mathbb{R}^2$ that satisfy all n constraints at the same time. In other words, we want to find all the points lying in the common intersection of the half-planes in H. (In the previous section we reduced the casting problem to finding *some* point in the intersection of a set of half-planes. The problem we study now is more general.)

The shape of the intersection of a set of half-planes is easy to determine: a half-plane is convex, and the intersection of convex sets is again a convex set, so the intersection of a set of half-planes is a convex region in the plane. Every point on the intersection boundary must lie on the bounding line of some half-plane. Hence, the boundary of the region consists of edges contained in these bounding lines. Since the intersection is convex, every bounding line can contribute at most one edge. It follows that the intersection of n half-planes is a convex polygonal region bounded by at most n edges. Figure 4.2 shows a few examples of intersections of half-planes. To which side of its bounding line a half-plane lies is indicated by dark shading in the figure; the common intersection is shaded lightly. As you can see in Figures 4.2 (ii) and (iii), the

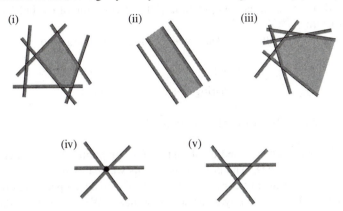

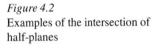

Figure 4.2
Examples of the intersection of
half-planes

intersection does not have to be bounded. The intersection can also degenerate to a line segment or a point, as in (iv), or it can be empty, as in (v).

We give a rather straightforward divide-and-conquer algorithm to compute the intersection of a set of n half-planes. It is based on a routine INTERSECTCON-VEXREGIONS to compute the intersection of two convex polygonal regions. We first give the overall algorithm.

Algorithm INTERSECTHALFPLANES(H)
Input. A set H of n half-planes in the plane.
Output. The convex polygonal region $C := \bigcap_{h \in H} h$.
1. **if** card(H) = 1
2.   **then** $C \leftarrow$ the unique half-plane $h \in H$
3.   **else** Split H into sets H_1 and H_2 of size $\lceil n/2 \rceil$ and $\lfloor n/2 \rfloor$.
4. $C_1 \leftarrow$ INTERSECTHALFPLANES(H_1)
5. $C_2 \leftarrow$ INTERSECTHALFPLANES(H_2)
6. $C \leftarrow$ INTERSECTCONVEXREGIONS(C_1, C_2)

What remains is to describe the procedure INTERSECTCONVEXREGIONS. But wait—didn't we see this problem before, in Chapter 2? Indeed, Corollary 2.7 states that we can compute the intersection of two polygons in $O(n \log n + k \log n)$ time, where n is the total number of vertices in the two polygons. We must be a bit careful in applying this result to our problem, because the regions we have can be unbounded, or degenerate to a segment or a point. Hence, the regions are not necessarily polygons. But it is not difficult to modify the algorithm from Chapter 2 so that it still works.

Let's analyze this approach. Assume we have already computed the two re-gions C_1 and C_2 by recursion. Since they are both defined by at most $n/2 + 1$

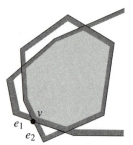

half-planes, they both have at most $n/2+1$ edges. The algorithm from Chapter 2 computes their overlay in time $O((n+k)\log n)$, where k is the number of intersection points between edges of C_1 and edges of C_2. What is k? Look at an intersection point v between an edge e_1 of C_1 and an edge e_2 of C_2. No matter how e_1 and e_2 intersect, v must be a vertex of $C_1 \cap C_2$. But $C_1 \cap C_2$ is the intersection of n half-planes, and therefore has at most n edges and vertices. It follows that $k \leqslant n$, so the computation of the intersection of C_1 and C_2 takes $O(n\log n)$ time.

This gives the following recurrence for the total running time:

$$T(n) = \begin{cases} O(1), & \text{if } n = 1, \\ O(n\log n) + 2T(n/2), & \text{if } n > 1. \end{cases}$$

This recurrence solves to $T(n) = O(n\log^2 n)$.

To obtain this result we used a subroutine for computing the intersection of two arbitrary polygons. The polygonal regions we deal with in INTERSECT-HALFPLANES are always convex. Can we use this to develop a more efficient algorithm? The answer is yes, as we show next. We will assume that the regions we want to intersect are 2-dimensional; the case where one or both of them is a segment or a point is easier and left as an exercise.

First, let's specify more precisely how we represent a convex polygonal region C. We will store the left and the right boundary of C separately, as sorted lists of half-planes. The lists are sorted in the order in which the bounding lines of the half-planes occur when the (left or right) boundary is traversed from top to bottom. We denote the left boundary list by $\mathcal{L}_{\text{left}}(C)$, and the right boundary list by $\mathcal{L}_{\text{right}}(C)$. Vertices are not stored explicitly; they can be computed by intersecting consecutive bounding lines.

To simplify the description of the algorithm, we shall assume that there are no horizontal edges. (To adapt the algorithm to deal with horizontal edges, one can define such edges to belong to the left boundary if they bound C from above, and to the right boundary if they bound C from below. With this convention only a few adaptations are needed to the algorithm stated below.)

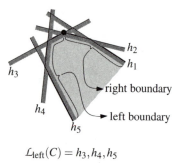

$$\mathcal{L}_{\text{left}}(C) = h_3, h_4, h_5$$
$$\mathcal{L}_{\text{right}}(C) = h_2, h_1$$

The new algorithm is a plane sweep algorithm, like the one in Chapter 2: we move a sweep line downward over the plane, and we maintain the edges of C_1 and C_2 intersecting the sweep line. Since C_1 and C_2 are convex, there are at most four such edges. Hence, there is no need to store these edges in a complicated data structure; instead we simply have pointers *left_edge_C1*, *right_edge_C1*, *left_edge_C2*, and *right_edge_C2* to them. If the sweep line does not intersect the right or left boundary of a region, then the corresponding pointer is nil. Figure 4.3 illustrates the definitions.

How are these pointers initialized? Let y_1 be the y-coordinate of the topmost vertex of C_1; if C_1 has an unbounded edge extending upward to infinity then we define $y_1 = \infty$. Define y_2 similarly for C_2, and let $y_{\text{start}} = \min(y_1, y_2)$. To compute the intersection of C_1 and C_2 we can restrict our attention to

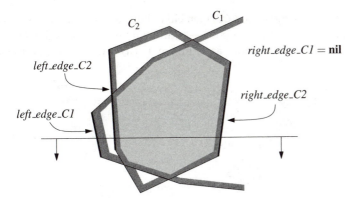

$right\_edge\_C1 = \textbf{nil}$

Figure 4.3
The edges maintained by the sweep
line algorithm

the part of the plane with y-coordinate less than or equal to y_{start}. Hence, we let the sweep line start at y_{start}, and we initialize the edges *left_edge_C1*, *right_edge_C1*, *left_edge_C2*, and *right_edge_C2* as the ones intersecting the line $y = y_{start}$.

In a plane sweep algorithm one normally also needs a queue to store the events. In our case the events are the points where edges of C_1 or of C_2 start or stop to intersect the sweep line. This implies that the next event point, which determines the next edge to be handled, is the highest of the lower endpoints of the edges intersecting the sweep line. (Endpoints with the same y-coordinate are handled from left to right. If two endpoints coincide then the leftmost edge is treated first.) Hence, we don't need an event queue; the next event can be found in constant time using the pointers *left_edge_C1*, *right_edge_C1*, *left_edge_C2*, and *right_edge_C2*.

At each event point some new edge e appears on the boundary. To handle the edge e we first check whether e belongs to C_1 or to C_2, and whether it is on the left or the right boundary, and then call the appropriate procedure. We shall only describe the procedure that is called when e is on the left boundary of C_1. The other procedures are similar.

Let p be the upper endpoint of e. The procedure that handles e will discover three possible edges that C might have: the edge with p as upper endpoint, the edge with $e \cap left\_edge\_C2$ as upper endpoint, and the edge with $e \cap right\_edge\_C2$ as upper endpoint.

The intersection C has an edge with p as its upper endpoint when p lies inside C_2. We can test this by checking whether p lies in between *left_edge_C2* and *right_edge_C2*. If this is the case, then e contributes an edge to C, which starts at p. We then add the half-plane whose bounding line contains e to the list $\mathcal{L}_{left}(C)$.

Now suppose that e intersects *left_edge_C2*. The intersection point is a vertex of C, and the edge of C starting at that vertex is either a part of e or it is a part of *left_edge_C2*. We can decide between these possibilities in constant time. After we decided whether e or *left_edge_C2* contributes the edge to C, we add the appropriate half-plane to $\mathcal{L}_{left}(C)$.

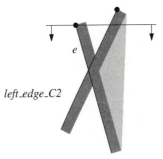

If e intersects $right\_edge\_C2$, then either both edges contribute an edge to C starting at the intersection point—as in Figure 4.4(i)—or both edges contribute an edge ending there—as in Figure 4.4(ii). Again, we can decide in

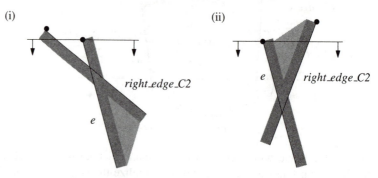

(i)   (ii)

$right\_edge\_C2$

e $right\_edge\_C2$

e

Figure 4.4
The two possibilities when e intersects
$right\_edge\_C2$

constant time between these two possibilities. If both edges contribute an edge starting at the intersection point, then we have to add the half-plane defining e to $\mathcal{L}_{\text{left}}(C)$ and the half-plane defining $right\_edge\_C2$ to $\mathcal{L}_{\text{right}}(C)$. If they contribute an edge ending at the intersection point we do nothing; these edges have already been discovered in some other way.

We now know which half-planes we may have to add to the lists, and when. Notice that we may have to add two half-planes to $\mathcal{L}_{\text{left}}(C)$: the half-plane bounding e and the half-plane bounding $left\_edge\_C2$. In which order should we add them? We add $left\_edge\_C2$ only if it defines an edge of C starting at the intersection point of $left\_edge\_C2$ and e. If we also decide to add the half-plane of e, it must be because e defines an edge of C starting at its upper endpoint. Hence, we should add the half-plane bounding e first.

We conclude that it takes constant time to handle an edge, so the intersection of two convex polygons can be computed in time $O(n)$. To show that the algorithm is correct, we have to prove that it adds the half-planes defining the edges of C in the right order. Consider an edge of C, and let p be its upper endpoint. Then p is either an upper endpoint of an edge in C_1 or C_2, or its is the intersection of two edges e and e' of C_1 and C_2, respectively. In the former case we discover the edge of C when p is reached, and in the latter case when the lower of the upper endpoints of e and e' is reached. Hence, all half-planes defining the edges of C are added. It is not difficult to prove that they are added in the correct order.

We get the following result:

Theorem 4.3 *The intersection of two convex polygonal regions in the plane can be computed in $O(n)$ time.*

This theorem shows that we can do the merge step in INTERSECTHALFPLANES in linear time. Hence, the recurrence for the running time of the algorithm becomes

$$T(n) = \begin{cases} O(1), & \text{if } n = 1, \\ O(n) + 2T(n/2), & \text{if } n > 1, \end{cases}$$

leading to the following result:

Corollary 4.4 *The common intersection of a set of n half-planes in the plane can be computed in $O(n \log n)$ time and linear storage.*

The problem of computing the intersection of half-planes is intimately related to the computation of convex hulls, and an alternative algorithm can be given that is almost identical to algorithm CONVEXHULL from Chapter 1. The relationship between convex hulls and intersections of half-planes is discussed in detail in Sections 8.2 and 11.4. Those sections are independent of the rest of their chapters, so if you are curious you can already have a look.

4.3 Incremental Linear Programming

In the previous section we showed how to compute the intersection of a set of n half-planes. In other words, we computed all solutions to a set of n linear constraints. The running time of our algorithm was $O(n \log n)$. One can prove that this is optimal: as for the sorting problem, any algorithm that solves the half-plane intersection problem must take $\Omega(n \log n)$ time in the worst case. In our application to the casting problem, however, we don't need to know *all* solutions to the set of linear constraints; just one solution will do fine. It turns out that this allows for a faster algorithm.

Finding a solution to a set of linear constraints is closely related to a well-known problem in operations research, called *linear optimization* or *linear programming*. (This term was coined before "programming" came to mean "giving instructions to a computer".) The only difference is that linear programming involves finding one specific solution to the set of constraints, namely the one that maximizes a given linear function of the variables. More precisely, a linear optimization problem is described as follows:

$$\text{Maximize} \quad c_1 x_1 + c_2 x_2 + \cdots + c_d x_d$$

$$\text{Subject to} \quad a_{1,1} x_1 + \cdots + a_{1,d} x_d \leqslant b_1$$
$$a_{2,1} x_1 + \cdots + a_{2,d} x_d \leqslant b_2$$
$$\vdots$$
$$a_{n,1} x_1 + \cdots + a_{n,d} x_d \leqslant b_n$$

where the c_i, and $a_{i,j}$, and b_i are real numbers, which form the input to the problem. The function to be maximized is called the *objective function*, and the set of constraints together with the objective function is a *linear program*. The number of variables, d, is the *dimension* of the linear program. We already saw that linear constraints can be viewed as half-spaces in $\mathbb{R}^d$. The intersection of these half-spaces, which is the set points satisfying all constraints, is called the *feasible region* of the linear program. Points (solutions) in this region are called *feasible*, points outside are *infeasible*. Recall from Figure 4.2 that the feasible region can be unbounded, and that it can be empty. In the latter case,

feasible region

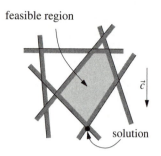

$\vec{c}$

solution

the linear program is called *infeasible*. The objective function can be viewed as a direction in $\mathbb{R}^d$; maximizing $c_1x_1 + c_2x_2 + \cdots + c_dx_d$ means finding a point $(x_1, \ldots, x_d)$ that is extreme in the direction $\vec{c} = (c_1, \ldots, c_d)$. Hence, the solution to the linear program is a point in the feasible region that is extreme in direction $\vec{c}$. We let $f_{\vec{c}}$ denote the objective function defined by a direction vector $\vec{c}$.

Many problems in operations research can be described by linear programs, and a lot of work has been dedicated to linear optimization. This has resulted in many different linear programming algorithms, several of which—the famous *simplex algorithm* for instance—perform well in practice.

Let's go back to our problem. We have n linear constraints in two variables and we want to find one solution to the set of constraints. We can do this by taking an arbitrary objective function, and then solving the linear program defined by the objective function and the linear constraints. For the latter step we can use the simplex algorithm, or any other linear programming algorithm developed in operations research. However, this particular linear program is quite different from the ones usually studied: in operations research both the number of constraints and the number of variables are large, but in our case the number of variables is only two. The traditional linear programming methods are not very efficient in such *low-dimensional linear programming problems*; methods developed in computational geometry, like the one described below, do better.

We denote the set of n linear constraints in our 2-dimensional linear programming problem by H. The vector defining the objective function is $\vec{c} = (c_x, c_y)$; thus the objective function is $f_{\vec{c}}(p) = c_xp_x + c_yp_y$. Our goal is to find a point $p \in \mathbb{R}^2$ such that $p \in \bigcap H$ and $f_{\vec{c}}(p)$ is maximized. We denote the linear program by $(H, \vec{c})$, and we use C to denote its feasible region. We can distinguish four cases for the solution of a linear program $(H, \vec{c})$. The four cases are illustrated in Figure 4.5; the vector defining the objective function is vertically downward in the examples.

Figure 4.5
Different types of solutions to a linear program.

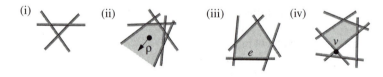

(i) The linear program is infeasible, that is, there is no solution to the set of constraints.

(ii) The feasible region is unbounded in direction $\vec{c}$. In this case there is a ray ρ completely contained in the feasible region C, such that the function $f_{\vec{c}}$ takes arbitrarily large values along ρ. We require the description of such a ray in this case.

(iii) The feasible region has an edge e whose outward normal points in the direction $\vec{c}$. In this case, there is a solution to the linear program, but it is not unique: any point on e is a feasible point that maximizes $f_{\vec{c}}(p)$.

(iv) If none of the preceding three cases applies, then there is a unique solution, which is the vertex v of C that is extreme in the direction $\vec{c}$.

A simple convention allows us to say that case (iii) also has a unique solution: if there are several optimal points, then we want the lexicographically smallest one. Conceptually, this is equivalent to rotating $\vec{c}$ a little in clockwise direction, such that it is no longer normal to any half-plane. With this convention, any bounded linear program that is feasible has a unique solution, which is a vertex of the feasible region. We call this vertex the *optimal vertex*.

Our algorithm for 2-dimensional linear programming is incremental. It adds the constraints one by one, and maintains the optimal vertex of the intermediate feasible regions. However, this requires that such a vertex exists, which is not the case when the linear program is unbounded. Therefore we will develop a separate subroutine UNBOUNDEDLP for dealing with unbounded linear programs. If the linear program $(H,\vec{c})$ is unbounded, then UNBOUNDEDLP$(H,\vec{c})$ returns a ray that is completely contained in the feasible region, and we are done. If the linear program is not unbounded, then UNBOUNDEDLP returns two half-planes h_1, h_2 from H, such that $(\{h_1,h_2\},\vec{c})$ is bounded. We call these half-planes *certificates*: they 'prove' that $(H,\vec{c})$ is really bounded. We shall look at the subroutine UNBOUNDEDLP in more detail later. For now we concentrate on bounded linear programs.

Let $(H,\vec{c})$ be a bounded linear program, and let h_1 and h_2 be the two certificates returned by UNBOUNDEDLP. We number the remaining half-planes h_3, h_4, $\ldots$, h_n. Let C_i be the feasible region when the first i half-planes have been added:

$$C_i := h_1 \cap h_2 \cap \cdots \cap h_i.$$

We denote the optimal vertex of C_i by v_i. Clearly, we have

$$C_2 \supseteq C_3 \supseteq C_4 \cdots \supseteq C_n = C.$$

This implies that if $C_i = \emptyset$ for some i, then $C_j = \emptyset$ for all $j \geq i$, and the linear program is infeasible. So our algorithm can stop once the linear program becomes infeasible.

The next lemma investigates how the optimal vertex changes when we add a half-plane h_i. It is the basis of our algorithm.

Lemma 4.5 *Let $2 < i \leq n$, and let C_i and v_i be defined as above. Then we have*
(i)   *If $v_{i-1} \in h_i$, then $v_i = v_{i-1}$.*
(ii) *If $v_{i-1} \notin h_i$, then either $C_i = \emptyset$ or $v_i \in \ell_i$, where ℓ_i is the line bounding h_i.*

Proof. (i)   Let $v_{i-1} \in h_i$. Because $C_i = C_{i-1} \cap h_i$ and $v_{i-1} \in C_{i-1}$ this means that $v_{i-1} \in C_i$. Furthermore, the optimal point in C_i cannot be better than the optimal point in C_{i-1}, since $C_i \subseteq C_{i-1}$. Hence, v_{i-1} is the optimal vertex in C_i as well.
(ii) Let $v_{i-1} \notin h_i$. Suppose for a contradiction that C_i is not empty and that v_i does not lie on ℓ_i. Consider the line segment $\overline{v_{i-1}v_i}$. We have $v_{i-1} \in C_{i-1}$ and, since $C_i \subset C_{i-1}$, also $v_i \in C_{i-1}$. Together with the convexity of C_{i-1},

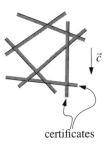

certificates

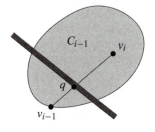

this implies that the segment $\overline{v_{i-1}v_i}$ is contained in C_{i-1}. Since v_{i-1} is the optimal point in C_{i-1} and the objective function $f_{\vec{c}}$ is linear, it follows that $f_{\vec{c}}(p)$ increases monotonically along $\overline{v_{i-1}v_i}$ as p moves from v_i to v_{i-1}. Now consider the intersection point q of $\overline{v_{i-1}v_i}$ and ℓ_i. This intersection point exists, because $v_{i-1} \notin h_i$ and $v_i \in C_i$. Since $\overline{v_{i-1}v_i}$ is contained in C_{i-1}, the point q must be in C_i. But the value of the objective function increases along $\overline{v_{i-1}v_i}$, so $f_{\vec{c}}(q) > f_{\vec{c}}(v_i)$. This contradicts the definition of v_i.

$\square$

Figure 4.6 illustrates the two cases that arise when adding a half-plane. In Figure 4.6(i), the optimal vertex v_4 that we have after adding the first four half-planes is contained in h_5, the next half-plane that we add. Therefore the optimal vertex remains the same. The optimal vertex is not contained in h_6, however, so when we add h_6 we must find a new optimal vertex. According

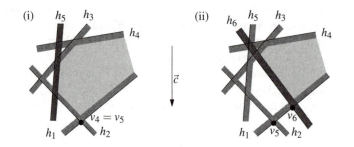

Figure 4.6
Adding a half-plane

to Lemma 4.5, this vertex v_6 is contained in the line bounding h_6, as is shown in Figure 4.6(ii). But Lemma 4.5 does not tell us how to find the new optimal vertex. Fortunately, this is not so difficult, as we show next.

Assume that the current optimal vertex v_{i-1} is not contained in the next half-plane h_i. The problem we have to solve can be stated as follows:

> Find the point p on ℓ_i that maximizes $f_{\vec{c}}(p)$, subject to the constraints $p \in h_j$, for $1 \leqslant j < i$.

To simplify the terminology, we assume that ℓ_i is the x-axis. Let x_j denote the x-coordinate of the intersection point of ℓ_j and ℓ_i. (If ℓ_j and ℓ_i are parallel, then either the constraint h_j is satisfied by any point on ℓ_i, or by no point on ℓ_i. In the former case we can ignore the constraint, in the latter case we can report the linear program infeasible.) Depending on whether $\ell_i \cap h_j$ is bounded to the left or to the right, we get a constraint on the x-coordinate of the solution of the form $x \geqslant x_j$ or of the form $x \leqslant x_j$. We can thus restate our problem as follows:

> Maximize $\quad f_{\vec{c}}((x,0))$
>
> subject to $\quad x \geqslant x_j, \quad 1 \leqslant j < i$ and $\ell_i \cap h_j$ is bounded to the left
> $\qquad\qquad x \leqslant x_k, \quad 1 \leqslant k < i$ and $\ell_i \cap h_k$ is bounded to the right

This is a 1-dimensional linear program. Solving it is very easy. Let

$$x_{\text{left}} = \max_{1 \leqslant j < i} \{x_j : \ell_i \cap h_j \text{ is bounded to the left}\}$$

and

$$x_{\text{right}} = \min_{1 \leqslant k < i} \{x_k : \ell_i \cap h_k \text{ is bounded to the right}\}.$$

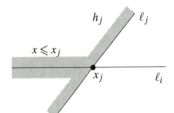

The interval $[x_{\text{left}} : x_{\text{right}}]$ is the feasible region of the 1-dimensional linear program. Hence, the linear program is infeasible if $x_{\text{left}} > x_{\text{right}}$, and otherwise the optimal point is either x_{left} or x_{right}, depending on the objective function.

We get the following lemma:

Lemma 4.6 *A 1-dimensional linear program can be solved in linear time. Hence, if case (ii) of Lemma 4.5 arises, then we can compute the new optimal vertex v_i, or decide that the linear program is infeasible, in $O(i)$ time.*

We can now describe the linear programming algorithm in more detail. As above, we use ℓ_i to denote the line that bounds the half-plane h_i.

Algorithm 2DLINEARPROGRAMMING$(H, \vec{c})$
Input. A linear program $(H, \vec{c})$, where H is a set of n half-planes and $\vec{c} \in \mathbb{R}^2$.
Output. If $(H, \vec{c})$ is unbounded, a ray ρ completely contained in the feasible region is reported. If $(H, \vec{c})$ is infeasible, then this fact is reported. Otherwise, a feasible point p that maximizes $f_{\vec{c}}(p)$ is reported.
1.   **if** UNBOUNDEDLP$(H, \vec{c})$ reports that $(H, \vec{c})$ is unbounded or infeasible
2.   **then** Report that information and, in the unbounded case, a ray along which $(H, \vec{c})$ is unbounded.
3.   **else** Let $h_1, h_2 \in H$ be the two certificate half-planes returned by UN-BOUNDEDLP, and let v_2 be the intersection point of ℓ_1 and ℓ_2.
4.   Let $h_3, \ldots, h_n$ be the remaining half-planes of H.
5.   **for** $i \leftarrow 3$ **to** n
6.   **do if** $v_{i-1} \in h_i$
7.   **then** $v_i \leftarrow v_{i-1}$
8.   **else** $v_i \leftarrow$ the point p on ℓ_i that maximizes $f_{\vec{c}}(p)$, subject to the constraints $h_1, \ldots, h_{i-1}$
9.   **if** p does not exist
10.   **then** Report that the linear program is infeasible and quit.
11.   **return** v_n

We now analyze the performance of our algorithm.

Lemma 4.7 *Algorithm* 2DLINEARPROGRAMMING *computes the solution to a linear program with n constraints and two variables in $O(n^2)$ time and linear storage.*

Proof. To prove that the algorithm correctly finds the solution, we have to show that after every stage—whenever we have added a new half-plane h_i—the point v_i is still the optimum point for C_i. This follows immediately from Lemma 4.5. If the 1-dimensional linear program on ℓ_i is infeasible, then C_i is empty, and consequently $C = C_n \subset C_i$ is empty, which means that the linear program is infeasible.

It is easy to see that the algorithm requires only linear storage, so we address the time it takes. We shall show later that the procedure UNBOUNDEDLP takes $O(n)$ time—see Lemma 4.10. The remaining part of the algorithm, where we add the half-planes one by one, has $n - 2$ stages. The time spent in stage i is dominated by the solving of a 1-dimensional linear program in line 8, which takes time $O(i)$. Hence, the total time is bounded by

$$\sum_{i=3}^{n} O(i) = O(n^2),$$

as stated in the lemma. $\qquad \square$

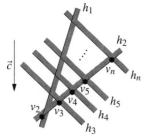

Although our linear programming algorithm is nice and simple, its running time is disappointing—the algorithm is much slower than the previous algorithm, which computed the whole feasible region. Is our analysis too crude? We bounded the cost of every stage i by $O(i)$. This may be too crude for some stages: Stage i takes $\Theta(i)$ time only when $v_{i-1} \notin h_i$; when $v_{i-1} \in h_i$ then stage i takes constant time. So if we could bound the number of times the optimal vertex changes, we might be able to prove a better running time. Unfortunately the optimum vertex can change $n - 2$ times: there are orders for some configurations where every new half-plane makes the previous optimum illegal. This means that the algorithm will really spend $\Theta(n^2)$ time. How can we avoid such nasty cases?

4.4 Randomized Linear Programming

If we have a second look at the case where the optimum changes $n - 2$ times, we see that the problem is not so much that the set of half-planes is bad. If we had added them in the order $h_n, h_{n-1}, \ldots, h_3$, then the optimal vertex would not change anymore after the addition of h_n. In this case the running time would be $O(n)$. Is this a general phenomenon? Is it true that, for any set H of half-planes, there is a good order to treat them? The answer to this question is "yes," but that doesn't seem to help us much. Even if such a good order exists, there seems to be no easy way to actually *find* it. Remember that we have to find the order at the beginning of the algorithm, when we don't know anything about the intersection of the half-planes yet.

We now meet a quite intriguing phenomenon. Although we have no way to determine an ordering of H that is guaranteed to lead to a good running time, we have a very simple way out of our problem. We simply pick a *random* ordering of H. Of course, we could have bad luck and pick an order that leads to a quadratic running time. But with some luck, we pick an order that makes it run much faster. Indeed, we shall prove below that most orders lead to a fast algorithm. For completeness, we first repeat the algorithm.

Algorithm 2DRANDOMIZEDLP$(H, \vec{c})$
Input. A linear program $(H, \vec{c})$, where H is a set of n half-planes and $\vec{c} \in \mathbb{R}^2$.
Output. If $(H, \vec{c})$ is unbounded, a ray ρ completely contained in the feasible region is reported. If $(H, \vec{c})$ is infeasible, then this fact is reported. Otherwise, a feasible point p that maximizes $f_{\vec{c}}(p)$ is reported.
1. **if** UNBOUNDEDLP$(H, \vec{c})$ reports that $(H, \vec{c})$ is unbounded or infeasible
2.   **then** Report that information and, in the unbounded case, a ray along which $(H, \vec{c})$ is unbounded.
3.   **else** Let $h_1, h_2 \in H$ be the two certificate half-planes returned by UNBOUNDEDLP, and let v_2 be the intersection point of ℓ_1 and ℓ_2.
4.   Compute a *random* permutation $h_3, \ldots, h_n$ of the remaining half-planes by calling RANDOMPERMUTATION$(H[3 \cdots n])$.
5.   **for** $i \leftarrow 3$ **to** n
6.   **do if** $v_{i-1} \in h_i$
7.   **then** $v_i \leftarrow v_{i-1}$
8.   **else** $v_i \leftarrow$ the point p on ℓ_i that maximizes $f_{\vec{c}}(p)$, subject to the constraints $h_1, \ldots, h_{i-1}$
9.   **if** p does not exist
10.   **then** Report that the linear program is infeasible and quit.
11.   **return** v_n

The only difference from the previous algorithm is in line 4, where we put the half-planes in random order before we start adding them one by one. To be able to do this, we assume that we have a random number generator, RANDOM(k), which has an integer k as input and generates a random integer between 1 and k in constant time. Computing a random permutation can then be done with the following linear time algorithm.

Algorithm RANDOMPERMUTATION(A)
Input. An array $A[1 \cdots n]$.
Output. The array $A[1 \cdots n]$ with the same elements, but rearranged into a random permutation.
1. **for** $k \leftarrow n$ **downto** 2
2.   **do** $rndindex \leftarrow$ RANDOM(k)
3.   Exchange $A[k]$ and $A[rndindex]$.

The new linear programming algorithm is called a *randomized algorithm*; its running time depends on certain random choices made by the algorithm. (In the linear programming algorithm, these random choices were made in the subroutine RANDOMPERMUTATION.)

What is the running time of this randomized version of our incremental linear programming algorithm? There is no easy answer to that. It all depends on the order that is computed in line 4. Consider a fixed set H of n half-planes. 2DRANDOMIZEDLP treats them depending on the permutation chosen in line 4. Since there are $(n-2)!$ possible permutations of $n-2$ objects, there are $(n-2)!$ possible ways in which the algorithm can proceed, each with its own running time. Because the permutation is random, each of these running times is equally likely. So what we do is analyze the *expected running time* of the algorithm, which is the *average running time over all* $(n-2)!$ *possible permutations*. The theorem below states that the expected running time of our randomized linear programming algorithm is $O(n)$. It is important to realize that we do not make any assumptions about the input: the expectancy is with respect to the random order in which the half-planes are treated and holds for any set of half-planes.

Theorem 4.8 *The 2-dimensional linear programming problem with n constraints can be solved in $O(n)$ randomized expected time using worst-case linear storage.*

Proof. As we observed before, the storage needed by the algorithm is linear.

The running times of UNBOUNDEDLP and of RANDOMPERMUTATION are $O(n)$, so what remains is to analyze the time needed to add the half-planes $h_3, \ldots, h_n$. Adding a half-plane takes constant time when the optimal vertex does not change. When the optimal vertex does change we need to solve a 1-dimensional linear program. We now bound the time needed for all these 1-dimensional linear programs.

Let X_i be a random variable, which is 1 if $v_{i-1} \notin h_i$, and 0 otherwise. Recall that a 1-dimensional linear program on i constraints can be solved in $O(i)$ time. The total time spent in line 8 over all half-planes $h_3, \ldots, h_n$ is therefore

$$\sum_{i=3}^{n} O(i) \cdot X_i.$$

To bound the expected value of this sum we will use *linearity of expectation*: the expected value of a sum of random variables is the sum of the expected values of the random variables. This holds even if the random variables are dependent. Hence, the expected time for solving all 1-dimensional linear programs is

$$E[\sum_{i=3}^{n} O(i) \cdot X_i] = \sum_{i=3}^{n} O(i) \cdot E[X_i].$$

But what is $E[X_i]$? It is exactly the probability that $v_{i-1} \notin h_i$. Let's analyze this probability.

We will do this with a technique called *backwards analysis*: we look at the algorithm "backwards." Assume that it has already finished, and that it has computed the optimum vertex v_n. Since v_n is a vertex of C_n, it is defined by at least two of the half-planes. Now we make one step backwards in time, and look at C_{n-1}. Note that C_{n-1} is obtained from C_n by removing the half-plane h_n. When does the optimum point change? This happens exactly if v_n

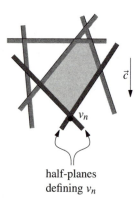

half-planes
defining v_n

lies in the interior of C_{n-1}, which is only possible if h_n is one of the half-planes that define v_n. But the half-planes are added in random order, so h_n is a random element of $\{h_3, h_4, \ldots, h_n\}$. Hence, the probability that h_n is one of the half-planes defining v_n is at most $2/(n-2)$. Why do we say "at most"? Because it is possible that the boundaries of more than two half-planes pass through v_n. In that case, removing one of the two half-planes containing the edges incident to $\vec{x}_n$ may fail to change v_n, in which case the probability is even less than $2/(n-2)$.

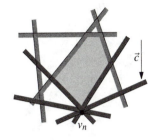

The same argument works in general: to bound $E[X_i]$, we fix the subset of the first i half-planes. This determines C_i. To analyze what happened in the last step, when we added h_i, we think backwards. The probability that we had to compute a new optimal vertex when adding h_i is the same as the probability that the optimal vertex changes when we remove a half-plane from C_i. The latter event only takes place for at most two half-planes of our fixed set $\{h_3, \ldots, h_i\}$. Since the half-planes are added in random order, the probability that h_i is one of the special half-planes is at most $2/(i-2)$. We derived this probability under the condition that the first i half-planes are some fixed subset of H. But since the derived bound holds for *any* fixed subset, it holds unconditionally. Hence, $E[X_i] \leqslant 2/(i-2)$. We can now bound the expected total time for solving all 1-dimensional linear programs by

$$\sum_{i=3}^{n} O(i) \cdot \frac{2}{i-2} = O(n).$$

We already noted that the time spent in the rest of the algorithm is $O(n)$ as well. $\quad\square$

Note again that the expectancy here is solely with respect to the random choices made by the algorithm. We do not average over possible choices for the input. For *any* input set of n half-planes, the expected running time of the algorithm is $O(n)$; there are no bad inputs.

4.5 Unbounded Linear Programs

In the preceding sections we assumed that we had a subroutine UNBOUNDED-LP available for dealing with unbounded linear programs. This subroutine should not only be able to solve unbounded linear programs—that is, return a ray contained in the feasible region—but it should also return some useful information when the linear program is bounded, namely two certificate half-planes $h_1, h_2 \in H$ such that $(\{h_1, h_2\}, \vec{c})$ is bounded.

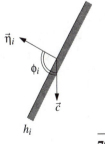

Let $H = \{h_1, h_2, \ldots, h_n\}$, and let $\vec{\eta}_i$ be the outward normal of h_i. Recall that ℓ_i denotes the line bounding h_i. Let ϕ_i denote the smaller angle that $\vec{\eta}_i$ makes with $\vec{c}$, with $0° \leqslant \phi_i \leqslant 180°$. It turns out that half-planes h_i with a minimal value of ϕ_i are crucial in deciding whether the linear program is bounded or

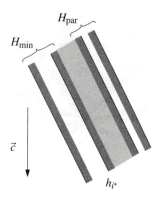

not. Let h_i be a half-plane such that $\phi_i = \min_{1 \leqslant j \leqslant n} \phi_j$, and define

$$H_{\min} := \{h_j \in H \mid \vec{\eta}_j = \vec{\eta}_i\}.$$

(Note that $H_{\min}$ does not necessarily contain all half-planes h_j with $\phi_j = \phi_i$, because if ϕ_j is a clockwise angle and ϕ_i is a counterclockwise angle then we need not have $\vec{\eta}_j = \vec{\eta}_i$.) Finally, let $H_{\mathrm{par}} \subset H \setminus H_{\min}$ denote the half-planes whose bounding lines are parallel to the bounding lines of the half-planes in $H_{\min}$, but that lie on the opposite side of their bounding line:

$$H_{\mathrm{par}} := \{h_j \in H \mid \vec{\eta}_j = -\vec{\eta}_i\}.$$

Note that H_{par} can be empty. We first compute the intersection of the half-planes in $H_{\min} \cup H_{\mathrm{par}}$. Because the bounding lines of these half-planes are parallel, their intersection is a slab, which can be computed in linear time. There are two cases to consider.

First of all, the slab can be empty. This implies that the linear program $(H, \vec{c})$ is infeasible, and so we are done.

The more interesting case is when the slab is not empty. Now the slab is bounded on one side by a line ℓ_{i^*} with $h_{i^*} \in H_{\min}$. The slab may or may not be bounded by another line, but the line ℓ_{i^*} is the one we are interested in: as we will show, we can decide whether the linear program is unbounded by looking at the intersections of the other half-planes with ℓ_{i^*}. Such an intersection $\ell_{i^*} \cap h_j$ is a ray ρ_j. This ray can either be bounded in direction $\vec{c}$ or unbounded. In the first case the linear program $(\{h_{i^*}, h_j\}, \vec{c})$ is bounded. If the second case occurs for all the other half-planes, then the whole linear program is unbounded. The next lemma states this more formally.

Lemma 4.9 *Let $H = \{h_1, h_2, \ldots, h_n\}$ be a set of half-planes, and let $H_{\min}, H_{\mathrm{par}}, \ell_{i^*}$ be defined as above. Assuming that $\bigcap (H_{\min} \cup H_{\mathrm{par}})$ is not empty, we have*
(i) *If $\ell_{i^*} \cap h_j$ is unbounded in the direction $\vec{c}$ for every half-plane h_j in the set $H \setminus (H_{\min} \cup H_{\mathrm{par}})$, then $(H, \vec{c})$ is unbounded along a ray contained in ℓ_{i^*}. Moreover, such a ray can be computed in linear time.*
(ii) *If $\ell_{i^*} \cap h_{j^*}$ is bounded in the direction $\vec{c}$ for some h_{j^*} in $H \setminus (H_{\min} \cup H_{\mathrm{par}})$, then the linear program $(\{h_{i^*}, h_{j^*}\}, \vec{c})$ is bounded.*

Proof. (i) Let $\rho_j := \ell_{i^*} \cap h_j$. Since all these rays ρ_j are unbounded in direction $\vec{c}$, their intersection is a ray that is unbounded in direction $\vec{c}$ as well. This ray can easily be computed in linear time.

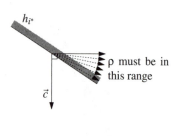

(ii) Suppose for a contradiction that $\ell_{i^*} \cap h_{j^*}$ is bounded in direction $\vec{c}$, but that $(\{h_{i^*}, h_{j^*}\}, \vec{c})$ is unbounded. Then the ray $\rho := \ell_{j^*} \cap h_{i^*}$ must be unbounded in direction $\vec{c}$. This means that the direction of ρ must lie in the smaller range defined by the ℓ_{i^*}, which we assume is directed downward, and the vector perpendicular to $\vec{c}$ and directed into the interior of h_{i^*}. Because $\ell_{i^*} \cap h_{j^*}$ is bounded in direction $\vec{c}$, this implies that the angle that $\vec{c}$ makes with $\vec{\eta}_{j^*}$ is smaller than the angle $\vec{c}$ makes with $\vec{\eta}_{i^*}$, which is impossible since $h_{i^*} \in H_{\min}$ and $h_{j^*} \notin H_{\min}$. ☐

The above leads to the following subroutine to check for unbounded linear programs.

Algorithm UNBOUNDEDLP($H, \vec{c}$)

Input. A linear program $(H, \vec{c})$, where H is a set of n half-planes and $\vec{c} \in \mathbb{R}^2$ is the vector defining the objective function.

Output. If $(H, \vec{c})$ is unbounded, the output is a ray that is contained in the feasible region. If $(H, \vec{c})$ is bounded, the output either consists of two half-planes h_{i^*} and h_{j^*} from H such that $(\{h_{i^*}, h_{j^*}\}, \vec{c})$ is bounded, or it is reported that the linear program is infeasible.

1.   For each half-plane $h_i \in H$, compute the angle ϕ_i defined above.
2.   Let h_i be a half-plane with $\phi_i = \min_{1 \leqslant j \leqslant n} \phi_j$.
3. $H_{\min} \leftarrow \{h_j \in H \mid \vec{\eta}_j = \vec{\eta}_i\}$
4. $H_{\mathrm{par}} \leftarrow \{h_j \in H \mid \vec{\eta}_j = -\vec{\eta}_i\}$
5. $\widehat{H} \leftarrow H \setminus (H_{\min} \cup H_{\mathrm{par}})$
6.   Compute the intersection of the half-planes in $H_{\min} \cup H_{\mathrm{par}}$.
7.   **if** the intersection is empty
8.   **then** Report that $(H, \vec{c})$ is infeasible.
9.   **else** Let $h_{i^*} \in H_{\min}$ be the half-plane whose bounding line bounds the intersection.
10.   **if** there is a half-plane $h_{j^*} \in \widehat{H}$ such that $\ell_{i^*} \cap h_{j^*}$ is bounded in direction $\vec{c}$
11.   **then** Report that $(\{h_{i^*}, h_{j^*}\}, \vec{c})$ is bounded.
12.   **else** Report that $(H, \vec{c})$ is unbounded along the ray $\ell_{i^*} \cap (\bigcap \widehat{H})$.

When $\bigcap(H_{\min} \cup H_{\mathrm{par}})$ is empty, the algorithm reports that the linear program is infeasible. But it does not detect other causes of infeasibility; it may report two half-planes h_{i^*} and h_{j^*} such that $(\{h_{i^*}, h_{j^*}\}, \vec{c})$ is bounded when $(H, \vec{c})$ is infeasible. This is not a problem: in such a case our main routine 2DRANDOMIZEDLP (or 2DLINEARPROGRAMMING) will detect the infeasibility.

UNBOUNDEDLP clearly runs in linear time. We get the following result.

Lemma 4.10 *Algorithm* UNBOUNDEDLP *decides in* $O(n)$ *time whether a 2-dimensional linear program with n constraints is unbounded. If it is, it finds a ray contained in the feasible region along which the objective function is unbounded.*

In some situations we do not need the subroutine for dealing with unbounded linear programs. One such case is when we only want to know whether there is a feasible point, like in our molding application. This means that we can choose the direction $\vec{c}$ ourselves. Hence, we can take any two non-parallel half-planes h_{i^*}, h_{j^*} from H and choose the direction $\vec{c}$ such that $(\{h_{i^*}, h_{j^*}\}, \vec{c})$ is bounded. In other applications we may have an a priori bound A on the absolute value of the solution. In that case we can add four constraints of the form $x \leqslant A$, $x \geqslant -A, y \leqslant A, y \geqslant -A$.

4.6* Linear Programming in Higher Dimensions

The linear programming algorithm presented in the previous sections can be generalized to higher dimensions. When the dimension is not too high, then the resulting algorithm compares favorably with traditional algorithms, such as the simplex algorithm.

Let H be a set of n closed half-spaces in $\mathbb{R}^d$. Given a vector $\vec{c} = (c_1, \ldots, c_d)$, we want to find the point $p = (p_1, \ldots, p_d) \in \mathbb{R}^d$ that maximizes the linear function $f_{\vec{c}}(p) := c_1 p_1 + \cdots + c_d p_d$, subject to the constraint that p lies in h for all $h \in H$. To make sure that the solution is unique when the linear program is bounded, we agree to look for the lexicographically smallest point that maximizes $f_{\vec{c}}(p)$.

As in the planar version, we need a subroutine UNBOUNDEDLP to deal with unbounded linear programs. UNBOUNDEDLP$(H, \vec{c})$ should return a ray that is completely contained in the feasible region if $(H, \vec{c})$ is unbounded, and otherwise it should return d certificate half-spaces $h_1, \ldots, h_d$ such that $(\{h_1, \ldots, h_d\}, \vec{c})$ is bounded. (In the latter case it may also report that $(H, \vec{c})$ is infeasible, if this is the case.) It turns out that the linear time subroutine we developed for unbounded 2-dimensional linear programs can be generalized to higher dimensions. We leave this as an exercise, and concentrate on bounded linear programs.

Let $h_1, \ldots, h_d \in H$ be the d certificate half-spaces that UNBOUNDEDLP returns. Because of our convention that we look for the lexicographically smallest point maximizing $f_{\vec{c}}$, we know that the solution to the linear program $(\{h_1, \ldots, h_d\}, \vec{c})$ is the vertex of the feasible region $\bigcap_{1 \leqslant i \leqslant d} h_i$. We now continue as in the 2-dimensional case: we add the remaining half-spaces one by one in random order, and maintain the optimal vertex during the process. Let's study this process in more detail.

Let $h_{d+1}, h_{d+2}, \ldots, h_n$ be a random permutation of $H \setminus \{h_1, \ldots, h_d\}$. Furthermore, define C_i to be the feasible region when the first i half-spaces have been added:

$$C_i := h_1 \cap h_2 \cap \cdots \cap h_i.$$

Let v_i denote the optimal vertex of C_i, that is, the vertex that maximizes $f_{\vec{c}}$. Lemma 4.5 gave us an easy way to maintain the optimal vertex in the 2-dimensional case: either the optimal vertex doesn't change, or the new optimal vertex is contained in the line that bounds the half-plane h_i that we are adding. The following lemma generalizes this result to higher dimensions; its proof is a straightforward generalization of the proof of Lemma 4.5.

Lemma 4.11 *Let $d < i \leqslant n$, and let C_i and v_i be defined as above. Then we have*
(i) *If $v_{i-1} \in h_i$, then $v_i = v_{i-1}$.*
(ii) *If $v_{i-1} \notin h_i$, then either $C_i = \emptyset$ or $v_i \in g_i$, where g_i is the hyperplane that bounds h_i.*

The global algorithm now looks as follows. We use g_i to denote the hyperplane that bounds the half-space h_i.

Algorithm RANDOMIZEDLP($H, \vec{c}$)

Input. A linear program $(H, \vec{c})$, where H is a set of n half-spaces in $\mathbb{R}^d$ and $\vec{c} \in \mathbb{R}^d$.

Output. If $(H, \vec{c})$ is unbounded, a ray ρ completely contained in the feasible region is reported. If $(H, \vec{c})$ is infeasible, then this fact is reported. Otherwise, a feasible point p which maximizes $f_{\vec{c}}(p)$ is reported.

1.   **if** UNBOUNDEDLP($H, \vec{c}$) reports that $(H, \vec{c})$ is unbounded or infeasible
2.   **then** Report that information and, in the unbounded case, a ray along which $(H, \vec{c})$ is unbounded.
3.   **else** Let $h_1, \dots, h_d \in H$ be the certificate half-planes returned by UN-BOUNDEDLP, and let v_d be the vertex of their common intersection.
4.   Compute a random permutation $h_{d+1}, \dots, h_n$ of the remaining half-spaces.
5.   **for** $i \leftarrow d+1$ **to** n
6.   **do if** $v_{i-1} \in h_i$
7.   **then** $v_i \leftarrow v_{i-1}$
8.   **else** $v_i \leftarrow$ the point p on g_i that maximizes $f_{\vec{c}}(p)$, subject to the constraints $h_1, \dots, h_{i-1}$.
9.   **if** p does not exist
10.   **then** Report that the linear program is infeasible and quit.
11.   **return** v_n

But how do we implement step 8 now? In two dimensions this was easy to do in linear time, because everything was restricted to a line. Let's look at the 3-dimensional case. In three dimensions, g_i is a plane, and $g_i \cap C_{i-1}$ is a 2-dimensional convex polygonal region. So what do we have to do to find the optimum in $g_i \cap C_{i-1}$? We have to solve a 2-dimensional linear program! So in the 3-dimensional case we implement line 8 as follows: we compute the intersection of all $i-1$ half-spaces with g_i, and project the vector $\vec{c}$ on g_i. This results in a linear program in two dimensions, which we solve using algorithm 2DRANDOMIZEDLP.

Theorem 4.12 *The 3-dimensional linear programming problem with n constraints can be solved in $O(n)$ expected time using worst-case linear storage.*

Proof. The bound on the storage follows immediately from the storage bound for 2DRANDOMIZEDLP. The running time of the algorithm is linear as long as the time spent in line 8 is not counted. So it suffices to bound the time spent in calls to 2DRANDOMIZEDLP.

As in the analysis of the 2-dimensional algorithm, we define a random variable X_i, which is 1 if $v_{i-1} \notin h_i$, and 0 otherwise. The total expected time spent in all executions of line 8 is

$$\sum_{i=4}^{n} O(i) \cdot X_i,$$

since the expected running time of 2DRANDOMIZEDLP for a linear program with i constraints is $O(i)$. But what is $E[X_i]$ in three dimensions?

Again we apply backwards analysis. Fix the subset of the first i half-spaces, and consider the situation after they have been added. The optimum point is a vertex v_i of C_i, so it is defined by three of the half-spaces. Now we make one step backwards in time. The optimum point changes only if we remove one of the half-spaces defining v_i. Since $h_4, \ldots, h_i$ is a random permutation, the probability that this happens is at most $3/(i-3)$. (As before, we have to say "at most", because there could be more than three half-spaces whose bounding plane passes through v_i. In this case the number of half-spaces whose removal changes the optimum may be smaller than three.) we derived the $3/(i-3)$ bound for a fixed subset of the first i half-spaces; since the derived bound holds for any fixed subset, it holds unconditionally. Therefore, the expected running time of the algorithm is

$$O(n) + \sum_{i=4}^{n} O(i) \frac{3}{i-3} = O(n).$$

The general strategy to implement line 8 of RANDOMIZEDLP is this: in the d-dimensional version, we make a recursive call to the $(d-1)$-dimensional version. The recursion bottoms out when we get to a 1-dimensional linear program, which can be solved directly in linear time.

The following theorem states the performance of RANDOMIZEDLP. Although we consider d a constant, it is useful to give the dependency of the running time on d.

Theorem 4.13 *The d-dimensional linear programming problem with n constraints can be solved in $O(d!n)$ expected time using linear storage.*

Proof. The proof is basically identical to the proofs of the 3- and 2-dimensional cases we have seen before. It is easy to see that the amount of storage is linear, so we concentrate on the running time.

Denote the expected running time of the algorithm by $T(d,n)$. The running time is clearly $O(dn)$ as long as we do not count the recursive calls, so we just have to bound the expected time spent in line 8. Define a random variable X_i, which is 1 if $v_{i-1} \notin h_i$, and 0 otherwise. Now the total expected time spent in line 8 is

$$\sum_{i=d+1}^{n} (O(di) + T(d-1, i-1)) \cdot X_i,$$

since it takes $O(di)$ time to generate the $(d-1)$-dimensional linear program in i constraints, and $T(d-1,i)$ expected time to solve it recursively.

To bound $E[X_i]$, we apply backwards analysis. Consider the situation after adding $h_1, \ldots, h_i$. The optimum point is a vertex v_i of C_i, so it is defined by d of the half-spaces. Now we make one step backwards in time. The optimum point changes only if we remove one of the half-spaces defining v_i. Since $h_{d+1}, \ldots, h_i$ is a random permutation, the probability that this happens is at most $d/(i-d)$. Consequently, we get the following recursion for the expected running time of the algorithm:

$$T(d,n) \leqslant O(dn) + \sum_{i=d+1}^{n} \frac{d}{i-d} T(d-1, i-1).$$

This recurrence solves to $T(d,n) = O(d!n)$. ⌑

It should be clear that the dependence on d makes this algorithm useful only for rather small dimensions.

4.7* Smallest Enclosing Discs

The simple randomized technique we used above turns out to be surprisingly powerful. It can be applied not only to linear programming but to a variety of other optimization problems as well. In this section we shall look at one such problem.

Consider a robot armwhose base is fixed to the work floor. The arm has to pick up items at various points and place them at other points. What would be a good position for the base of the arm? This would be somewhere "in the middle" of the points it must be able to reach. More precisely, a good position is at the center of the smallest disc that encloses all the points. This point minimizes the maximum distance between the base of the arm and any point it has to reach. We arrive at the following problem: given a set P of n points in the plane (the points on the work floor that the arm must be able to reach), find the *smallest enclosing disc* for P, that is, the smallest disc that contains all the points of P. This smallest enclosing disc is unique—see Lemma 4.15(i) below, which is a generalization of this statement.

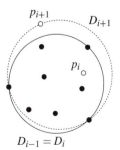

As in the previous sections, we will give a randomized incremental algorithm for the problem: First we generate a random permutation $p_1, \ldots, p_n$ of the points in P. Let $P_i := \{p_1, \ldots, p_i\}$. We add the points one by one while we maintain D_i, the smallest enclosing disc of P_i.

In the case of linear programming, there was a nice fact that helped us to maintain the optimal vertex: when the current optimal vertex is contained in the next half-plane then it does not change, and otherwise the new optimal vertex lies on the boundary of the half-plane. Is a similar statement true for smallest enclosing discs? The answer is yes:

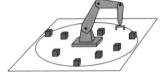

Lemma 4.14 *Let* $2 < i < n$, *and let* P_i *and* D_i *be defined as above. Then we have*

(i) *If* $p_i \in D_{i-1}$, *then* $D_i = D_{i-1}$.
(ii) *If* $p_i \notin D_{i-1}$, *then* p_i *lies on the boundary of* D_i.

We shall prove this lemma later, after we have seen how we can use it to design a randomized incremental algorithm that is quite similar to the linear programming algorithm.

Algorithm MINIDISC(P)
Input. A set P of n points in the plane.
Output. The smallest enclosing disc for P.
1. Compute a random permutation $p_1, \ldots, p_n$ of P.
2. Let D_2 be the smallest enclosing disc for $\{p_1, p_2\}$.
3. **for** $i \leftarrow 3$ **to** n
4.   **do if** $p_i \in D_{i-1}$
5.   **then** $D_i \leftarrow D_{i-1}$
6.   **else** $D_i \leftarrow$ MINIDISCWITHPOINT($\{p_1, \ldots, p_{i-1}\}, p_i$)
7. **return** D_n

The critical step occurs when $p_i \notin D_{i-1}$. We need a subroutine that finds the smallest disc enclosing P_i, using the knowledge that p_i must lie on the boundary of that disc. How do we implement this routine? Let $q := p_i$. We use the same framework once more: we add the points of P_{i-1} in random order, and maintain the smallest enclosing disc of $P_{i-1} \cup \{q\}$ under the extra constraint that it should have q on its boundary. The addition of a point p_j will be facilitated by the following fact: when p_j is contained in the currently smallest enclosing disc then this disc remains the same, and otherwise it must have p_j on its boundary. So in the latter case, the disc has both q and p_j and its boundary. We get the following subroutine.

MINIDISCWITHPOINT(P, q)
Input. A set P of n points in the plane, and a point q such that there exists an enclosing disc for P with q on its boundary.
Output. The smallest enclosing disc for P with q on its boundary.
1. Compute a random permutation $p_1, \ldots, p_n$ of P.
2. Let D_1 be the smallest disc with q and p_1 on its boundary.
3. **for** $j \leftarrow 2$ **to** n
4.   **do if** $p_j \in D_{j-1}$
5.   **then** $D_j \leftarrow D_{j-1}$
6.   **else** $D_j \leftarrow$ MINIDISCWITH2POINTS($\{p_1, \ldots, p_{j-1}\}, p_j, q$)
7. **return** D_n

How do we find the smallest enclosing disc for a set under the restriction that two given points q_1 and q_2 are on its boundary? We simply apply the same approach one more time. Thus we add the points in random order and maintain the optimal disc; when the point p_k we add is inside the current disc we don't have to do anything, and when p_k is not inside the current disc it must be on the boundary of the new disc. In the latter case we have three points on the disc boundary: q_1, q_2, and p_k. This means there is only one disc left: the unique disc with q_1, q_2, and p_k on its boundary. This following routine describes this in more detail.

MINIDISCWITH2POINTS(P, q_1, q_2)

Input. A set P of n points in the plane, and two points q_1 and q_2 such that there exists an enclosing disc for P with q_1 and q_2 on its boundary.

Output. The smallest enclosing disc for P with q_1 and q_2 on its boundary.

1.   Let D_0 be the smallest disc with q_1 and q_2 on its boundary.
2.   **for** $k \leftarrow 1$ **to** n
3.   **do if** $p_k \in D_{k-1}$
4.   **then** $D_k \leftarrow D_{k-1}$
5.   **else** $D_k \leftarrow$ the disc with q_1, q_2, and p_k on its boundary
6.   **return** D_n

This finally completes the algorithm for computing the smallest enclosing disc of a set of points. Before we analyze it, we must validate its correctness by proving some facts that we used in the algorithms. For instance, we used the fact that when we added a new point and this point was outside the current optimal disc, then the new optimal disc must have this point on its boundary.

Lemma 4.15 *Let P be a set of points in the plane, let R be a possibly empty set of points with $P \cap R = \emptyset$, and let $p \in P$. Then the following holds:*

(i)   *If there is a disc that encloses P and has all points of R on its boundary, then the smallest such disc is unique. We denote it by $md(P, R)$.*

(ii) *If $p \in md(P \setminus \{p\}, R)$, then $md(P, R) = md(P \setminus \{p\}, R)$.*

(iii) *If $p \notin md(P \setminus \{p\}, R)$, then $md(P, R) = md(P \setminus \{p\}, R \cup \{p\})$.*

Proof. (i)   Assume that there are two distinct enclosing discs D_0 and D_1 with centers x_0 and x_1, respectively, and with the same radius. Clearly, all points of P must lie in the intersection $D_0 \cap D_1$. We define a continuous family $\{D(\lambda) \mid 0 \leqslant \lambda \leqslant 1\}$ of discs as follows. Let z be an intersection point of ∂D_0 and ∂D_1, the boundaries of D_0 and D_1. The center of $D(\lambda)$ is the point $x(\lambda) := (1 - \lambda)x_0 + \lambda x_1$, and the radius of $D(\lambda)$ is $r(\lambda) := d(x(\lambda), z)$. We have $D_0 \cap D_1 \subset D(\lambda)$ for all λ with $0 \leqslant \lambda \leqslant 1$ and, in particular, for $\lambda = 1/2$. Hence, since both D_0 and D_1 enclose all points of P, so must $D(1/2)$. Moreover, $\partial D(1/2)$ passes through the intersection points of ∂D_0 and ∂D_1. Because $R \subset \partial D_0 \cap \partial D_1$, this implies that $R \subset \partial D(1/2)$. In other words, $D(1/2)$ is an enclosing disc for P with R on its boundary. But the radius of $D(1/2)$ is strictly less than the radii of D_0 and D_1. So whenever there are two distinct enclosing discs of the same radius with R on their boundary, then there is a smaller enclosing disc with R on its boundary. Hence, the smallest enclosing disc $md(P, R)$ is unique.

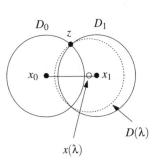

(ii) Let $D := md(P \setminus \{p\}, R)$. If $p \in D$, then D contains P and has R on its boundary. There cannot be any smaller disc containing P with R on its boundary, because such a disc would also be a containing disc for $P \setminus \{p\}$ with R on its boundary, contradicting the definition of D. It follows that $D = md(P, R)$.

(iii) Let $D_0 := md(P \setminus \{p\}, R)$ and let $D_1 := md(P, R)$. Consider the family $D(\lambda)$ of discs defined above. Note that $D(0) = D_0$ and $D(1) = D_1$, so the family defines a continous deformation of D_0 to D_1. By assumption we have $p \notin D_0$. We also have $p \in D_1$, so by continuity there must be some

$0 < \lambda^* \leqslant 1$ such that p lies on the boundary of $D(\lambda^*)$. As in the proof of (i), we have $P \subset D(\lambda^*)$ and $R \subset \partial D(\lambda^*)$. Since the radius of any $D(\lambda)$ with $0 < \lambda < 1$ is strictly less than the radius of D_1, and D_1 is by definition the smallest enclosing disc for P, we must have $\lambda^* = 1$. In other words, D_1 has p on its boundary.

☐

Lemma 4.15 implies that MINIDISC correctly computes the smallest enclosing disc of a set of points. The analysis of the running time is given in the proof of the following theorem.

Theorem 4.16 *The smallest enclosing disc for a set of n points in the plane can be computed in $O(n)$ expected time using worst-case linear storage.*

Proof. MINIDISCWITH2POINTS runs in $O(n)$ time because every iteration of the loop takes constant time, and it uses linear storage. MINIDISCWITHPOINT and MINIDISC also need linear storage, so what remains is to analyze their expected running time.

The running time of MINIDISCWITHPOINT is $O(n)$ as long as we don't count the time spent in calls to MINIDISCWITH2POINTS. What is the probability of having to make such a call? Again we use backwards analysis to bound this probability: Fix a subset $\{p_1, \dots, p_i\}$, and let D_i be the smallest disc enclosing $\{p_1, \dots, p_i\}$ and having q on its boundary. Imagine that we re-move one of the points $\{p_1, \dots, p_i\}$. When does the smallest enclosing circle change? That happens only when we remove one of the three points on the boundary. (When there are more points on the boundary, then the smallest disc cannot change after removing one point.) One of the points on the boundary is q, so there are at most two points that cause the smallest enclosing circle to shrink. The probability that p_i is one of those points is $2/i$. So we can bound the total expected running time of MINIDISCWITHPOINT by

$$O(n) + \sum_{i=2}^{n} O(i)\frac{2}{i} = O(n).$$

Applying the same argument once more, we find that the expected running time of MINIDISC is $O(n)$ as well. ☐

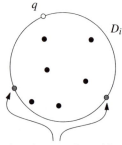

points that together with q define D_i

Algorithm MINIDISC can be improved in various ways. First of all, it is not necessary to use a fresh random permutation in every instance of subroutine MINIDISCWITHPOINT. Instead, one can compute a permutation once, at the start of MINIDISC, and pass the permutation to MINIDISCWITHPOINT. Fur-thermore, instead of writing three different routines, one could write a single algorithm MINIDISCWITHPOINTS(P,R) that computes $md(P,R)$ as defined in Lemma 4.15.

4.8 Notes and Comments

It is only recently that the geometric computations needed in computer aided manufacturing have received consideration from computational geometry. We have only scratched the surface of the area of manufacturing processes; Bose and Toussaint [52] give an extensive overview.

The computation of the common intersection of half-planes is an old and well-studied problem. As we will explain in Chapter 11, the problem is dual to the computation of the convex hull of points in the plane. Both problems have a long history in the field, and Preparata and Shamos [286] already list a number of solutions. More information on the computation of 2-dimensional convex hulls can be found in the notes and comments of Chapter 1.

Computing the common intersection of half-spaces, which can be done in $O(n \log n)$ time in the plane, becomes a more computationally demanding problem when the dimension increases. The reason is that the number of (lower-dimensional) faces of the convex polytope formed as the common intersection can be as large as $\Theta(n^{\lfloor d/2 \rfloor})$ [129]. So if the only goal is to find a feasible point, computing the common intersection explicitly soon becomes an unattractive approach.

Linear programming is one of the basic problems in numerical analysis and combinatorial optimization. It goes beyond the scope of this chapter to survey this literature, and we restrict ourselves to mentioning the simplex algorithm and its variants [117], and the polynomial-time solutions of Khachiyan [198] and Karmarkar [193]. More information on linear programming can be found in books by Chvátal [106] and Schrijver [302].

Linear programming as a problem in computational geometry was first considered by Megiddo [239], who showed that the problem of testing whether the intersection of half-spaces is empty is strictly simpler than the computation of the intersection. He gave the first deterministic algorithm for linear programming whose running time is of the form $O(C_d n)$, where C_d is a factor depending on the dimension only. His algorithm is linear in n for any fixed dimension. The factor C_d in his algorithm is 2^{2^d}. This was later improved to 3^{d^2} [125, 108]. More recently, a number of simpler and more practical randomized algorithms have been given [110, 309, 314]. There are a number of randomized algorithms whose running time is subexponential, but still not polynomial in the dimension [188, 233]. Finding a strongly polynomial algorithm, that is of combinatorial polynomial complexity, for linear programming is one of the major open problems in the area.

The simple randomized incremental algorithm for two and higher dimensions given here is due to Seidel [309]. The generalization to the computation of smallest enclosing discs is due to Welzl [333], who also showed how to find the smallest enclosing ball of a set of points in higher dimensions, and the smallest enclosing ellipse or ellipsoid. Sharir and Welzl further generalized the technique and introduced the notion of *LP-type problems*, which can be solved efficiently with an algorithm similar to the ones described here [314]. Generally speaking, the technique is applicable to optimization problems where the solution either does not change when a new constraint is added, or the solution is partially defined by the new constraint so that the dimension of the problem is reduced. It has also been shown that the special properties of LP-type problems give rise to so-called Helly-type theorems [7].

Randomization is a technique that often produces algorithms that are simple and efficient. We will see more examples in the following chapters. The price we pay is that the running time is only an expected bound and—as we observed—there is a certain chance that the algorithm takes much longer. Some people take this as a reason to say that randomized algorithms cannot be trusted and shouldn't be used (think of a computer in an intensive care station in a hospital, or in a nuclear power plant).

On the other hand, deterministic algorithms are only perfect in theory. In practice, any non-trivial algorithm may contain bugs, and even if we neglect this, there is the risk of hardware malfunction or "soft errors": single bits in core memory flipping under the influence of ambient α-radiation. Because randomized algorithms are often much simpler and have shorter code, the probability of such a mishap is smaller. Therefore the total probability that a randomized algorithm fails to produce the correct answer in time need not be larger than the probability that a deterministic algorithms fails. Moreover, we can always reduce the probability that the actual running time of a randomized algorithm exceeds its expected running time by allowing a larger constant in the expected running time.

4.9 Exercises

4.1 In this chapter we studied the casting problem for molds of one piece. A sphere cannot be manufactured in this manner, but it can be manufactured if we use a two-piece mold. Give an example of an object that cannot be manufactured with a two-piece mold, but that can be manufactured with a three-piece mold.

4.2 Consider the casting problem in the plane: we are given polygon P and a 2-dimensional mold for it. Describe a linear time algorithm that decides whether P can be removed from the mold by a single translation.

4.3 Suppose that, in the 3-dimensional casting problem, we do not want the object to slide along a facet of the mold when we remove it. How does this affect the geometric problem (computing a point in the intersection of half-planes) that we derived?

4.4 Let P be a castable simple polyhedron with top facet f. Let $\vec{d}$ be a removal direction for P. Show that any line with direction $\vec{d}$ intersects P if and only if it intersects f. Also show that for any line ℓ with direction $\vec{d}$, the intersection $\ell \cap P$ is connected.

4.5 Let $\mathcal{P}$ be a simple polyhedron $\mathcal{P}$ with n vertices. If $\mathcal{P}$ is castable with some facet f as top facet, then a necessary condition is that the facets adjacent to f must lie completely to one side of h_f, the plane through f. (The reverse is not necessarily true, of course: if all adjacent facets lie to one side of h_f then $\mathcal{P}$ is not necessarily castable with f as top facet.) Give a linear time algorithm to compute all facets of $\mathcal{P}$ for which this condition holds.

4.6* Consider the restricted version of the casting problem in which we insist that the object is removed from its mold using a vertical translation (perpendicular to the top facet).

    a.   Prove that in this case there is always only a constant number of possible top facets.

    b.   Give a linear time algorithm that determines whether for a given object a mold exists under this restricted model.

4.7 Instead of removing the object from the mold by a single translation, we can also try to remove it by a single rotation. For simplicity, let's consider the planar variant of this version of the casting problem, and let's only look at clockwise rotations.

    a.   Give an example of a simple polygon $\mathcal{P}$ with top facet f that is not castable when we require that $\mathcal{P}$ should be removed from the mold by a single translation, but that is castable using rotation around a point.

    b.   Show that the problem of finding a center of rotation that allows us to remove $\mathcal{P}$ with a single rotation from its mold can be reduced to the problem of finding a point in the common intersection of a set of half-planes.

4.8 The plane $z = 1$ can be used to represent all directions of vectors in 3-dimensional space that have a positive z-value. How can we represent all directions of vectors in 3-dimensional space that have a non-negative z-value? And how can we represent the directions of all vectors in 3-dimensional space?

4.9 Let H be a set of at least three half-planes. We call a half-plane $h \in H$ *redundant* if it does not contribute an edge to $\bigcap H$. Prove that for any redundant half-plane $h \in H$ there are two half-planes $h', h'' \in H$ such that $h' \cap h'' \subset h$.

4.10 Prove that RANDOMPERMUTATION(A) is correct, that is, prove that every possible permutation of A is equally likely to be the output.

4.11 In the text we gave a linear time algorithm for computing a random permutation. The algorithm needed a random number generator that can produce a random integer between 1 and n in constant time. Now assume we have a restricted random number generator available that can

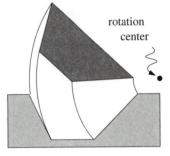

rotation
center

only generate a random bit (0 or 1) in constant time. How can we generate a random permutation with this restricted random number generator? What is the running time of your procedure?

4.12 Here is a paranoid algorithm to compute the maximum of a set A of n real numbers:

Algorithm PARANOIDMAXIMUM(A)
1.   **if** card(A) = 1
2.   **then return** the unique element $x \in A$
3.   **else** Pick a random element x from A.
4. $x' \leftarrow$ PARANOIDMAXIMUM($A \setminus \{x\}$)
5.   **if** $x \leqslant x'$
6.   **then return** x'
7.   **else** Now we suspect that x is the maximum, but to be absolutely sure, we compare x with all card(A) $- 1$ other elements of A.
8.   **return** x

What is the worst-case running time of this algorithm? What is the expected running time (with respect to the random choice in line 3)?

4.13 a.   Write a recursive version of algorithm 2DRANDOMIZEDLP.
    b.* Write a recursive version of algorithm RANDOMIZEDLP. The algorithm should use recursion in two ways: there are recursive calls where the dimension of the problem reduces by one, and there are recursive calls where the number of half-spaces reduces by one.

4.14 A simple polygon $\mathcal{P}$ is called *star-shaped* if it contains a point q such that for any point p in $\mathcal{P}$ the line segment $\overline{pq}$ is contained in $\mathcal{P}$. Give a linear time algorithm to decide whether a simple polygon is star-shaped.

4.15 On n parallel railway tracks n trains are going with constant speeds $v_1, v_2, \ldots, v_n$. At time $t = 0$ the trains are at positions $k_1, k_2, \ldots, k_n$. Give an $O(n \log n)$ algorithm that detects all trains that at some moment in time are leading. To this end, use the algorithm for computing the intersection of half-planes.

4.16* a.   Give an algorithm 3DUNBOUNDEDLP($H, \vec{c}$) to determine whether the 3-dimensional linear program $(H, \vec{c})$ is unbounded. If so, it should return a ray ρ along which $(H, \vec{c})$ is unbounded; otherwise it should return three certificates, that is, half-spaces h_1, h_2, h_3 such that $(\{h_1, h_2, h_3\}, \vec{c})$ is bounded.
     b.   Generalize the algorithm to arbitrary dimensions.

4.17* Show how to implement MINIDISC using a single routine MINIDISC-WITHPOINTS(P, R) that computes $md(P, R)$ as defined in Lemma 4.15. Your algorithm should compute only a single random permutation during the whole computation.

5 Orthogonal Range Searching

Querying a Database

At first sight it seems that databases have little to do with geometry. Nevertheless, many types of questions—from now on called *queries*—about data in a database can be interpreted geometrically. To this end we transform records in a database into points in a multi-dimensional space, and we transform the queries about the records into queries on this set of points. Let's demonstrate this with an example.

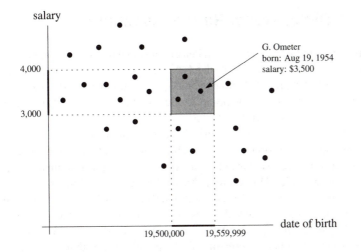

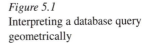

Figure 5.1
Interpreting a database query geometrically

Consider a database for personnel administration. In such a database the name, address, date of birth, salary, and so on, of each employee are stored. A typical query one may want to perform is to report all employees born between 1950 and 1955 who earn between $3,000 and $4,000 a month. To formulate this as a geometric problem we represent each employee by a point in the plane. The first coordinate of the point is the date of birth, represented by the integer $10,000 \times year + 100 \times month + day$, and the second coordinate is the monthly salary. With the point we also store the other information we have about the employee, such as name and address. The database query asking for all employees born between 1950 and 1955 who earn between $3,000 and

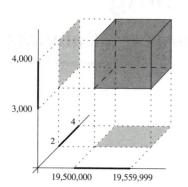

$4,000 transforms into the following geometric query: report all points whose first coordinate lies between 19,500,000 and 19,559,999, and whose second coordinate lies between 3,000 and 4,000. In other words, we want to report all the points inside an axis-parallel query rectangle—see Figure 5.1.

What if we also have information about the number of children of each employee, and we would like to be able to ask queries like "report all employees born between 1950 and 1955 who earn between $3,000 and $4,000 a month and have between two and four children"? In this case we represent each employee by a point in 3-dimensional space: the first coordinate represents the date of birth, the second coordinate the salary, and the third coordinate the number of children. To answer the query we now have to report all points inside the axis-parallel box $[19,500,000 : 19,559,999] \times [3,000 : 4,000] \times [2 : 4]$. In general, if we are interested in answering queries on d fields of the records in our database, we transform the records to points in d-dimensional space. A query asking to report all records whose fields lie between specified values then transforms to a query asking for all points inside a d-dimensional axis-parallel box. Such a query is called a *rectangular range query*, or an *orthogonal range query*, in computational geometry. In this chapter we shall study data structures for such queries.

5.1 1-Dimensional Range Searching

Before we try to tackle the 2- or higher-dimensional rectangular range searching problem, let's have a look at the 1-dimensional version. The data we are given is a set of points in 1-dimensional space—in other words, a set of real numbers. A query asks for the points inside a 1-dimensional query rectangle—in other words, an interval $[x : x']$.

Let $P := \{p_1, p_2, \ldots, p_n\}$ be the given set of points on the real line. We can solve the 1-dimensional range searching problem efficiently using a well-known data structure: a balanced binary search tree $\mathcal{T}$. (A solution that uses an array is also possible. This solution does not generalize to higher dimensions, however, nor does it allow for efficient updates on P.) The leaves of $\mathcal{T}$ store the points of P and the internal nodes of $\mathcal{T}$ store splitting values to guide the search. We denote the splitting value stored at a node v by x_v. We assume that the left subtree of a node v contains all the points smaller than or equal to x_v, and that the right subtree contains all the points strictly greater than x_v.

To report the points in a query range $[x : x']$ we proceed as follows. We search with x and x' in $\mathcal{T}$. Let μ and μ' be the two leaves where the searches end, respectively. Then the points in the interval $[x : x']$ are the ones stored in the leaves in between μ and μ' plus, possibly, the point stored at μ and the point stored at μ'. When we search with the interval $[18 : 77]$ in the tree of Figure 5.2, for instance, we have to report all the points stored in the dark grey leaves, plus the point stored in the leaf μ. How can we find the leaves in between μ and μ'? As Figure 5.2 already suggests, they are the leaves of certain subtrees in between the search paths to μ and μ'. (In Figure 5.2, these subtrees are dark

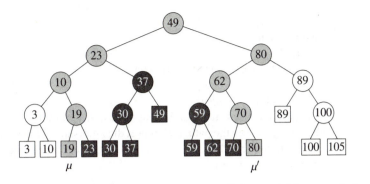

Figure 5.2
A 1-dimensional range query in a
binary search tree

grey, whereas the nodes on the search paths are light grey.) More precisely, the
subtrees that we select are rooted at nodes v in between the two search paths
whose parents are on the search path. To find these nodes we first search for
the node v_{split} where the paths to x and x' split. This is done with the following
subroutine. Let $lc(v)$ and $rc(v)$ denote the left and right child, respectively, of
a node v.

FINDSPLITNODE(T, x, x')
Input. A tree T and two values x and x' with $x \leqslant x'$.
Output. The node v where the paths to x and x' split, or the leaf where both
paths end.
1.　　$v \leftarrow root(T)$
2.　　**while** v is not a leaf **and** ($x' \leqslant x_v$ or $x > x_v$)
3.　　　**do if** $x' \leqslant x_v$
4.　　　　**then** $v \leftarrow lc(v)$
5.　　　　**else** $v \leftarrow rc(v)$
6.　　**return** v

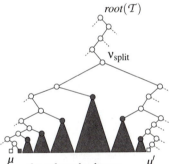

the selected subtrees

Starting from v_{split} we then follow the search path of x. At each node where the
path goes left, we report all the leaves in the right subtree, because this subtree
is in between the two search paths. Similarly, we follow the path of x' and we
report the leaves in the left subtree of nodes where the path goes right. Finally,
we have to check the points stored at the leaves where the paths end; they may
or may not lie in the range $[x : x']$.
　　Next we describe the query algorithm in more detail. It uses a subroutine
REPORTSUBTREE, which traverses the subtree rooted at a given node and reports the points stored at its leaves. Since the number of internal nodes of any
binary tree is less than its number of leaves, this subroutine takes an amount of
time that is linear in the number of reported points.

Algorithm 1DRANGEQUERY($T, [x : x']$)
Input. A range tree T and a range $[x : x']$.
Output. All points that lie in the range.
1.　　$v_{split} \leftarrow$FINDSPLITNODE(T, x, x')
2.　　**if** v_{split} is a leaf
3.　　　**then** Check if the point stored at v_{split} must be reported.
4.　　　**else** (∗ Follow the path to x and report the points in subtrees right of
　　　　　　the path. ∗)

5. $v \leftarrow lc(v_{\text{split}})$
6.   **while** v is not a leaf
7.   **do if** $x \leqslant x_v$
8.   **then** REPORTSUBTREE($rc(v)$)
9. $v \leftarrow lc(v)$
10.   **else** $v \leftarrow rc(v)$
11.   Check if the point stored at the leaf v must be reported.
12.   Similarly, follow the path to x', report the points in subtrees left of the path, and check if the point stored at the leaf where the path ends must be reported.

We first prove the correctness of the algorithm.

Lemma 5.1 *Algorithm* 1DRANGEQUERY *reports exactly those points that lie in the query range.*

Proof. We first show that any reported point p lies in the query range. If p is stored at the leaf where the path to x or to x' ends, then p is tested explicitly for inclusion in the query range. Otherwise, p is reported in a call to REPORTSUB-TREE. Assume this call was made when we followed the path to x. Let v be the node on the path such that p was reported in the call REPORTSUBTREE($rc(v)$). Since v and, hence, $rc(v)$ lie in the left subtree of v_{split}, we have $p < x_{v_{\text{split}}}$. Because the search path of x' goes right at v_{split} this means that $p < x'$. On the other hand, the search path of x goes left at v and p is in the right subtree of v, so $x < p$. It follows that $p \in [x : x']$. The proof that p lies in the range when it is reported while following the path to x' is symmetrical.

It remains to prove that any point p in the range is reported. Let μ be the leaf where p is stored, and let v be the lowest ancestor of μ that is visited by the query algorithm. We claim that $v = \mu$, which implies that p is reported. Assume for a contradiction that $v \neq \mu$. Observe that v cannot be a node visited in a call to REPORTSUBTREE, because all descendants of such a node are visited. Hence, v is either on the search path to x, or on the search path to x', or both. Because all three cases are similar, we only consider the third case. Assume first that μ is in the left subtree of v. Then the search path of x goes right at v (otherwise v would not be the lowest visited ancestor). But this implies that $p < x$. Similarly, if μ is in the right subtree of v, then the path of x' goes left at v, and $p > x'$. In both cases, the assumption that p lies in the range is contradicted. $\qquad\qquad\Box$

We now turn our attention to the performance of the data structure. Because it is a balanced binary search tree, it uses $O(n)$ storage and it can be built in $O(n \log n)$ time. What about the query time? In the worst case all the points could be in the query range. In this case the query time will be $\Theta(n)$, which seems bad. Indeed, we do not need any data structure to achieve $\Theta(n)$ query time; simply checking all the points against the query range leads to the same result. On the other hand, a query time of $\Theta(n)$ cannot be avoided when we have to report all the points. Therefore we shall give a more refined analysis of the query time. The refined analysis takes not only n, the number of points in the set P, into account, but also k, the number of reported points. In other words, we will show that the query algorithm is *output-sensitive*, a concept we already encountered in Chapter 2.

Recall that the time spent in a call to REPORTSUBTREE is linear in the number of reported points. Hence, the total time spent in all such calls is $O(k)$. The remaining nodes that are visited are nodes on the search path of x or x'. Because $\mathcal{T}$ is balanced, these paths have length $O(\log n)$. The time we spend at each node is $O(1)$, so the total time spent in these nodes is $O(\log n)$, which gives a query time of $O(\log n + k)$.

The following theorem summarizes the results for 1-dimensional range searching:

Theorem 5.2 *Let P be a set of n points in 1-dimensional space. The set P can be stored in a balanced binary search tree, which uses $O(n)$ storage and has $O(n \log n)$ construction time, such that the points in a query range can be reported in time $O(k + \log n)$, where k is the number of reported points.*

5.2 Kd-Trees

Now let's go to the 2-dimensional rectangular range searching problem. Let P be a set of n points in the plane. In the remainder of this section we assume that no two points in P have the same x-coordinate, and no two points have the same y-coordinate. This restriction is not very realistic, especially not if the points represent employees and the coordinates are things like salary or number of children. Fortunately, the restriction can be overcome with a nice trick that we describe in Section 5.5.

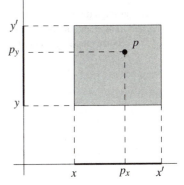

A 2-dimensional rectangular range query on P asks for the points from P lying inside a query rectangle $[x : x'] \times [y : y']$. A point $p := (p_x, p_y)$ lies inside this rectangle if and only if

$$p_x \in [x : x'] \qquad \text{and} \qquad p_y \in [y : y'].$$

We could say that a 2-dimensional rectangular range query is composed of two 1-dimensional sub-queries, one on the x-coordinate of the points and one on the y-coordinate.

In the previous section we saw a data structure for 1-dimensional range queries. How can we generalize this structure—which was just a binary search tree—to 2-dimensional range queries? Let's consider the following recursive definition of the binary search tree: the set of (1-dimensional) points is split into two subsets of roughly equal size; one subset contains the points smaller than or equal to the splitting value, the other subset contains the points larger than the splitting value. The splitting value is stored at the root, and the two subsets are stored recursively in the two subtrees.

In the 2-dimensional case each point has two values that are important: its x- and its y-coordinate. Therefore we first split on x-coordinate, next on y-coordinate, then again on x-coordinate, and so on. More precisely, the process

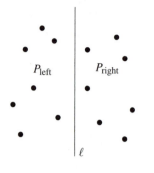

is as follows. At the root we split the set P with a vertical line ℓ into two subsets of roughly equal size. The splitting line is stored at the root. P_{left}, the subset of points to the left or on the splitting line, is stored in the left subtree, and P_{right}, the subset to the right of it, is stored in the right subtree. At the left child of the root we split P_{left} into two subsets with a horizontal line; the points below or on it are stored in the left subtree of the left child, and the points above it are stored in the right subtree. The left child itself stores the splitting line. Similarly, the set P_{right} is split with a horizontal line into two subsets, which are stored in the left and right subtree of the right child. At the grandchildren of the root, we split again with a vertical line. In general, we split with a vertical line at nodes whose depth is even, and we split with a horizontal line at nodes whose depth is odd. Figure 5.3 illustrates how the splitting is done and what the corresponding binary tree looks like. A tree like this is called a *kd-tree*. Originally, the name

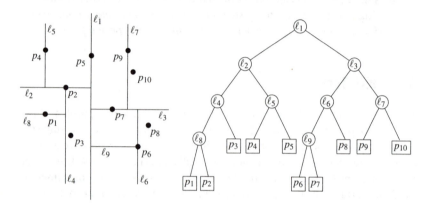

Figure 5.3
A kd-tree: on the left the way the plane is subdivided and on the right the corresponding binary tree

stood for k-dimensional tree; the tree we described above would be a 2d-tree. Nowadays, the original meaning is lost, and what used to be called a 2d-tree is now called a 2-dimensional kd-tree.

We can construct a kd-tree with the recursive procedure described below. This procedure has two parameters: a set of points and an integer. The first parameter is the set for which we want to build the kd-tree; initially this is the set P. The second parameter is depth of recursion or, in other words, the depth of the root of the subtree that the recursive call constructs. The depth parameter is zero at the first call. The depth is important because, as explained above, it determines whether we must split with a vertical or a horizontal line. The procedure returns the root of the kd-tree.

Algorithm BUILDKDTREE($P, depth$)
Input. A set of points P and the current depth $depth$.
Output. The root of a kd-tree storing P.
1. **if** P contains only one point
2.   **then return** a leaf storing this point
3.   **else if** $depth$ is even
4.   **then** Split P into two subsets with a vertical line ℓ through the
            median x-coordinate of the points in P. Let P_1 be the set of

points to the left of ℓ or on ℓ, and let P_2 be the set of points to the right of ℓ.

5.   **else** Split P into two subsets with a horizontal line ℓ through the median y-coordinate of the points in P. Let P_1 be the set of points below ℓ or on ℓ, and let P_2 be the set of points above ℓ.

6. $v_{\text{left}} \leftarrow$ BUILDKDTREE($P_1, depth + 1$)

7. $v_{\text{right}} \leftarrow$ BUILDKDTREE($P_2, depth + 1$)

8.   Create a node v storing ℓ, make v_{left} the left child of v, and make v_{right} the right child of v.

9.   **return** v

The algorithm uses the convention that the point on the splitting line—the one determining the median x- or y-coordinate—belongs to the subset to the left of, or below, the splitting line. For this to work correctly, the median of a set of n numbers should be defined as the $\lfloor n/2 \rfloor$-th smallest number. This means that the median of two values is the smaller one, which ensures that the algorithm terminates.

Before we come to the query algorithm, let's analyze the construction time of a 2-dimensional kd-tree. The most expensive step that is performed at every recursive call is finding the splitting line. This requires determining the median x-coordinate or the median y-coordinate, depending on whether the depth is even or odd. Median finding can be done in linear time. Linear time median finding algorithms, however, are rather complicated. A better approach is to presort the set of points both on x- and on y-coordinate. The parameter set P is now passed to the procedure in the form of two sorted lists, one on x-coordinate and one on y-coordinate. Given the two sorted lists, it is easy to find the median x-coordinate (when the depth is even) or the median y-coordinate (when the depth is odd) in linear time. It is also easy to construct the sorted lists for the two recursive calls in linear time from the given lists. Hence, the building time $T(n)$ satisfies the recurrence

$$T(n) = \begin{cases} O(1), & \text{if } n = 1, \\ O(n) + 2T(\lceil n/2 \rceil), & \text{if } n > 1, \end{cases}$$

which solves to $O(n \log n)$. This bound subsumes the time we spend for pre-sorting the points on x- and y-coordinate.

To bound the amount of storage we note that each leaf in the kd-tree stores a distinct point of P. Hence, there are n leaves. Because a kd-tree is a binary tree, and every leaf and internal node uses $O(1)$ storage, this implies that the total amount of storage is $O(n)$. This leads to the following lemma.

Lemma 5.3 *A kd-tree for a set of n points uses $O(n)$ storage and can be constructed in $O(n \log n)$ time.*

We now turn to the query algorithm. The splitting line stored at the root partitions the plane into two half-planes. The points in the left half-plane are stored in the left subtree, and the points in the right half-plane are stored in the right subtree. In a sense, the left child of the root corresponds to the left half-plane and the right child corresponds to the right half-plane. (The convention used in BUILDKDTREE that the point on the splitting line belongs to the left subset implies that the left half-plane is closed to the right and the right half-plane is open to the left.) The other nodes in a kd-tree correspond to a region of the plane as well. The left child of the left child of the root, for instance, corresponds to the region bounded to the right by the splitting line stored at the root and bounded from above by the line stored at the left child of the root. In general, the region corresponding to a node v is a rectangle, which can be unbounded on one or more sides. It is bounded by splitting lines stored at ancestors of v—see Figure 5.4. We denote the region corresponding to a node

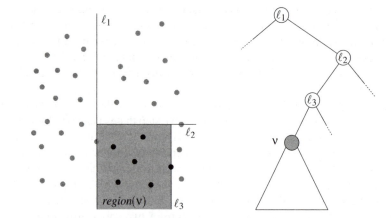

Figure 5.4
Correspondence between nodes in a
kd-tree and regions in the plane

v by *region*(v). The region of the root of a kd-tree is simply the whole plane. Observe that a point is stored in the subtree rooted at a node v if and only if it lies in *region*(v). For instance, the subtree of the node v in Figure 5.4 stores the points indicated as black dots. Therefore we have to search the subtree rooted at v only if the query rectangle intersects *region*(v). This observation leads to the following query algorithm: we traverse the kd-tree, but visit only nodes whose region is intersected by the query rectangle. When a region is fully contained in the query rectangle, we can report all the points stored in its subtree. When the traversal reaches a leaf, we have to check whether the point stored at the leaf is contained in the query region and, if so, report it. Figure 5.5 illustrates the query algorithm. (Note that the kd-tree of Figure 5.5 could not have been constructed by Algorithm BUILDKDTREE; the median wasn't always chosen as the split value.) The grey nodes are visited when we query with the grey rectangle. The node marked with a star corresponds to a region that is completely contained in the query rectangle; in the figure this rectangular region is shown darker. Hence, the dark grey subtree rooted at this node is traversed and all points stored in it are reported. The other leaves that are visited correspond to regions that are only partially inside the query rectangle. Hence, the points stored in them must be tested for inclusion in the query range; this results in points p_6 and p_{11} being reported, and points p_3, p_{12}, and p_{13} not being reported. The query algorithm is described by the following re-

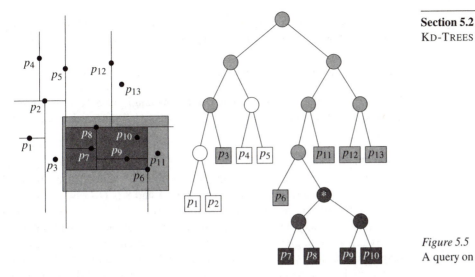

Figure 5.5
A query on a kd-tree

cursive procedure, which takes as arguments the root of a kd-tree and the query range R. It uses a subroutine REPORTSUBTREE(ν), which traverses the subtree rooted at a node ν and reports all the points stored at its leaves. Recall that $lc(\nu)$ and $rc(\nu)$ denote the left and right child of a node ν, respectively.

Algorithm SEARCHKDTREE(ν, R)
Input. The root of a (subtree of a) kd-tree and a range R.
Output. All points at leaves below ν that lie in the range.
1. **if** ν is a leaf
2.   **then** Report the point stored at ν if it lies in R.
3.   **else if** *region*$(lc(\nu))$ is fully contained in R
4.   **then** REPORTSUBTREE($lc(\nu)$)
5.   **else if** *region*$(lc(\nu))$ intersects R
6.   **then** SEARCHKDTREE($lc(\nu), R$)
7.   **if** *region*$(rc(\nu))$ is fully contained in R
8.   **then** REPORTSUBTREE($rc(\nu)$)
9.   **else if** *region*$(rc(\nu))$ intersects R
10.   **then** SEARCHKDTREE($rc(\nu), R$)

The main test the query algorithm performs is whether the query range R intersects the region corresponding to some node ν. To be able to do this test we can compute *region*(ν) for all nodes ν during the preprocessing phase and store it, but this is not necessary: one can maintain the current region through the recursive calls using the lines stored in the internal nodes. For instance, the region corresponding to the left child of a node ν at even depth can be computed from *region*(ν) as follows:

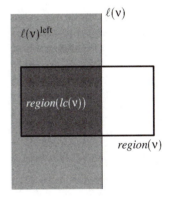

$$region(lc(\nu)) = region(\nu) \cap \ell(\nu)^{\text{left}},$$

where $\ell(\nu)$ is the splitting line stored at ν, and $\ell(\nu)^{\text{left}}$ is the half-plane to the left of and including $\ell(\nu)$.

Observe that the query algorithm above never assumes that the query range R is a rectangle. Indeed, it works for any other query range as well.

We now analyze the time a query with a rectangular range takes.

Lemma 5.4 *A query with an axis-parallel rectangle in a kd-tree storing n points can be performed in $O(\sqrt{n}+k)$ time, where k is the number of reported points.*

Proof. First of all, note that the time to traverse a subtree and report the points stored in its leaves is linear in the number of reported points. Hence, the total time required for traversing subtrees in steps 4 and 8 is $O(k)$, where k is the total number of reported points. It remains to bound the number of nodes visited by the query algorithm that are not in one of the traversed subtrees. (These are the light grey nodes in Figure 5.5.) For each such node v, the query range properly intersects $region(v)$, that is, $region(v)$ is intersected by, but not fully contained in the range. In other words, the boundary of the query range intersects $region(v)$. To analyze the number of such nodes, we shall bound the number of regions intersected by any vertical line. This will give us an upper bound on the number of regions intersected by the left and right edge of the query rectangle. The number of regions intersected by the bottom and top edges of the query range can be bounded in the same way.

Let ℓ be a vertical line, and let $\mathcal{T}$ be a kd-tree. Let $\ell(root(\mathcal{T}))$ be the splitting line stored at the root of the kd-tree. The line ℓ intersects either the region to the left of $\ell(root(\mathcal{T}))$ or the region to the right of $\ell(root(\mathcal{T}))$, but not both. This observation seems to imply that $Q(n)$, the number of intersected regions in a kd-tree storing a set of n points, satisfies the recurrence $Q(n) = 1 + Q(n/2)$. But this is not true, because the splitting lines are horizontal at the children of the root. This means that if the line ℓ intersects for instance $region(lc(root(\mathcal{T})))$, then it will always intersect the regions corresponding to both children of $lc(root(\mathcal{T}))$. Hence, the recursive situation we get is not the same as the original situation, and the recurrence above is incorrect. To overcome this problem we have to make sure that the recursive situation is exactly the same as the original situation: the root of the subtree must contain a vertical splitting line. This leads us to redefine $Q(n)$ as the number of intersected regions in a kd-tree storing n points whose root contains a vertical splitting line. To write a recurrence for $Q(n)$ we now have to go down two steps in the tree. Each of the four nodes at depth two in the tree corresponds to a region containing $n/4$ points. (To be precise, a region can contain at most $\lceil \lceil n/2 \rceil / 2 \rceil = \lceil n/4 \rceil$ points, but asymptotically this does not influence the outcome of the recurrence below.) Two of the four nodes correspond to intersected regions, so we have to count the number of intersected regions in these subtrees recursively. Moreover, ℓ intersects the region of the root and of one of its children. Hence, $Q(n)$ satisfies the recurrence

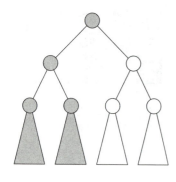

$$Q(n) = \begin{cases} O(1), & \text{if } n = 1, \\ 2 + 2Q(n/4), & \text{if } n > 1. \end{cases}$$

This recurrence solves to $Q(n) = O(\sqrt{n})$. In other words, any vertical line intersects $O(\sqrt{n})$ regions in a kd-tree. In a similar way one can prove that the total number of regions intersected by a horizontal line is $O(\sqrt{n})$. The total number of regions intersected by the boundary of a rectangular query range is bounded by $O(\sqrt{n})$ as well. ⊟

The analysis of the query time that we gave above is rather pessimistic: we bounded the number of regions intersecting an edge of the query rectangle by the number of regions intersecting the line through it. In many practical situations the range will be small. As a result, the edges are short and will intersect much fewer regions. For example, when we search with a range $[x : x] \times [y : y]$—this query effectively asks whether the point (x, y) is in the set—the query time is bounded by $O(\log n)$.

The following theorem summarizes the performance of kd-trees.

Theorem 5.5 *A kd-tree for a set P of n points in the plane uses $O(n)$ storage and can be built in $O(n \log n)$ time. A rectangular range query on the kd-tree takes $O(\sqrt{n} + k)$ time, where k is the number of reported points.*

Kd-trees can be also be used for point sets in 3- or higher-dimensional space. The construction algorithm is very similar to the planar case: At the root, we split the set of points into two subsets of roughly the same size by a hyperplane perpendicular to the x_1-axis. In other words, at the root the point set is partitioned based on the first coordinate of the points. At the children of the root the partition is based on the second coordinate, at nodes at depth two on the third coordinate, and so on, until at depth $d - 1$ we partition on the last coordinate. At depth d we start all over again, partitioning on first coordinate. The recursion stops when there is only one point left, which is then stored at a leaf. Because a d-dimensional kd-tree for a set of n points is a binary tree with n leaves, it uses $O(n)$ storage. The construction time is $O(n \log n)$. (Here we assume d to be a constant. The precise bounds are $O(d \cdot n)$ for storage and $O(d \cdot n \log n)$ for construction.)

Nodes in a d-dimensional kd-tree correspond to regions, as in the plane. The query algorithm visits those nodes whose regions are properly intersected by the query range, and traverses subtrees (to report the points stored in the leaves) that are rooted at nodes whose region is fully contained in the query range. It can be shown that the query time is bounded by $O(n^{1-1/d} + k)$.

5.3 Range Trees

Kd-trees, which were described in the previous section, have $O(\sqrt{n} + k)$ query time. So when the number of reported points is small, the query time is relatively high. In this section we shall describe another data structure for rectangular range queries, the *range tree*, which has a better query time, namely

$O(\log^2 n + k)$. The price we have to pay for this improvement is an increase in storage from $O(n)$ for kd-trees to $O(n\log n)$ for range trees.

As we observed before, a 2-dimensional range query is essentially composed of two 1-dimensional sub-queries, one on the x-coordinate of the points and one on the y-coordinate. This gave us the idea to split the given point set alternately on x- and y-coordinate, leading to the kd-tree. To obtain the range tree, we shall use this observation in a different way.

Let P be a set of n points in the plane that we want to preprocess for rectangular range queries. Let $[x : x'] \times [y : y']$ be the query range. We first concentrate on finding the points whose x-coordinate lies in $[x : x']$, the x-interval of the query rectangle, and worry about the y-coordinate later. If we only care about the x-coordinate then the query is a 1-dimensional range query. In Section 5.1 we have seen how to answer such a query: with a binary search tree on the x-coordinate of the points. The query algorithm was roughly as follows. We search with x and x' in the tree until we get to a node v_{split} where the search paths split. From the left child of v_{split} we continue the search with x, and at every node v where the search path of x goes left, we report all points in the right subtree of v. Similarly, we continue the search with x' at the right child of v_{split}, and at every node v where the search path of x' goes right we report all points in the left subtree of v. Finally, we check the leaves μ and μ' where the two paths end to see if they contain a point in the range. In effect, we select a collection of subtrees that together contain exactly the points whose x-coordinate lies in the x-interval of the query rectangle. For any range, there is only a logarithmic number of subtrees selected.

Let's call the subset of points stored in the leaves of the subtree rooted at a node v the *canonical subset* of v. The canonical subset of the root of the tree, for instance, is the whole set P. The canonical subset of a leaf is simply the point stored at that leaf. We denote the canonical subset of node v by $P(v)$. We have just seen that the subset of points whose x-coordinate lies in a query range can be expressed as the disjoint union of $O(\log n)$ canonical subsets; these are the sets $P(v)$ of the nodes v that are the roots of the selected subtrees. We are not interested in all the points in such a canonical subset $P(v)$, but only want to report the ones whose y-coordinate lies in the interval $[y : y']$. This is another 1-dimensional query, which we can solve, provided we have a binary search tree on the y-coordinate of the points in $P(v)$ available. This leads to the following data structure for rectangular range queries on a set P of n points in the plane.

- The main tree is a balanced binary search tree $\mathcal{T}$ built on the x-coordinate of the points in P.

- For any internal or leaf node v in $\mathcal{T}$, the canonical subset $P(v)$ is stored in a balanced binary search tree $\mathcal{T}_{\text{assoc}}(v)$ on the y-coordinate of the points. The node v stores a pointer to the root of $\mathcal{T}_{\text{assoc}}(v)$, which is called the *associated structure* of v.

This data structure is called a range tree. Figure 5.6 shows the structure of a range tree. Data structures where nodes have pointers to associated structures

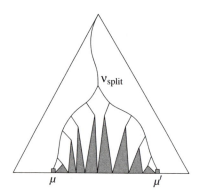

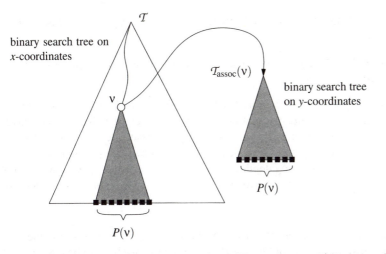

binary search tree on
x-coordinates

$\mathcal{T}$

$\mathcal{T}_{assoc}(v)$

v

binary search tree
on y-coordinates

$P(v)$

$P(v)$

Figure 5.6
A 2-dimensional range tree

are often called *multi-level data structures*. The main tree $\mathcal{T}$ is then called the *first-level tree*, and the associated structures are *second-level trees*. Multi-level data structures play an important role in computational geometry; more examples can be found in Chapters 10 and 16.

A range tree can be constructed with the following recursive algorithm, which receives as input the set $P := \{p_1, ..., p_n\}$ of points sorted on x-coordinate and returns the root of a 2-dimensional range tree $\mathcal{T}$ of P. As in the previous section, we assume that no two points have the same x- or y-coordinate. We shall get rid of this assumption in Section 5.5.

Algorithm BUILD2DRANGETREE(P)
Input. A set P of points in the plane.
Output. The root of a 2-dimensional range tree.
1. Construct the associated structure: Build a binary search tree $\mathcal{T}_{assoc}$ on the set P_y of y-coordinates of the points in P. Store at the leaves of $\mathcal{T}_{assoc}$ not just the y-coordinate of the points in P_y, but the points themselves.
2. **if** P contains only one point
3. **then** Create a leaf v storing this point, and make $\mathcal{T}_{assoc}$ the associated structure of v.
4. **else** Split P into two subsets; one subset P_{left} contains the points with x-coordinate less than or equal to x_{mid}, the median x-coordinate, and the other subset P_{right} contains the points with x-coordinate larger than x_{mid}.
5. $v_{left} \leftarrow$ BUILD2DRANGETREE(P_{left})
6. $v_{right} \leftarrow$ BUILD2DRANGETREE(P_{right})
7. Create a node v storing x_{mid}, make v_{left} the left child of v, make v_{right} the right child of v, and make $\mathcal{T}_{assoc}$ the associated structure of v.
8. **return** v

Note that in the leaves of the associated structures we do not just store the y-coordinate of the points but the points themselves. This is important because, when searching the associated structures, we need to report the points and not just the y-coordinates.

Lemma 5.6 *A range tree on a set of n points in the plane requires $O(n \log n)$ storage.*

Proof. A point p in P is stored only in the associated structure of nodes on the path in T towards the leaf containing p. Hence, for all nodes at a given depth of T, the point p is stored in exactly one associated structure. Because 1-dimensional range trees use linear storage it follows that the associated structures of all nodes at any depth of T together use $O(n)$ storage. The depth of T is $O(\log n)$. Hence, the total amount of storage required is bounded by $O(n \log n)$. ☐

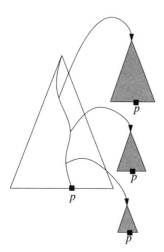

Algorithm BUILD2DRANGETREE as it is described will not result in the optimal construction time of $O(n \log n)$. To obtain this we have to be a bit careful. Constructing a binary search tree on an unsorted set of n keys takes $O(n \log n)$ time. This means that constructing the associated structure in line 1 would take $O(n \log n)$ time. But we can do better if the points in P_y are presorted on y-coordinate; then the binary search tree can be constructed bottom-up in linear time. During the construction algorithm we therefore maintain the set of points in two lists, one sorted on x-coordinate and one sorted on y-coordinate. This way the time we spend at a node in the main tree T is linear in the size of its canonical subset. This implies that the total construction time is the same as the amount of storage, namely $O(n \log n)$. Since the presorting takes $O(n \log n)$ time as well, the total construction time is again $O(n \log n)$.

The query algorithm first selects $O(\log n)$ canonical subsets that together contain the points whose x-coordinate lie in the range $[x : x']$. This can be done with the 1-dimensional query algorithm. Of those subsets, we then report the points whose y-coordinate lie in the range $[y : y']$. For this we also use the 1-dimensional query algorithm; this time it is applied to the associated structures that store the selected canonical subsets. Thus the query algorithm is virtually the same as the 1-dimensional query algorithm 1DRANGEQUERY; the only difference is that calls to REPORTSUBTREE are replaced by calls to 1DRANGEQUERY.

Algorithm 2DRANGEQUERY$(T, [x : x'] \times [y : y'])$
Input. A 2-dimensional range tree T and a range $[x : x'] \times [y : y']$.
Output. All points in T that lie in the range.
1. $v_{split} \leftarrow$ FINDSPLITNODE(T, x, x')
2. **if** v_{split} is a leaf
3.   **then** Check if the point stored at v_{split} must be reported.
4.   **else** (∗ Follow the path to x and call 1DRANGEQUERY on the subtrees right of the path. ∗)
5. $v \leftarrow lc(v_{split})$
6.   **while** v is not a leaf
7.   **do if** $x \leqslant x_v$

8.   **then** 1DRANGEQUERY($\mathcal{T}_{assoc}(rc(v)), [y : y']$)
9. $v \leftarrow lc(v)$
10.   **else** $v \leftarrow rc(v)$
11.   Check if the point stored at v must be reported.
12.   Similarly, follow the path from $rc(v_{split})$ to x', call 1DRANGE-QUERY with the range $[y : y']$ on the associated structures of subtrees left of the path, and check if the point stored at the leaf where the path ends must be reported.

Lemma 5.7 *A query with an axis-parallel rectangle in a range tree storing n points takes $O(\log^2 n + k)$ time, where k is the number of reported points.*

Proof. At each node v in the main tree $\mathcal{T}$ we spend constant time to decide where the search path continues, and we possibly call 1DRANGEQUERY. Theorem 5.2 states that the time we spend in this recursive call is $O(\log n + k_v)$, where k_v is the number of points reported in this call. Hence, the total time we spend is

$$\sum_v O(\log n + k_v),$$

where the summation is over all nodes in the main tree $\mathcal{T}$ that are visited. Notice that the sum $\sum_v k_v$ equals k, the total number of reported points. Furthermore, the search paths of x and x' in the main tree $\mathcal{T}$ have length $O(\log n)$. Hence, $\sum_v O(\log n) = O(\log^2 n)$. The lemma follows. ☐

The following theorem summarizes the performance of 2-dimensional range trees.

Theorem 5.8 *Let P be a set of n points in the plane. A range tree for P uses $O(n \log n)$ storage and can be constructed in $O(n \log n)$ time. By querying this range tree one can report the points in P that lie in a rectangular query range in $O(\log^2 n + k)$ time, where k is the number of reported points.*

The query time stated in Theorem 5.8 can be improved to $O(\log n + k)$ by a technique called *fractional cascading*. This is described in Section 5.6.

5.4 Higher-Dimensional Range Trees

It is fairly straightforward to generalize 2-dimensional range trees to higher-dimensional range trees. We only describe the global approach.

Let P be a set of points in d-dimensional space. We construct a balanced binary search tree on the first coordinate of the points. The canonical subset $P(v)$ of a node v in this first-level tree, the main tree, consists of the points stored in the leaves of the subtree rooted at v. For each node v we construct an associated structure $\mathcal{T}_{assoc}(v)$; the second-level tree $\mathcal{T}_{assoc}(v)$ is a $(d-1)$-dimensional range tree for the points in $P(v)$, restricted to their last $d-1$ coordinates. This

$(d-1)$-dimensional range tree is constructed recursively in the same way: it is a balanced binary search tree on the second coordinate of the points, in which each node has a pointer to a $(d-2)$-dimensional range tree of the points in its subtree, restricted to the last $(d-2)$ coordinates. The recursion stops when we are left with points restricted to their last coordinate; these are stored in a 1-dimensional range tree—a balanced binary search tree.

The query algorithm is also very similar to the 2-dimensional case. We use the first-level tree to locate $O(\log n)$ nodes whose canonical subsets together contain all the points whose first coordinates are in the correct range. These canonical subsets are queried further by performing a range query on the corresponding second-level structures. In each second-level structure we select $O(\log n)$ canonical subsets. This means there are $O(\log^2 n)$ canonical subsets in the second-level structures in total. Together, they contain all points whose first and second coordinate lie in the correct ranges. The third-level structures storing these canonical subsets are then queried with the range for the third coordinate, and so on, until we reach the 1-dimensional trees. In these trees we find the points whose last coordinate lies in the correct range and report them. This approach leads to the following result.

Theorem 5.9 *Let P be a set of n points in d-dimensional space, where $d \geqslant 2$. A range tree for P uses $O(n \log^{d-1} n)$ storage and it can be constructed in $O(n \log^{d-1} n)$ time. One can report the points in P that lie in a rectangular query range in $O(\log^d n + k)$ time, where k is the number of reported points.*

Proof. Let $T_d(n)$ denote the construction time for a range tree on a set of n points in d-dimensional space. By Theorem 5.8 we know that $T_2(n) = O(n \log n)$. The construction of a d-dimensional range tree consists of building a balanced binary search tree, which takes time $O(n \log n)$, and the construction of associated structures. At the nodes at any depth of the first-level tree, each point is stored in exactly one associated structure. The time required to build all associated structures of the nodes at some depth is $O(T_{d-1}(n))$, the time required to build the associated structure of the root. This follows because the building time is at least linear. Hence, the total construction time satisfies

$$T_d(n) = O(n \log n) + O(\log n) \cdot T_{d-1}(n).$$

Since $T_2(n) = O(n \log n)$, this recurrence solves to $O(n \log^{d-1} n)$. The bound on the amount of storage follows in the same way.

Let $Q_d(n)$ denote the time spent in querying a d-dimensional range tree on n points, not counting the time to report points. Querying the d-dimensional range tree involves searching in a first-level tree, which takes time $O(\log n)$, and querying a logarithmic number of $(d-1)$-dimensional range trees. Hence,

$$Q_d(n) = O(\log n) + O(\log n) \cdot Q_{d-1}(n),$$

where $Q_2(n) = O(\log^2 n)$. This recurrence easily solves to $Q_d(n) = O(\log^d n)$. We still have to add the time needed to report points, which is bounded by $O(k)$. The bound on the query time follows. $\qed$

As in the 2-dimensional case, the query time can be improved by a logarithmic factor—see Section 5.6.

5.5 General Sets of Points

Until now we imposed the restriction that no two points have equal x- or y-coordinate, which is highly unrealistic. Fortunately, this is easy to remedy. The crucial observation is that we never assumed the coordinate values to be real numbers. We only need that they come from a totally ordered universe, so that we can compare any two coordinates and compute medians. Therefore we can use the trick described next.

We replace the coordinates, which are real numbers, by elements of the so-called *composite-number space*. The elements of this space are pairs of reals. The *composite number* of two reals a and b is denoted by $(a|b)$. We define a total order on the composite-number space by using a lexicographic order. So, for two composite numbers $(a|b)$ and $(a'|b')$, we have

$$(a|b) < (a'|b') \quad \Leftrightarrow \quad a < a' \text{ or } (a = a' \text{ and } b < b').$$

Now assume we are given a set P of n points in the plane. The points are distinct, but many points can have the same x- or y-coordinate. We replace each point $p := (p_x, p_y)$ by a new point $p' := ((p_x|p_y), (p_y|p_x))$ that has composite numbers as coordinate values. This way we obtain a new set P' of n points. The first coordinate of any two points in P' are distinct; the same holds true for the second coordinate. Using the order defined above one can now construct kd-trees and 2-dimensional range trees for P'.

Now suppose we want to report the points of P that lie in a range $R := [x : x'] \times [y : y']$. To this end we must query the tree we have constructed for P'. This means that we must also transform the query range to our new composite space. The transformed range R' is defined as follows:

$$R' := [(x| - \infty) : (x'| + \infty)] \times [(y| - \infty) : (y'| + \infty)].$$

It remains to prove that our approach is correct, that is, that the points of P' that we report when we query with R' correspond exactly to the points of P that lie in R.

Lemma 5.10 *Let p be a point and R a rectangular range. Then*

$$p \in R \quad \Leftrightarrow \quad p' \in R'.$$

Proof. Let $R := [x : x'] \times [y : y']$. Assume that $p := (p_x, p_y)$ lies inside R, that is, that $x \leqslant p_x \leqslant x'$ and $y \leqslant p_y \leqslant y'$. This is easily seen to imply that $(x| - \infty) \leqslant (p_x|p_y) \leqslant (x'| + \infty)$ and $(y| - \infty) \leqslant (p_y|p_x) \leqslant (y'| + \infty)$. Hence, p' lies in R'. Now assume that p does not lie in R. So either $p_x < x$, or $p_x > x'$, or $p_y < y$, or $p_y > y'$. In the first case we have $(p_x|p_y) < (x| - \infty)$, so p' does not lie in R'. Also in the other three cases one can show that p' does not lie in R'. $\square$

We can conclude that our approach is indeed correct: we will get the correct answer to a query. Observe that there is no need to actually store the transformed points: we can just store the original points, provided that we do comparisons between two x-coordinates or two y-coordinates in the composite space.

The approach of using composite numbers can also be used in higher dimensions.

5.6* Fractional Cascading

In Section 5.3 we described a data structure for rectangular range queries in the plane, the range tree, whose query time is $O(\log^2 n + k)$. (Here n is the total number of points stored in the data structure, and k is the number of reported points.) In this section we describe a technique, called *fractional cascading*, to reduce the query time to $O(\log n + k)$.

Let's briefly recall how a range tree works. A range tree for a set P of points in the plane is a two-level data structure. The main tree is a binary search tree on the x-coordinate of the points. Each node v in the main tree has an associated structure $\mathcal{T}_{assoc}(v)$, which is a binary search tree on the y-coordinate of the points in $P(v)$, the canonical subset of v. A query with a rectangular range $[x : x'] \times [y : y']$ is performed as follows: First, a collection of $O(\log n)$ nodes in the main tree is identified whose canonical subsets together contain the points with x-coordinate in the range $[x : x']$. Second, the associated structures of these nodes are queried with the range $[y : y']$. Querying an associated structure $\mathcal{T}_{assoc}(v)$ is a 1-dimensional range query, so it takes $O(\log n + k_v)$ time, where k_v is the number of reported points. Hence, the total query time is $O(\log^2 n + k)$.

If we could perform the searches in the associated structures in $O(1 + k_v)$ time, then the total query time would reduce to $O(\log n + k)$. But how can we do this? In general, it is not possible to answer a 1-dimensional range query in $O(1 + k)$ time, with k the number of answers. What saves us is that we have to do *many* 1-dimensional searches with the *same range*, and that we can use the result of one search to speed up other searches.

We first illustrate the idea of fractional cascading with a simple example. Let S_1 and S_2 be two sets of objects, where each object has a key that is a real number. These sets are stored in sorted order in arrays A_1 and A_2. Suppose we want to report all objects in S_1 and in S_2 whose keys lie in a query interval $[y : y']$. We can do this as follows: we do a binary search with y in A_1 to find the smallest key larger than or equal to y. From there we walk through the array to the right, reporting the objects we pass, until a key larger than y' is encountered. The objects from S_2 can be reported in a similar fashion. If the total number of reported objects is k, then the query time will be $O(k)$ plus the time for two binary searches, one in A_1 and one in A_2. If, however, the keys of the objects in S_2 are a subset of the keys of the objects in S_1, then we can avoid the second binary search as follows. We add pointers from the entries in A_1 to the entries in A_2: if $A_1[i]$ stores an object with key y_i, then we store a pointer to the entry in A_2 with the smallest key larger than or equal to y_i. If there is no such key then the pointer from $A_1[i]$ is **nil**. Figure 5.7 illustrates this. (Only the keys are shown in this figure, not the corresponding objects.)

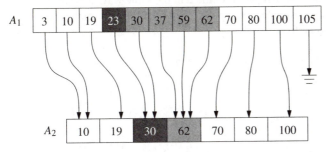

Figure 5.7
Speeding up the search by adding
pointers

How can we use this structure to report the objects in S_1 and S_2 whose keys are in a query interval $[y : y']$? Reporting the objects in S_1 is still done as before: do a binary search with y in A_1, and walk through A_1 to the right until a key larger than y' is encountered. To report the points from S_2 we proceed as follows. Let the search for y in A_1 end at $A[i]$. Hence, the key of $A[i]$ is the smallest one in S_1 that is larger than or equal to y. Since the keys from S_2 form a subset of the keys from S_1, this means that the pointer from $A[i]$ must point to the smallest key from S_2 larger than or equal to y. Hence, we can follow this pointer, and from there start to walk to the right through A_2. This way the binary search in A_2 is avoided, and reporting the objects from S_2 takes only $O(1+k)$ time, with k the number of reported answers.

Figure 5.7 shows an example of a query. We query with the range $[20 : 65]$. First we use binary search in A_1 to find 23, the smallest key larger than or equal to 20. From there we walk to the right until we encounter a key larger than 65. The objects that we pass have their keys in the range, so they are reported. Then we follow the pointer from 23 into A_2. We get to the key 30, which is the smallest one larger than or equal to 20 in A_2. From there we also walk to the right until we reach a key larger than 65, and report the objects from S_2 whose keys are in the range.

Now let's return to range trees. The crucial observation here is that the canonical subsets $P(lc(v))$ and $P(rc(v))$ both are subsets of $P(v)$. As a result we can use the same idea to speed up the query time. The details are slightly more complicated, because we now have two subsets of $P(v)$ to which we need fast access rather than only one. Let $\mathcal{T}$ be a range tree on a set P of n points in the plane. Each canonical subset $P(v)$ is stored in an associated structure. But instead of using a binary search tree as associated structure, as we did in Section 5.3, we now store it in an array $A(v)$. The array is sorted on the y-coordinate of the points. Furthermore, each entry in an array $A(v)$ stores two pointers: a pointer into $A(lc(v))$ and a pointer into $A(rc(v))$. More precisely, we add the following pointers. Suppose that $A(v)[i]$ stores a point p. Then we store a pointer from $A(v)[i]$ to the entry of $A(lc(v))$ such that the y-coordinate of the point p' stored there is the smallest one larger than or equal to p_y. As noted above, $P(lc(v))$ is a subset of $P(v)$. Hence, if p has the smallest y-coordinate larger than or equal to some value y of any point in $P(v)$, then p' has the smallest y-coordinate larger than or equal to y of any point in $P(lc(v))$. The pointer into $A(rc(v))$ is defined in the same way: it points to the entry such that the y-coordinate of the point stored there is the smallest one that is larger than or equal to p_y.

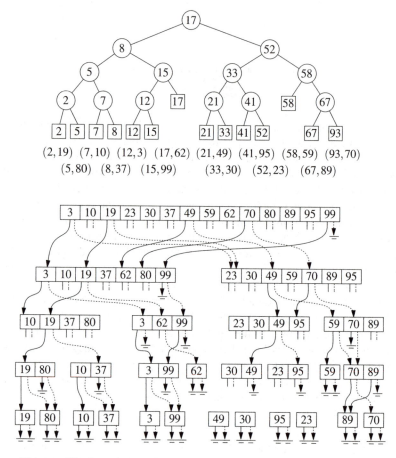

Figure 5.8
The main tree of a layered range tree: the leaves show only the x-coordinates; the points stored are given below

Figure 5.9
The arrays associated with the nodes in the main tree, with the y-coordinate of the points of the canonical subsets in sorted order (not all pointers are shown)

This modified version of the range tree is called a *layered range tree*; Figures 5.8 and 5.9 show an example. (The positions in which the arrays are drawn corresponds to the positions of the nodes in the tree they are associated with: the topmost array is associated with the root, the left array below it is associated with the left child of the root, and so on. Not all pointers are shown in the figure.)

Let's see how to answer a query with a range $[x : x'] \times [y : y']$ in a layered range tree. As before we search with x and x' in the main tree $\mathcal{T}$ to determine $O(\log n)$ nodes whose canonical subsets together contain the points with x-coordinate in the range $[x : x']$. These nodes are found as follows. Let v_{split} be the node where the two search paths split. The nodes that we are looking for are the ones below v_{split} that are the right child of a node on the search path to x where the path goes left, or the left child of a node on the search path to x' where the path goes right. At v_{split} we find the entry in $A(v_{split})$ whose y-coordinate is the smallest one larger than or equal to y. This can be done in $O(\log n)$ time by binary search. While we search further with x and x' in the main tree, we keep track of the entry in the associated arrays whose y-coordinate is the smallest one larger than or equal to y. They can be maintained in constant time by following the pointers stored in the arrays. Now let v be one of the $O(\log n)$ nodes we selected. We must report the points stored in $A(v)$ whose y-coordinate is in the range $[y : y']$. For this it suffices to be able to find the point with smallest y-coordinate larger than or equal to y; from there we can just walk through the array, reporting points as long as their y-coordinate is less than or equal to y'. This point can be found in constant time, because $parent(v)$ is on the search path, and we kept track of the points with smallest y-coordinate larger than or equal to y in the arrays on the search path. Hence, we can report the points of $A(v)$ whose y-coordinate is in the range $[y : y']$ in $O(1 + k_v)$ time, where k_v is the number of reported answers at node v. The total query time now becomes $O(\log n + k)$.

Fractional cascading also improves the query time of higher-dimensional range trees by a logarithmic factor. Recall that a d-dimensional range query was solved by first selecting the points whose d-th coordinate is in the correct range in $O(\log n)$ canonical subsets, and then solving a $(d-1)$-dimensional query on these subsets. The $(d-1)$-dimensional query is solved recursively in the same way. This continues until we arrive at a 2-dimensional query, which can be solved as described above. This leads to the following theorem.

Theorem 5.11 *Let P be a set of n points in d-dimensional space, with $d \geqslant 2$. A layered range tree for P uses $O(n \log^{d-1} n)$ storage and it can be constructed in $O(n \log^{d-1} n)$ time. With this range tree one can report the points in P that lie in a rectangular query range in $O(\log^{d-1} n + k)$ time, where k is the number of reported points.*

5.7 Notes and Comments

In the 1970s—the early days of computational geometry—orthogonal range searching was one of the most important problems in the field, and many people worked on it. This resulted in a large number of results, of which we discuss a few below.

One of the first data structures for orthogonal range searching was the quadtree, which is discussed in Chapter 14 in the context of meshing. Unfortunately, the worst-case behavior of quadtrees is quite bad. Kd-trees, described first by Bentley [32] in 1975, are an improvement over quadtrees. Samet's books [296, 297] discuss quadtrees, kd-trees, and their applications in great detail. A few years later, the range tree was discovered independently by several people [34, 216, 224, 335]. The improvement in query time to $O(\log n + k)$ by fractional cascading was described by Lueker [224] and

Willard [334]. Fractional cascading applies in fact not only to range trees, but in many situations where one does many searches with the same key. Chazelle and Guibas [84, 85] discuss this technique in its full generality. Fractional cascading can also be used in a dynamic setting [241]. The most efficient data structure for 2-dimensional range queries is a modified version of the layered range tree, described by Chazelle [66]; he succeeded in improving the storage to $O(n \log n / \log \log n)$ while keeping the query time $O(\log n + k)$. Chazelle [69, 70] also proved that this is optimal. If the query range is unbounded to one side, for instance when it is of the form $[x : x'] \times [y : +\infty]$, then one can achieve $O(\log n)$ query time with only linear space, using a priority search tree—see Chapter 10. In higher dimensions the best result for orthogonal range searching is also due to Chazelle [69]: he gave a structure for d-dimensional queries with $O(n(\log n / \log \log n)^{d-1})$ storage and polylogarithmic query time. Again, this result is optimal. Trade-offs between storage and query time are also possible [301, 339].

The lower-bound result is only valid under certain models of computation. This allows for improved results in specific cases. In particular Overmars [264] describes more efficient data structures for range searching when the points lie on a $U \times U$ grid, yielding query time bounds of $O(\log \log U + k)$ or $O(\sqrt{U} + k)$, depending on the preprocessing time allowed. The results use earlier data structures by Willard [337, 338]. When compared to the general case, better time bounds can be obtained for many problems in computational geometry once the coordinates of the objects are restricted to lie on grid points. Examples are the nearest neighbor searching problem [190, 191], point location [253], and line segment intersection [192].

In databases, range queries are considered the most general of three basic types of multi-dimensional queries. The two other types are *exact match queries* and *partial match queries*. Exact match queries are simply queries of the type: Is the object (point) with attribute values (coordinates) such and such present in the database? The obvious data structure for exact match queries is the balanced binary tree that uses, for instance, a lexicographical order on the coordinates. With this structure exact match queries can be answered in $O(\log n)$ time. If the dimension—the number of attributes—increases, it can be useful to express the efficiency of queries not only in terms of n, but also in the dimension d. If a binary tree is used for exact match queries, the query time is $O(d \log n)$ because it takes $O(d)$ time to compare two points. This can easily be reduced to $O(d + \log n)$ time, which is optimal. A partial match query specifies only a value for a subset of the coordinates and asks for all points with the specified coordinate values. In the plane, for instance, a partial match query specifies only an x-coordinate, or only a y-coordinate. Interpreted geometrically, a partial match query in the plane asks for the points on a horizontal line, or on a vertical line. With a d-dimensional kd-tree, a partial match query specifying s coordinates (with $s < d$) can be answered in $O(n^{1-s/d} + k)$ time, where k is the number of reported points [32].

In many applications the data that we are given are not a set of points, but a set of certain objects such as polygons. If we want to report the objects that are completely contained in a query range $[x : x'] \times [y : y']$, then it is possible to transform the query to a query on point data in higher dimensions—see Exercise 5.12. Often one also wants to find the objects that are only partially in the range. This specific problem is called the windowing problem and is discussed in Chapter 10.

Other variants of the range searching problem are obtained by allowing other types of query ranges, such as circles or triangles. Many of these variants can be solved using so-called partition trees, which are discussed in Chapter 16.

5.8 Exercises

5.1 In the proof of the query time of the kd-tree we found the following recurrence:

$$Q(n) = \begin{cases} O(1), & \text{if } n = 1, \\ 2 + 2Q(n/4), & \text{if } n > 1. \end{cases}$$

Prove that this recurrence solves to $Q(n) = O(\sqrt{n})$.

5.2 Describe algorithms to insert and delete points from a kd-tree. In your algorithm you do not need to take care of rebalancing the structure.

5.3 In Section 5.2 it was indicated that kd-trees can also be used to store sets of points in higher-dimensional space. Let P be a set of n points in d-dimensional space.

a. Describe the procedure to construct a d-dimensional kd-tree of the points in P. Prove that the tree uses linear storage and that your algorithm takes $O(n \log n)$ time. (You may consider d to be a constant here.)

b. Describe the query algorithm for performing a d-dimensional range query. Prove that the query time is bounded by $O(n^{1-1/d} + k)$.

c. Assuming that d is not a constant, give the dependence on d of the construction time, amount of storage required, and query time.

5.4 Kd-trees can be used for partial match queries. A 2-dimensional partial match query specifies one of the coordinates and asks for all point that have the specified coordinate value. In higher dimensions we specify values for a subset of the coordinates. Here we allow multiple points to have equal values for coordinates.

a. Show that 2-dimensional kd-trees can answer partial match queries in $O(\sqrt{n} + k)$ time, where k is the number of reported answers.

b. Describe a data structure that uses linear storage and solves 2-dimensional partial match queries in $O(\log n + k)$ time.

c. Show that with a d-dimensional kd-tree we can solve a partial match query in $O(n^{1-s/d} + k)$ time, where s is the number of specified coordinates.

d. Show that, when we allow for $O(d2^d n)$ storage, d-dimensional partial match queries can be solved in time $O(d \log n + k)$.

5.5 Describe algorithms to insert and delete points from a range tree. You don't have to take care of rebalancing the structure.

5.6 Algorithm SEARCHKDTREE can also be used when querying with other ranges than rectangles. For example, a query is answered correctly if the range is a triangle.

 a. Show that the query time for range queries with triangles is linear in the worst case, even if no answers are reported at all. *Hint:* Choose all points to be stored in the kd-tree on the line $y = x$.
 b. Suppose that a data structure is needed that can answer triangular range queries, but only for triangles whose edges are horizontal, vertical, or have slope $+1$ or -1. Develop a linear size data structure that answers such range queries in $O(n^{3/4} + k)$ time, where k is the number of points reported. *Hint:* Choose 4 coordinate axes in the plane and use a 4-dimensional kd-tree.
 c. Improve the query time to $O(n^{2/3} + k)$. *Hint:* Solve Exercise 5.4 first.

5.7 In the proof of Lemma 5.7 we made a rather rough estimate of the query time in the associated structures by stating that this was bounded by $O(\log n)$. In reality the query time is dependent on the actual number of points in the associated structure. Let n_v denote the number of points in the canonical subset $P(v)$. Then the total time spend is

$$\sum_v O(\log n_v + k_v),$$

where the summation is over all nodes in the main tree $\mathcal{T}$ that are visited. Show that this bound is still $\Theta(\log^2 n + k)$. (That is, our more careful analysis only improves the constant in the bound, not the order of magnitude.)

5.8 Theorem 5.8 showed that a range tree on a set of n points in the plane requires $O(n \log n)$ storage. One could bring down the storage requirements of the range tree by storing associated structures only with a subset of the nodes in the main tree.

 a. Suppose that only the nodes with depth 0, 2, 4, ... have an associated structure. Show how the query algorithm can be adapted to answer queries correctly.
 b. Analyze the storage requirements and query time of such a data structure.
 c. Suppose that only the nodes with depth 0, $\lfloor \frac{1}{j} \log n \rfloor$, $\lfloor \frac{2}{j} \log n \rfloor$, ... have an associated structure, where $j \geqslant 2$ is a fixed integer. Analyze the storage requirements and query time of this data structure. Express the bounds in n and j.

5.9 One can use the data structures described in this chapter to determine whether a particular point (a,b) is in the set by performing a range query with range $[a : a] \times [b : b]$.

    a.   Prove that performing such a range query on a kd-tree takes time $O(\log n)$.

    b.   What is the time bound for such a query on a range tree? Prove your answer.

5.10 In some applications one is interested only in the number of points that lie in a range rather than in reporting all of them. Such queries are often referred to as *range counting queries*. In this case one would like to avoid paying the $O(k)$ in the query time.

    a.   Describe how a 1-dimensional range tree can be adapted such that a range counting query can be performed in $O(\log n)$ time. Prove the query time bound.

    b.   Describe how d-dimensional range counting queries can be answered in $O(\log^d n)$ time, using the solution to the 1-dimensional problem. Prove the query time.

    c.*   Describe how fractional cascading can be used to improve the running time with a factor of $O(\log n)$ for 2- and higher-dimensional range counting queries.

5.11 In Section 5.5 it was shown that one can treat sets of points in the plane that contain equal coordinate values using composite numbers. Extend this notion to points in d-dimensional space. To this end define the composite number of d numbers and define an appropriate order on them. Next transform the points and the range according to this order.

5.12 In many application one wants to do range searching among objects other than points.

    a.   Let P be a set of n axis-parallel rectangles in the plane. We want to be able to report all rectangles in P that are completely contained in a query rectangle $[x : x'] \times [y : y']$. Describe a data structure for this problem that uses $O(n \log^3 n)$ storage and has $O(\log^4 n + k)$ query time, where k is the number of reported answers. *Hint:* Transform the problem to an orthogonal range searching problem in some higher-dimensional space.

    b.   Let P now consist of a set of n polygons in the plane. Again describe a data structure that uses $O(n \log^3 n)$ storage and has $O(\log^4 n + k)$ query time, where k is the number of reported answers, to report all polygons completely contained in the query rectangle.

    c.*   Improve the query time to $O(\log^3 n + k)$.

5.13* Prove the $O(n \log^{d-1} n)$ bounds on the storage and construction time in Theorem 5.11.

6 Point Location

Knowing Where You Are

This book has, for the most part, been written in Europe. More precisely, it has been written very close to a point at longitude 5°6′ east and latitude 52°3′ north. Where that is? You can find that out yourself from a map of Europe: using the scales on the sides of the map, you will find that the point with the coordinates stated above is located in a little country named "the Netherlands".

In this way you would have answered a *point location query*: Given a map and a query point q specified by its coordinates, find the region of the map containing q. A map, of course, is nothing more than a subdivision of the plane into regions, a *planar subdivision*, as defined in Chapter 2.

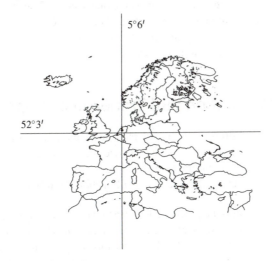

Figure 6.1
Point location in a map

Point location queries arise in various settings. Suppose that you are sailing on a sea with sand banks and dangerous currents in various parts of it. To be able to navigate safely, you will have to know the current at your present position. Fortunately there are maps that indicate the kind of current in the various parts of the sea. To use such a map, you will have to do the following. First, you must determine your position. Until not so long ago, you would have to rely for this on the stars or the sun, and a good chronometer.

Nowadays it is much easier to determine your position: there are little boxes on the market that compute your position for you, using information from various satellites. Once you have determined the coordinates of your position, you will have to locate the point on the map showing the currents, or to find the region of the sea you are presently in.

One step further would be to automate this last step: store the map electronically, and let the computer do the point location for you. It could then display the current—or any other information for which you have a thematic map in electronic form—of the region you are in continuously. In this situation we have a set of presumably rather detailed thematic maps and we want to answer point location queries frequently, to update the displayed information while the ship is moving. This means that we will want to *preprocess* the maps, and to store them in a data structure that makes it possible to answer point location queries fast.

Point location problems arise on a quite different scale as well. Assume that we want to implement an interactive geographic information system that displays a map on a screen. By clicking with the mouse on a country, the user can retrieve information about that country. While the mouse is moved the system should display the name of the country underneath the mouse pointer somewhere on the screen. Every time the mouse is moved, the system has to recompute which name to display. Clearly this is a point location problem in the map displayed on the screen, with the mouse position as the query point. These queries occur with high frequency—after all, we want to update the screen information in real time—and therefore have to be answered fast. Again, we need a data structure that supports fast point location queries.

6.1 Point Location and Trapezoidal Maps

Let S be a planar subdivision with n edges. The *planar point location* problem is to store S in such a way that we can answer queries of the following type: Given a query point q, report the face f of S that contains q. If q lies on an edge or coincides with a vertex, the query algorithm should return this information.

To get some insight into the problem, let's first give a very simple data structure to perform point location queries. We draw vertical lines through all vertices of the subdivision, as in Figure 6.2. This partitions the plane into vertical *slabs*. We store the x-coordinates of the vertices in sorted order in an array. This makes it possible to determine in $O(\log n)$ time the slab that contains a query point q. Within a slab, there are no vertices of S. This means that the part of the subdivision lying inside the slab has a special form: all edges intersecting a slab completely cross it—they have no endpoint in the slab—and they don't cross each other. This means that they can be ordered from top to bottom. Notice that every region in the slab between two consecutive edges belongs to a unique face of S. The lowest and highest region of the slab are unbounded,

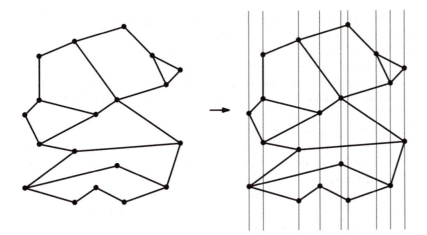

Figure 6.2
Partition into slabs

and are part of the unbounded face of S. The special structure of the edges intersecting a slab implies that we can store them in sorted order in an array. We label each edge with the face of S that is immediately above it inside the slab.

The total query algorithm is now as follows. First, we do a binary search with the x-coordinate of the query point q in the array storing the x-coordinates of the vertices of the subdivision. This tells us the slab containing q. Then we do a binary search with q in the array for that slab. The elementary operation in this binary search is: Given a segment s and a point q such that the vertical line through q intersects s, determine whether q lies above s, below s, or on s. This tells us the segment directly below q, provided there is one. The label stored with that segment is the face of S containing q. If we find that there is no segment below q then q lies in the unbounded face.

The query time for the data structure is good: we only perform two binary searches, the first in an array of length at most $2n$ (the n edges of the subdivision have at most $2n$ vertices), and the second in an array of length at most n (a slab is crossed by at most n edges). Hence, the query time is $O(\log n)$.

What about the storage requirements? First of all, we have an array on the x-coordinates of the vertices, which uses $O(n)$ storage. But we also have an array for every slab. Such an array stores the edges intersecting its slab, so it uses $O(n)$ storage. Since there are $O(n)$ slabs, the total amount of storage is $O(n^2)$. It's easy to give an example where $n/4$ slabs are intersected by $n/4$ edges each, which shows that this worst-case bound can actually occur.

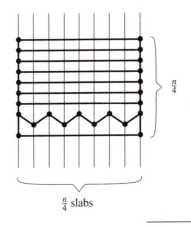

$\frac{n}{4}$ slabs

The amount of storage required makes this data structure rather uninteresting—a quadratic size structure is useless in most practical applications, even for moderately large values of n. (One may argue that in practice the quadratic behavior does not occur. But it is rather likely that the amount of storage is something like $O(n\sqrt{n})$.) Where does this quadratic behavior come from? Let's have a second look at Figure 6.2. The segments and the vertical lines through the endpoints define a new subdivision S', whose faces are trapezoids, triangles, and unbounded trapezoid-like faces. Furthermore, S' is a *refinement* of the original subdivision S: every face of S' lies completely in one face of S. The query algorithm described above is in fact an algorithm to do planar point location in this refined subdivision. This solves the original planar point location as well: because S' is a refinement of S, once we know the face $f' \in S'$ containing q, we know the face $f \in S$ containing q. Unfortunately, the refined subdivision can have quadratic complexity. It is therefore not surprising that the resulting data structure has quadratic size.

Perhaps we should look for a different refinement of S that—like the decomposition shown above—makes point location queries easier, and that—unlike the decomposition shown above—has a complexity that is not much larger than the complexity of the original subdivision S. Indeed such a refinement exists. In the rest of this section, we shall describe the *trapezoidal map*, a refinement that has the desirable properties just mentioned.

We call two line segments in the plane *non-crossing* if their intersection is either empty or a common endpoint. Notice that the edges of any planar subdivision are non-crossing.

Let S be a set of n non-crossing segments in the plane. Trapezoidal maps can be defined for such sets in general, but we shall make two simplifications that make life easier for us in this and the next sections.

First, it will be convenient to get rid of the unbounded trapezoid-like faces that occur at the boundary of the scene. This can be done by introducing a large, axis-parallel rectangle R that contains the whole scene, that is, that contains all segments of S. For our application—point location in subdivisions—this is not a problem: a query point outside R always lies in the unbounded face of S, so we can safely restrict our attention to what happens inside R.

The second simplification is more difficult to justify: we will assume that no two distinct endpoints of segments in the set S have the same x-coordinate. A consequence of this is that there cannot be any vertical segments. This assumption is not very realistic: vertical edges occur frequently in many applications, and the situation that two non-intersecting segments have an endpoint with the same x-coordinate is not so unusual either, because the precision in which the coordinates are given is often limited. We will make this assumption nevertheless, postponing the treatment of the general case to Section 6.3.

So we have a set S of n non-crossing line segments, enclosed in a bounding box R, and with the property that no two distinct endpoints lie on a common vertical line. We call such a set *a set of line segments in general position*. The *trapezoidal map* $\mathcal{T}(S)$ of S—also known as the *vertical decomposition* or *trapezoidal decomposition* of S—is obtained by drawing two *vertical extensions* from every endpoint p of a segment in S, one extension going upwards and one going downwards. The extensions stop when they meet another segment of S or the boundary of R. We call the two vertical extensions starting in an endpoint p the *upper vertical extension* and the *lower vertical extension*. The trapezoidal map of S is simply the subdivision induced by S, the rectangle R, and the upper and lower vertical extensions. Figure 6.3 shows an example.

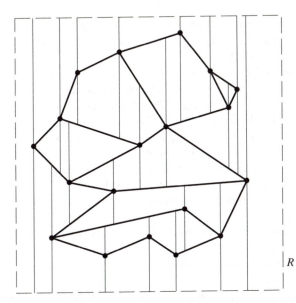

R

Figure 6.3
A trapezoidal map

A face in $\mathcal{T}(S)$ is bounded by a number of edges of $\mathcal{T}(S)$. Some of these edges may be adjacent and collinear. We call the union of such edges a *side* of the face. In other words, the sides of a face are the segments of maximal length that are contained in the boundary of a face.

Lemma 6.1 *Each face in a trapezoidal map of a set S of line segments in general position has one or two vertical sides and exactly two non-vertical sides.*

Proof. Let f be a face in $\mathcal{T}(S)$. We first prove that f is convex.

Because the segments in S are non-crossing, any corner of f is either an endpoint of a segment in S, a point where a vertical extension abuts a segment of S or an edge of R, or it is a corner of R. Due to the vertical extensions, no corner that is a segment endpoint can have an interior angle greater than 180°. Moreover, any angle at a point where a vertical extension abuts a segment must be less than or equal to 180° as well. Finally, the corners of R are 90°. Hence, f is convex—the vertical extensions have removed all non-convexities.

Because we are looking at sides of f, rather than at edges of $\mathcal{T}(S)$ on the boundary of f, the convexity of f implies that it can have at most two vertical sides. Now suppose for a contradiction that f has more than two non-vertical sides. Then there must be two such sides that are adjacent and either both bound f from above or both bound f from below. Because any non-vertical side must be contained in a segment of S or in an edge of R, and the segments are non-crossing, the two adjacent sides must meet in a segment endpoint. But then the vertical extensions for that endpoint prevent the two sides from being adjacent, a contradiction. Hence, f has at most two non-vertical sides.

Finally, we observe that f is bounded (since we have enclosed the whole scene in a bounding box R), which implies that it cannot have less than two non-vertical sides and that it must have at least one vertical side. ☐

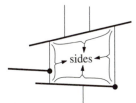

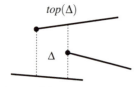

$top(\Delta)$

Δ

$bottom(\Delta)$

Lemma 6.1 shows that the trapezoidal map deserves its name: each face is either a trapezoid or a triangle, which we can view as a trapezoid with one degenerate edge of length zero.

In the proof of Lemma 6.1 we observed that a non-vertical side of a trapezoid is contained in a segment of S or in a horizontal edge of R. We denote the non-vertical segment of S, or edge of R, bounding a trapezoid Δ from above by $top(\Delta)$, and the one bounding it from below by $bottom(\Delta)$.

By the general position assumption, a vertical side of a trapezoid either consists of vertical extensions, or it is the vertical edge of R. More precisely, we can distinguish five different cases for the left side and the right side of a trapezoid Δ. The cases for the left side are as follows:

(a) It degenerates to a point, which is the common left endpoint of $top(\Delta)$ and $bottom(\Delta)$.

(b) It is the lower vertical extension of the left endpoint of $top(\Delta)$ that abuts on $bottom(\Delta)$.

(c) It is the upper vertical extension of the left endpoint of $bottom(\Delta)$ that abuts on $top(\Delta)$.

(d) It consists of the upper and lower extension of the right endpoint p of a third segment s. These extensions abut on $top(\Delta)$ and $bottom(\Delta)$, respectively.

(e) It is the left edge of R. This case occurs for a single trapezoid of $\mathcal{T}(S)$ only, namely the unique leftmost trapezoid of $\mathcal{T}(S)$.

The first four cases are illustrated in Figure 6.4. The five cases for the right vertical edge of Δ are symmetrical. You should verify for yourself that the listing above is indeed exhaustive.

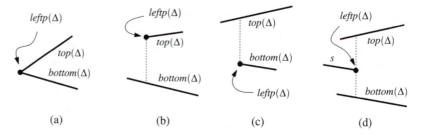

Figure 6.4
Four of the five cases for the left edge
of trapezoid Δ

(a)   (b)   (c)   (d)

For every trapezoid $\Delta \in \mathcal{T}(S)$, except the leftmost one, the left vertical edge of Δ is, in a sense, defined by a segment endpoint p: it is either contained in the vertical extensions of p, or—when it is degenerate—it is p itself. We will denote the endpoint defining the left edge of Δ by $leftp(\Delta)$. As shown above, $leftp(\Delta)$ is the left endpoint of $top(\Delta)$ or $bottom(\Delta)$, or it is the right endpoint of a third segment. For the unique trapezoid whose left side is the left edge of R, we define $leftp(\Delta)$ to be the lower left vertex of R. Similarly, we denote the endpoint that defines the right vertical edge of Δ by $rightp(\Delta)$. Notice that Δ is uniquely determined by $top(\Delta)$, $bottom(\Delta)$, $leftp(\Delta)$, and $rightp(\Delta)$. Therefore we will sometimes say that Δ is *defined* by these segments and endpoints.

The trapezoidal map of the edges of a subdivision is a refinement of that subdivision. It is not clear, however, why point location in a trapezoidal map should be any easier than point location in a general subdivision. But before we come to this in the next section, let's first verify that the complexity of the trapezoidal map is not too much larger than the number of segments in the set defining it.

Lemma 6.2 *The trapezoidal map $\mathcal{T}(S)$ of a set S of n line segments in general position contains at most $6n + 4$ vertices and at most $3n + 1$ trapezoids.*

Proof. A vertex of $\mathcal{T}(S)$ is either a vertex of R, an endpoint of a segment in S, or else the point where the vertical extension starting in an endpoint abuts on another segment or on the boundary of R. Since every endpoint of a segment induces two vertical extensions—one upwards, one downwards—this implies that the total number of vertices is bounded by $4 + 2n + 2(2n) = 6n + 4$.

A bound on the number of trapezoids follows from Euler's formula and the bound on the number of vertices. Here we give a direct proof, using the point $leftp(\Delta)$. Recall that each trapezoid has such a point $leftp(\Delta)$. This point is the endpoint of one of the n segments, or is the lower left corner of R. By looking at the five cases for the left side of a trapezoid, we find that the lower left corner of R plays this role for exactly one trapezoid, a right endpoint of a segment can play this role for at most one trapezoid, and a left endpoint of a segment can be the $leftp(\Delta)$ of at most two different trapezoids. (Since endpoints can coincide, a point in the plane can be $leftp(\Delta)$ for many trapezoids. However, if in case (a) we consider $leftp(\Delta)$ to be the left endpoint of $bottom(\Delta)$, then the left endpoint of a segment s can be $leftp(\Delta)$ for only two trapezoids, on above s and one below s.) It follows that the total number of trapezoids is at most $3n + 1$. $\Box$

We call two trapezoids Δ and Δ' *adjacent* if they meet along a vertical edge. In Figure 6.5(i), for example, trapezoid Δ is adjacent to Δ_1, Δ_2, and Δ_3, but not to Δ_4 and Δ_5. Because the set of line segments is in general position, a trapezoid has at most four adjacent trapezoids. If the set is not in general position, a trapezoid can have an arbitrary number of adjacent trapezoids, as illustrated in Figure 6.5(ii). Let Δ' be a trapezoid that is adjacent to Δ along the left vertical

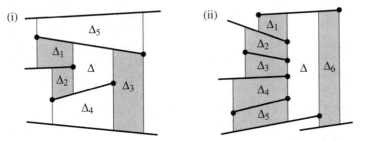

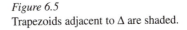

Figure 6.5
Trapezoids adjacent to Δ are shaded.

edge of Δ. Then either $top(\Delta) = top(\Delta')$ or $bottom(\Delta) = bottom(\Delta')$. In the first case we call Δ' the *upper left neighbor* of Δ, and in the second case Δ' is the *lower left neighbor* of Δ. So the trapezoid in Figure 6.4(b) has a bottom left neighbor but no top left neighbor, the trapezoid in Figure 6.4(c) has a top left neighbor but no bottom left neighbor, and the trapezoid in Figure 6.4(d) has both a top left neighbor and a bottom left neighbor. The trapezoid in Figure 6.4(a) and the single trapezoid whose left vertical edge is the left side of R

have no left neighbors. The *upper right neighbor* and *lower right neighbor* of a trapezoid are defined similarly.

To represent a trapezoidal map, we could use the doubly-connected edge list described in Chapter 2; after all, a trapezoidal map is a planar subdivision. However, the special shape of the trapezoidal map makes it more convenient to use a specialized structure. This structure uses the adjacency of trapezoids to link the subdivision as a whole. There are records for all line segments and endpoints of S, since they serve as $leftp(\Delta)$, $rightp(\Delta)$, $top(\Delta)$, and $bottom(\Delta)$. Furthermore, the structure contains records for the trapezoids of $T(S)$, but not for edges or vertices of $T(S)$. The record for a trapezoid Δ stores pointers to $top(\Delta)$ and $bottom(\Delta)$, pointers to $leftp(\Delta)$ and $rightp(\Delta)$, and finally, pointers to its at most four neighbors. Note that the geometry of a trapezoid Δ (that is, the coordinates of its vertices) is not available explicitly. However, Δ is uniquely defined by $top(\Delta)$, $bottom(\Delta)$, $leftp(\Delta)$, and $rightp(\Delta)$. This means that we can deduce the geometry of Δ in constant time from the information stored for Δ.

6.2 A Randomized Incremental Algorithm

In this section we will develop a randomized incremental algorithm that constructs the trapezoidal map $T(S)$ of a set S of n line segments in general position. During the construction of the trapezoidal map, the algorithm also builds a data structure $\mathcal{D}$ that can be used to perform point location queries in $T(S)$. This is the reason why a plane sweep algorithm isn't chosen to construct the trapezoidal map. It would construct it all right, but it wouldn't give us a data structure that supports point location queries, which is the main objective of this chapter.

Before discussing the algorithm, we first describe the point location data structure $\mathcal{D}$ that the algorithm constructs. This structure, which we call the *search structure*, is a directed acyclic graph with a single root and exactly one leaf for every trapezoid of the trapezoidal map of S. Its inner nodes have out-degree 2. There are two types of inner nodes: *x-nodes*, which are labeled with an endpoint of some segment in S, and *y-nodes*, which are labeled with a segment itself.

A query with a point q starts at the root and proceeds along a directed path to one of the leaves. This leaf corresponds to the trapezoid $\Delta \in T(S)$ containing q. At each node on the path, q has to be tested to determine in which of the two child nodes to proceed. At an x-node, the test is of the form: "Does q lie to the left or to the right of the vertical line through the endpoint stored at this node?" At a y-node, the test has the form: "Does q lie above or below the segment s stored here?" We will ensure that whenever we come to a y-node, the vertical line through q intersects the segment of that node, so that the test makes sense. The tests at the inner nodes only have two outcomes: left or right of an endpoint for an x-node, and above or below a segment for a y-node. What should we do if the query point lies exactly on the vertical line, or on the segment? For now, we shall simply make the assumption that this does not occur; Section 6.3, which shows how to deal with sets of segments that are not in general position, will also deal with this type of query point.

The search structure $\mathcal{D}$ and the trapezoidal map $\mathcal{T}(S)$ computed by the algorithm are interlinked: a trapezoid $\Delta \in \mathcal{T}(S)$ has a pointer to the leaf of $\mathcal{D}$ corresponding to it, and a leaf node of $\mathcal{D}$ has a pointer to the corresponding trapezoid in $\mathcal{T}(S)$. Figure 6.6 shows the trapezoidal map of a set of two line

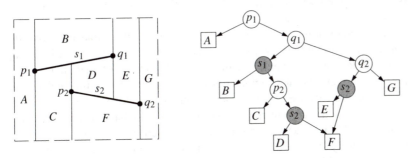

Figure 6.6
The trapezoidal map of two segments
and a search structure

segments s_1 and s_2, and a search structure for the trapezoidal map. The x-nodes are white, with the endpoint labeling it inside. The y-nodes are grey and have the segment labeling it inside. The leaves of the search structure are shown as squares, and are labeled with the corresponding trapezoid in the trapezoidal map.

The algorithm we give for the construction of the search structure is incremental: it adds the segments one at a time, and after each addition it updates the search structure and the trapezoidal map. The order in which the segments are added influences the search structure; some orders lead to a search structure with a good query time, while for others the query time will be bad. Instead of trying to be clever in finding a suitable order, we shall take the same approach we took in Chapter 4, where we studied linear programming: we use a random ordering. So the algorithm will be *randomized incremental*. Later we will prove that the search structure resulting from a randomized incremental algorithm is expected to be good. But first we describe the algorithm in more detail. We begin with its global structure; the various substeps will be explained after that.

Algorithm TRAPEZOIDALMAP(S)
Input. A set S of n non-crossing line segments.
Output. The trapezoidal map $\mathcal{T}(S)$ and a search structure $\mathcal{D}$ for $\mathcal{T}(S)$ in a bounding box.
1.   Determine a bounding box R that contains all segments of S, and initialize the trapezoidal map structure $\mathcal{T}$ and search structure $\mathcal{D}$ for it.
2.   Compute a random permutation $s_1, s_2, \ldots, s_n$ of the elements of S.
3.   **for** $i \leftarrow 1$ **to** n
4.   **do** Find the set $\Delta_0, \Delta_1, \ldots, \Delta_k$ of trapezoids in $\mathcal{T}$ properly intersected by s_i.

5.   Remove $\Delta_0, \Delta_1, \ldots, \Delta_k$ from $\mathcal{T}$ and replace them by the new trapezoids that appear because of the insertion of s_i.

6.   Remove the leaves for $\Delta_0, \Delta_1, \ldots, \Delta_k$ from $\mathcal{D}$, and create leaves for the new trapezoids. Link the new leaves to the existing inner nodes by adding some new inner nodes, as explained below.

We now describe the various steps of the algorithm in more detail. In the following, we let $S_i := \{s_1, s_2, \ldots, s_i\}$. The loop invariant of TRAPEZOIDALMAP is that $\mathcal{T}$ is the trapezoidal map for S_i, and that $\mathcal{D}$ is a valid search structure for $\mathcal{T}$.

The initialization of $\mathcal{T}$ as $\mathcal{T}(S_0) = \mathcal{T}(\emptyset)$ and of $\mathcal{D}$ in line 1 is easy: the trapezoidal map for the empty set consists of a single trapezoid—the bounding rectangle R—and the search structure for $\mathcal{T}(\emptyset)$ consists of a single leaf node for this trapezoid. For the computation of the random permutation in line 2 see Chapter 4. Now let's see how to insert a segment s_i in lines 4–6.

To modify the current trapezoidal map, we first have to know where it changes. This is exactly at the trapezoids that are intersected by s_i. Stated more precisely, a trapezoid of $\mathcal{T}(S_{i-1})$ is not present in $\mathcal{T}(S_i)$ if and only if it is intersected by s_i. Our first task is therefore to find the intersected trapezoids. Let $\Delta_0, \Delta_1, \ldots, \Delta_k$ denote these trapezoids, ordered from left to right along s_i. Observe that Δ_{j+1} must be one of the right neighbors of Δ_j. It is also easy to test which neighbor it is: if $\mathit{right}(\Delta_j)$ lies above s_i, then Δ_{j+1} is the lower right neighbor of Δ_j, otherwise it is the upper right neighbor. This means that once we know Δ_0, we can find $\Delta_1, \ldots, \Delta_k$ by traversing the representation of the trapezoidal map. So, to get started, we need to to find the trapezoid $\Delta_0 \in \mathcal{T}$ containing the left endpoint, p, of s_i. If p is not yet present in S_{i-1} as an endpoint, then, because of our general position assumption, it must lie in the interior of Δ_0. This means we can find Δ_0 by a point location with p in $\mathcal{T}(S_{i-1})$. And now comes the exciting part: at this stage in the algorithm $\mathcal{D}$ is a search structure for $\mathcal{T} = \mathcal{T}(S_{i-1})$, so all that we need to do is to perform a query on $\mathcal{D}$ with the point p.

If p is already an endpoint of a segment in S_{i-1}—remember that we allow different segments to share an endpoint—then we must be careful. To find Δ_0, we simply start to search in $\mathcal{D}$. If p happens not to be present yet, then the query algorithm will proceed without trouble, and end up in the leaf corresponding to Δ_0. If, however, p is already present, then the following will happen: at some point during the search, p will lie *on* the vertical line through the point in an x-node. Recall that we decided that such query points are illegal. To remedy this, we should imagine continuing the query with a point p' slightly to the right of p. Replacing p by p' is only done conceptually. What it actually means when implementing the search is this: whenever p lies on the vertical line of an x-node, we decide that it lies to the right. Similarly, whenever p lies on a segment s of a y-node (this can only happen if s_i shares its left endpoint, p, with s) we compare the slopes of s and s_i; if the slope of s_i is larger, we decide that p lies above s, otherwise we decide that it is below s. With this adaptation, the search will end in the first trapezoid Δ_0 intersected properly by s_i. In summary, we use the following algorithm to find $\Delta_0, \ldots, \Delta_k$.

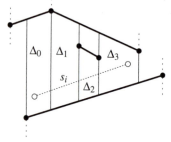

Algorithm FOLLOWSEGMENT($\mathcal{T}, s_i$)

Input. A trapezoidal map $\mathcal{T}$, and a new segment s_i.

Output. The sequence $\Delta_0, \ldots, \Delta_k$ of trapezoids intersected by s_i.

1.   Let p and q be the left and right endpoint of s_i.
2.   Search with p in the search structure to find Δ_0.
3. $j \leftarrow 0$;
4.   **while** q lies to the right of $rightp(\Delta_j)$
5.   **do if** $rightp(\Delta_j)$ lies above s_i
6.   **then** Let Δ_{j+1} be the lower right neighbor of Δ_j.
7.   **else** Let Δ_{j+1} be the upper right neighbor of Δ_j.
8. $j \leftarrow j + 1$
9.   **return** $\Delta_0, \Delta_1, \ldots, \Delta_j$

We have seen how to find the trapezoids intersecting s_i. The next step is to update $\mathcal{T}$ and $\mathcal{D}$. Let's start with the simple case that s_i is completely contained in a trapezoid $\Delta = \Delta_0$. We are in the situation depicted on the left hand side of Figure 6.7.

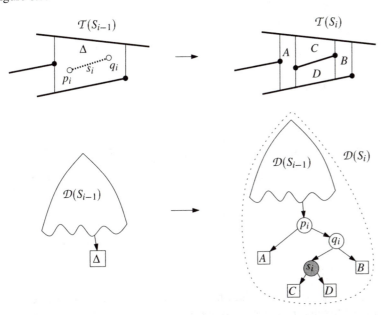

Figure 6.7
The new segment s_i lies completely in trapezoid Δ.

To update $\mathcal{T}$, we delete Δ from $\mathcal{T}$, and replace it by four new trapezoids A, B, C, and D. Notice that all the information we need to initialize the records for the new trapezoids correctly (their neighbors, the segments on the top and bottom, and the points defining their left and right vertical edges) are available: they can be determined in constant time from the segment s_i and the information stored for Δ.

It remains to update $\mathcal{D}$. What we must do is replace the leaf for Δ by a little tree with four leaves. The tree contains two x-nodes, testing with the left and right endpoint of s_i, and one y-node, testing with the segment s_i itself. This is sufficient to determine in which of the four new trapezoids A, B, C, or D a

129

query point lies, if we already know that it lies in Δ. The right hand side of Figure 6.7 illustrates the modifications to the search structure. Note that one or both endpoints of the segment s_i could be equal to $leftp(\Delta)$ or $rightp(\Delta)$. In that case there would be only two or three new trapezoids, but the modification is done in the same spirit.

The case where s_i intersects two or more trapezoids is only slightly more complicated. Let $\Delta_0, \Delta_1, \ldots, \Delta_k$ be the sequence of intersected trapezoids. To update $\mathcal{T}$, we first erect vertical extensions through the endpoints of s_i,

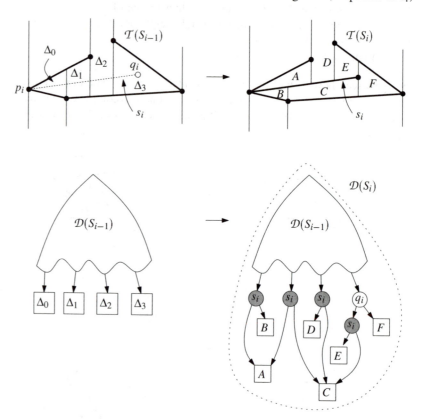

Figure 6.8
Segment s_i intersects four trapezoids.

partitioning Δ_0 and Δ_k into three new trapezoids each. This is only necessary for the endpoints of s_i that were not already present. Then we shorten the vertical extensions that now abut on s_i. This amounts to merging trapezoids along the segment s_i, see Figure 6.8. Using the information stored with the trapezoids $\Delta_0, \Delta_1, \ldots, \Delta_k$, this step can be done in time that is linear in the number of intersected trapezoids.

To update $\mathcal{D}$, we have to remove the leaves for $\Delta_0, \Delta_1, \ldots, \Delta_k$, we must create leaves for the new trapezoids, and we must introduce extra inner nodes. More precisely, we proceed as follows. If Δ_0 has the left endpoint of s_i in its interior (which means it has been partitioned into three new trapezoids) then we replace the leaf for Δ_0 with an x-node for the left endpoint of s_i and a y-node for the segment s_i. Similarly, if Δ_k has the right endpoint of s_i in its interior, we replace the leaf for Δ_k with an x-node for the right endpoint of s_i and a y-node for s_i. Finally, the leaves of Δ_1 to Δ_{k-1} are replaced with single y-nodes for the segment s_i. We make the outgoing edges of the new inner nodes point to the correct new leaves. Notice that, due to the fact that we have merged trapezoids stemming from different trapezoids of $\mathcal{T}$, there can be several incoming edges for a new trapezoid. Figure 6.8 illustrates this.

We have finished the description of the algorithm that constructs $\mathcal{T}(s)$ and, at the same time, builds a search structure $\mathcal{D}$ for it. The correctness of the algorithm follows directly from the loop invariant (stated just after Algorithm TRAPEZOIDALMAP), so it remains to analyze its performance.

The order in which we treat the segments has considerable influence on the resulting search structure $\mathcal{D}$ and on the running time of the algorithm itself. For some cases the resulting search structure has quadratic size and linear search time, but for other permutations of the same set of segments the resulting search structure is much better. As in Chapter 4, we haven't tried to determine a good sequence; we simply took a random insertion sequence. This means that the analysis will be probabilistic: we shall look at the *expected* performance of the algorithm and of the search structure. Perhaps it's not quite clear yet what the term "expected" means in this context. Consider a fixed set S of n non-crossing line segments. TRAPEZOIDALMAP computes a search structure $\mathcal{D}$ for $\mathcal{T}(S)$. This structure depends on the permutation chosen in line 2. Since there are $n!$ possible permutations of n objects, there are $n!$ possible ways in which the algorithm can proceed. The expected running time of the algorithm is the average of the running time, taken over all $n!$ permutations. For each permutation, a different search structure will result. The expected size of $\mathcal{D}$ is the average of the sizes of all these $n!$ search structures. Finally, the expected query time for a point q is the average of the query time for point q, over all $n!$ search structures. (Notice that this not the same as the average of the maximum query time for the search structure. Proving a bound on this quantity is a bit more technical, so we defer that to Section 6.4.)

Theorem 6.3 *Algorithm* TRAPEZOIDALMAP *computes the trapezoidal map* $\mathcal{T}(S)$ *of a set S of n line segments in general position and a search structure $\mathcal{D}$ for $\mathcal{T}(S)$ in $O(n \log n)$ expected time. The expected size of the search structure is $O(n)$ and for any query point q the expected query time is $O(\log n)$.*

Proof. As noted earlier, the correctness of the algorithm follows directly from the loop invariant, so we concentrate on the performance analysis.

We start with the query time of the search structure $\mathcal{D}$. Let q be a fixed query point. As the query time for q is linear in the length of the path in $\mathcal{D}$ that is traversed when querying with q, it suffices to bound the path length. A simple case analysis shows that the depth of $\mathcal{D}$ (that is, the maximum path length) increases by at most 3 in every iteration of the algorithm. Hence, $3n$ is an upper bound on the query time for q. This bound is the best possible *worst-case* bound over all possible insertion orders for S. We are not so much interested in the worst-case case behavior, however, but in the expected behavior: we want to bound the average query time for q with respect to the $n!$ possible insertion orders.

Consider the path traversed by the query for q in $\mathcal{D}$. Every node on this path was created at some iteration of the algorithm. Let X_i, for $1 \leqslant i \leqslant n$, denote the number of nodes on the path created in iteration i. Since we consider S and q to be fixed, X_i is a random variable—it depends on the random order of the segments only. We can now express the expected path length as

$$\mathrm{E}[\sum_{i=1}^{n} X_i] = \sum_{i=1}^{n} \mathrm{E}[X_i].$$

The equality here is *linearity of expectation*: the expected value of a sum is equal to the sum of the expected values.

We already observed that any iteration adds at most three nodes to the search path for any possible query point, so $X_i \leqslant 3$. In other words, if P_i denotes the probability that there exists a node on the search path of q that is created in iteration i, we have

$$\mathrm{E}[X_i] \leqslant 3P_i.$$

The central observation to derive a bound on P_i is this: iteration i contributes a node to the search path of q exactly if $\Delta_q(S_{i-1})$, the trapezoid containing q in $\mathcal{T}(S_{i-1})$, is not the same as $\Delta_q(S_i)$, the trapezoid containing q in $\mathcal{T}(S_i)$. In other words,

$$P_i = \Pr[\Delta_q(S_i) \neq \Delta_q(S_{i-1})].$$

If $\Delta_q(S_i)$ is not the same as $\Delta_q(S_{i-1})$, then $\Delta_q(S_i)$ must be one of the trapezoids created in iteration i. Note that all trapezoids Δ created in iteration i are adjacent to s_i, the segment that is inserted in that iteration: either $top(\Delta)$ or $bottom(\Delta)$ is s_i, or $leftp(\Delta)$ or $rightp(\Delta)$ is an endpoint of s_i.

Now consider a fixed set $S_i \subset S$. The trapezoidal map $\mathcal{T}(S_i)$, and therefore $\Delta_q(S_i)$, are uniquely defined as a function of S_i; $\Delta_q(S_i)$ does *not* depend on the order in which the segments in S_i have been inserted. To bound the probability that the trapezoid containing q has changed due to the insertion of s_i, we shall use a trick we also used in Chapter 4, called backwards analysis: we consider $\mathcal{T}(S_i)$ and look at the probability that $\Delta_q(S_i)$ disappears from the trapezoidal map when we remove the segment s_i. By what we said above, $\Delta_q(S_i)$ disappears if and only if one of $top(\Delta_q(S_i))$, $bottom(\Delta_q(S_i))$, $leftp(\Delta_q(S_i))$, or $rightp(\Delta_q(S_i))$ disappears with the removal of s_i. What is the probability that $top(\Delta_q(S_i))$ disappears? The segments of S_i have been inserted in random order, so every segment in S_i is equally likely to be s_i. This means that the probability that s_i happens to be $top(\Delta_q(S_i))$ is $1/i$. (If $top(\Delta_q(S_i))$ is the top edge of the rectangle R surrounding the scene, then the probability is even zero.) Similarly, the probability that s_i happens to be $bottom(\Delta_q(S_i))$ is at most $1/i$. There can be many segments sharing the point $leftp(\Delta_q(S_i))$. Hence, the probability that s_i is one of these segments can be large. But $leftp(\Delta_q(S_i))$ disappears only if s_i is the only segment in S_i with $leftp(\Delta_q(S_i))$ as an endpoint. Hence, the probability that $leftp(\Delta_q(S_i))$ disappears is at most $1/i$ as well. The same holds true for $rightp(\Delta_q(S_i))$. Hence, we can conclude that

$$P_i = \Pr[\Delta_q(S_i) \neq \Delta_q(S_{i-1})] = \Pr[\Delta_q(S_i) \notin T(S_{i-1})] \leqslant 4/i.$$

(A small technical point: In the argument above we fixed the set S_i. This means that the bound we derived on P_i holds under the condition that S_i is this fixed set. But since the bound does not depend on what the fixed set actually is, the bound holds unconditionally.)

Putting it all together we get the bound on the expected query time:

$$\mathrm{E}[\sum_{i=1}^{n} X_i] \leqslant \sum_{i=1}^{n} 3P_i \leqslant \sum_{i=1}^{n} \frac{12}{i} = 12 \sum_{i=1}^{n} \frac{1}{i} = 12 H_n.$$

Here, H_n is the n-th *harmonic number*, defined as

$$H_n := \frac{1}{1} + \frac{1}{2} + \frac{1}{3} + \cdots + \frac{1}{n}$$

Harmonic numbers arise quite often in the analysis of algorithms, so it's good to remember the following bound that holds for all $n > 1$:

$$\ln n < H_n < \ln n + 1.$$

(It can be derived by comparing H_n to the integral $\int_1^n 1/x\,dx = \ln n$.) We conclude that the expected query time for a query point q is $O(\log n)$, as claimed.

Let's now turn to the size of $\mathcal{D}$. To bound the size, it suffices to bound the number of nodes in $\mathcal{D}$. We first note that the leaves in $\mathcal{D}$ are in one-to-one correspondence with the trapezoids in Δ, of which there are $O(n)$ by Lemma 6.2. This means that the total number of nodes is bounded by

$$O(n) + \sum_{i=1}^{n} (\text{number of inner nodes created in iteration } i).$$

Let k_i be the number of new trapezoids that are created in iteration i, due to the insertion of segment s_i. In other words, k_i is the number of new leaves in $\mathcal{D}$. The number of inner nodes created in iteration i is exactly equal to $k_i - 1$. A simple worst case upper bound on k_i follows from the fact that the number of new trapezoids in $T(S_i)$ can obviously not be larger than the total number of trapezoids in $T(S_i)$, which is $O(i)$. This leads to a worst-case upper bound on the size of the structure of

$$O(n) + \sum_{i=1}^{n} O(i) = O(n^2).$$

Indeed, if we have bad luck and the order in which we insert the segments is very unfortunate, then the size of $\mathcal{D}$ can be quadratic. However, we are more interested in the expected size of the data structure, over all possible insertion orders. Using linearity of expectation, we find that this expected size is bounded by

$$O(n) + \mathrm{E}[\sum_{i=1}^{n}(k_i - 1)] = O(n) + \sum_{i=1}^{n} \mathrm{E}[k_i].$$

It remains to bound the expected value of k_i. We already prepared the necessary tools for this when we derived the bound on the query time. Consider a fixed set $S_i \subseteq S$. For a trapezoid $\Delta \in \mathcal{T}(S_i)$ and a segment $s \in S_i$, let

$$\delta(\Delta, s) := \begin{cases} 1 & \text{if } \Delta \text{ disappears from } \mathcal{T}(S_i) \text{ when } s \text{ is removed from } S_i, \\ 0 & \text{otherwise.} \end{cases}$$

In the analysis of the query time we observed that there are at most four segments that cause a given trapezoid to disappear. Hence,

$$\sum_{s \in S_i} \sum_{\Delta \in \mathcal{T}(S_i)} \delta(\Delta, s) \leqslant 4|\mathcal{T}(S_i)| = O(i).$$

Now, k_i is the number of trapezoids created by the insertion of s_i, or, equivalently, the number of trapezoids in $\mathcal{T}(S_i)$ that disappear when s_i is removed. Since s_i is a random element of S_i, we can find the expected value of k_i by taking the average over all $s \in S_i$:

$$\mathrm{E}[k_i] = \frac{1}{i} \sum_{s \in S_i} \sum_{\Delta \in \mathcal{T}(S_i)} \delta(\Delta, s) \leqslant \frac{O(i)}{i} = O(1).$$

We conclude that the expected number of newly created trapezoids is $O(1)$ in every iteration of the algorithm, from which the $O(n)$ bound on the expected amount of storage follows.

It remains to bound the expected running time of the construction algorithm. Given the analysis of the query time and storage, this is easy. We only need to observe that the time to insert segment s_i is $O(k_i)$ plus the time needed to locate the left endpoint of s_i in $\mathcal{T}(S_{i-1})$. Using the earlier derived bounds on k_i and the query time, we immediately get that the expected running time of the algorithm is

$$O(1) + \sum_{i=1}^{n} \left\{ O(\log i) + O(\mathrm{E}[k_i]) \right\} = O(n \log n).$$

This completes the proof. $\quad\square$

Note once again that the expectancy in Theorem 6.3 is solely with respect to the random choices made by the algorithm; we do not average over possible choices for the input. Hence, there are no bad inputs: for *any* input set of n line segments, the expected running time of the algorithm is $O(n \log n)$.

As discussed earlier, Theorem 6.3 does not guarantee anything about the expected maximum query time over all possible query points. In Section 6.4, however, it is proved that the expected maximum query time is $O(\log n)$ as well. Hence, we can build a data structure of expected size $O(n)$, whose expected query time is $O(\log n)$. This also proves the *existence* of a data structure with $O(n)$ size and $O(\log n)$ query time for any query point q—see Theorem 6.8.

Finally we go back to our original problem: point location in a planar subdivision S. We assume that S is given as a doubly-connected edge list with n edges. We use algorithm TRAPEZOIDALMAP to compute a search structure $\mathcal{D}$ for the trapezoidal map of the edges of S. To use this search structure for point location in S, however, we still need to attach to every leaf of $\mathcal{D}$ a pointer to the face f of S that contains the trapezoid of $\mathcal{T}(S)$ corresponding to that leaf. But this is rather easy: recall from Chapter 2 that the doubly-connected edge list of S stores with every half-edge a pointer to the face that is incident to the left. The face that is incident to the right can be found in constant time from $Twin(\vec{e})$. So for every trapezoid Δ of $\mathcal{T}(S)$ we simply look at the face of S incident to $top(\Delta)$ from below. If $top(\Delta)$ is the top edge of R, then Δ is contained in the unique unbounded face of S.

In the next section we show that the assumption that the segments be in general position may be dropped, which leads to a less restricted version of Theorem 6.3. It implies the following corollary.

Corollary 6.4 *Let S be a planar subdivision with n edges. In $O(n \log n)$ expected time one can construct a data structure that uses $O(n)$ expected storage, such that for any query point q, the expected time for a point location query is $O(\log n)$.*

6.3 Dealing with Degenerate Cases

In the previous sections we made two simplifying assumptions. First of all, we assumed that the set of line segments was in general position, meaning that no two distinct endpoints have the same x-coordinate. Secondly, we assumed that a query point never lies on the vertical line of an x-node on its search path, nor on the segment of a y-node. We now set out to get rid of these assumptions.

We first show how to avoid the assumption that no two distinct endpoints lie on a common vertical line. The crucial observation is that the vertical direction chosen to define the trapezoidal map of the set of line segments was immaterial. Therefore we can rotate the axis-system slightly. If the rotation angle is small enough, then no two distinct endpoints will lie on the same vertical line anymore. Rotations by very small angles, however, lead to numerical difficulties. Even if the input coordinates are integer, we need a significantly higher precision to do the calculations properly. A better approach is to do the rotation *symbolically*. In Chapter 5 we have seen another form of a symbolic transformation, composite numbers, which was used to deal with the case where data points have the same x- or y-coordinate. In this chapter, we will have another look at such a *symbolic perturbation*, and will try to interpret it geometrically.

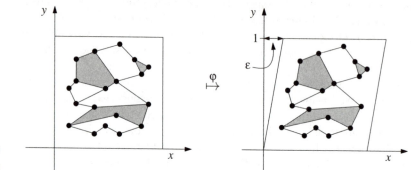

Figure 6.9
The shear transformation

It will be convenient not to use rotation, but to use an affine mapping called *shear transformation*. In particular, we shall use the shear transformation along the x-axis by some value $\varepsilon > 0$:

$$\varphi : \begin{pmatrix} x \\ y \end{pmatrix} \mapsto \begin{pmatrix} x + \varepsilon y \\ y \end{pmatrix}.$$

Figure 6.9 illustrates the effect of this shear transform. It maps a vertical line to a line with slope $1/\varepsilon$, so that any two distinct points on the same vertical line are mapped to points with distinct x-coordinates. Furthermore, if $\varepsilon > 0$ is small enough, then the transformation does not reverse the order in x-direction of the given input points. It is not difficult to compute an upper bound for ε to guarantee that this property holds. In the following, we will assume that we have a value of $\varepsilon > 0$ available that is small enough that the order of points is conserved by the shear transform. Surprisingly, we will later realize that we do not need to compute such an actual value for ε.

Given a set S of n arbitrary non-crossing line segments, we will run algorithm TRAPEZOIDALMAP on the set $\varphi S := \{\varphi s : s \in S\}$. As noted earlier, however, actually performing the transformation may lead to numerical problems, and therefore we use a trick: a point $\varphi p = (x + \varepsilon y, y)$ will simply be stored as (x, y). This is a unique representation. We only need to make sure that the algorithm treats the segments represented in this way correctly. Here it comes handy that the algorithm does not compute any geometric objects: we never actually compute the coordinates of the endpoints of vertical extensions, for instance. All it does is to apply two types of elementary operations to the input points. The first operation takes two distinct points p and q and decides whether q lies to the left, to the right, or on the vertical line through p. The second operation takes one of the input segments, specified by its two endpoints p_1 and p_2, and tests whether a third point q lies above, below, or on this segment. This second operation is only applied when we already know that a vertical line through q intersects the segment. All the points p, q, p_1, and p_2 are endpoints of segments in the input set S. (You may want to go through the description of the algorithm again, verifying that it can indeed be realized with these two operations alone.)

Let's look at how to apply the first operation to two transformed points φp and φq. These points have coordinates $(x_p + \varepsilon y_p, y_p)$ and $(x_q + \varepsilon y_q, y_q)$, respectively. If $x_q \neq x_p$, then the relation between x_q and x_p determines the outcome of the test—after all, we had chosen ε to have this property. If $x_q = x_p$, then the relation between y_q and y_p decides the horizontal order of the points. Therefore there is a strict horizontal order for any pair of distinct points. It follows that φq will *never* lie on the vertical line through φp, *except* when p and q coincide. But this is exactly what we need, since no two distinct input points should have the same x-coordinate.

For the second operation, we are given a segment φs with endpoints $\varphi p_1 = (x_1 + \varepsilon y_1, y_1)$ and $\varphi p_2 = (x_2 + \varepsilon y_2, y_2)$, and we want to test whether a point $\varphi q = (x + \varepsilon y, y)$ lies above, below, or on φs. The algorithm ensures that whenever we do this test, the vertical line through φq intersects φs. In other words,

$$x_1 + \varepsilon y_1 \leqslant x + \varepsilon y \leqslant x_2 + \varepsilon y_2.$$

This implies that $x_1 \leqslant x \leqslant x_2$. Moreover, if $x = x_1$ then $y \geqslant y_1$, and if $x = x_2$ then $y \leqslant y_2$. Let's now distinguish two cases.

If $x_1 = x_2$, the untransformed segment s is vertical. Since now $x_1 = x = x_2$, we have $y_1 \leqslant y \leqslant y_2$, which implies that q lies *on* s. Because the affine mapping φ is incidence preserving—if two points coincide before the transformation, they do so afterwards—we can conclude that φq lies on φs.

Now consider the case where $x_1 < x_2$. Since we already know that the vertical line through φq intersects φs, it is good enough to do the test with the line through φs. Now we observe that the mapping φ preserves the relation between points and lines: if a point is above (or on, or below) a given line, then the transformed point is above (or on, or below) the transformed line. Hence, we can simply perform the test with the untransformed point q and segment s.

This shows that to run the algorithm on φS instead of on S, the only modification we have to make is to compare points lexicographically, when we want to determine their horizontal order. Of course, the algorithm will compute the trapezoidal map for φS, and a search structure for $\mathcal{T}(\varphi S)$. Note that, as promised, we never actually needed the value of ε, so there is no need to compute such a value at the beginning. All we needed was that ε is small enough.

Using our shear transformation, we got rid of the assumption that any two distinct endpoints should have distinct x-coordinates. What about the restriction that a query point never lies on the vertical line of an x-node on the search path, nor on the segment of a y-node? Our approach solves this problem as well, as we show next.

Since the constructed search structure is for the transformed map $\mathcal{T}(\varphi S)$, we will also have to use the transformed query point φq when doing a query. In other words, all the comparisons we have to do during the search must be done in the transformed space. We already know how to do the tests in the transformed space:

At x-nodes we have to do the test lexicographically. As a result, no two distinct points will ever lie on a vertical line. (Trick question: How can this be true, if the transformation φ is bijective?) This does not mean that the outcome of the test at an x-node is always "to the right" or "to the left". The outcome can also be "on the line". This can only happen, however, when the query point coincides with the endpoint stored at the node—and this answers the query!

At y-nodes we have to test the transformed query point against a transformed segment. The test described above can have three outcomes: "above", "below", or "on". In the first two cases, there is no problem, and we can descend to the corresponding child of the y-node. If the outcome of the test is "on", then the untransformed point lies on the untransformed segment as well, and we can report that fact as the answer to the query.

We have generalized Theorem 6.3 to arbitrary sets of non-crossing line segments.

Theorem 6.5 *Algorithm* TRAPEZOIDALMAP *computes the trapezoidal map* $\mathcal{T}(S)$ *of a set* S *of* n *non-crossing line segments and a search structure* $\mathcal{D}$ *for* $\mathcal{T}(S)$ *in* $O(n\log n)$ *expected time. The expected size of the search structure is* $O(n)$ *and for any query point* q *the expected query time is* $O(\log n)$.

6.4* A Tail Estimate

Theorem 6.5 states that for any query point q, the expected query time is $O(\log n)$. This is a rather weak result. In fact, there is no reason to expect that the maximum query time of the search structure is small: it may be that for any permutation of the set of segments the resulting search structure has a bad query time for *some* query point. In this section we shall prove that there is no need to worry: the probability that the maximum query time is bad is very small. To this end we first prove the following *high-probability bound*.

Lemma 6.6 *Let* S *be a set of* n *non-crossing line segments, let* q *be a query point, and let* λ *be a parameter with* $\lambda > 0$. *Then the probability that the search path for* q *in the search structure computed by Algorithm* TRAPEZOIDALMAP *has more than* $3\lambda\ln(n+1)$ *nodes is at most* $1/(n+1)^{\lambda\ln 1.25-1}$.

Proof. We would like to define random variables X_i, with $1 \leqslant i \leqslant n$, such that X_i is 1 if at least one node on the search path for q has been created in iteration i of the algorithm, and 0 if no such node has been created. Unfortunately, the random variables defined this way are not independent. (We did not need independence in the proof of Theorem 6.3, but we shall need it now.) Therefore, we use a little trick.

We define a directed acyclic graph $\mathcal{G}$, with one source and one sink. Paths in $\mathcal{G}$ from the source to the sink correspond to the permutations of S. The graph $\mathcal{G}$ is defined as follows. There is a node for every subset of S, including one for the empty set. With a slight abuse of terminology, we shall often speak of "the subset S'" when we mean "the node corresponding to the subset S'". It is convenient to imagine the nodes as being grouped into $n+1$ layers, such that layer i contains the subsets of cardinality i. Notice that layers 0 and n both have exactly one node, which correspond to the empty set and to the set S, respectively. A node in layer i has outgoing arcs to some of the nodes in layer $i+1$. More precisely, a subset S' of cardinality i has outgoing arcs to the subsets S'' of cardinality $i+1$ if and only if that $S' \subset S''$. In other words, a subset S' has an outgoing arc to a subset S'' if S'' can be obtained by adding one segment of S to S'. We label the arc with this segment. Note that a subset S' in layer i has exactly i incoming arcs, each labeled with a segment in S', and exactly $n-i$ outgoing arcs, each labeled with a segment in $S \setminus S'$.

Directed paths from the source to the sink in G now correspond one-to-one to permutations of S, which correspond to possible executions of the algorithm TRAPEZOIDALMAP. Consider an arc of G from subset S' in layer i to subset S'' in layer $i+1$. Let the segment s be the label of the arc. The arc represents the insertion of s into the trapezoidal map of S'. We mark this arc if this insertion changes the trapezoid containing point q. To be able to say something about the number of marked arcs, we use the backwards analysis argument from the proof of Theorem 6.3: there are at most four segments that change the trapezoid containing q when they are *removed* from a subset S''. This means that any node in G has at most four marked incoming arcs. But it is possible that a node has less than four incoming arcs. In that case, we simply mark some other, arbitrary incoming arcs, so that the number of marked incoming arcs is exactly four. For nodes in the first three layers, which have less than four incoming arcs, we mark all incoming arcs.

We want to analyze the expected number of steps during which the trapezoid containing q changes. In other words, we want to analyze the expected number of marked edges on a source-to-sink path in G. To this end we define the random variable X_i as follows:

$$X_i := \begin{cases} 1 & \text{if the } i\text{-th arc on the sink-to-source path in } G \text{ is marked,} \\ 0 & \text{otherwise.} \end{cases}$$

Note the similarity of the definition of X_i and the definition of X_i on page 132. The i-th arc on the path is from a node in layer $i-1$ to a node in layer i, and each such arc is equally likely to be the i-th arc. Since each node in layer i has i incoming arcs, exactly four of which are marked (assuming $i \geqslant 4$), this implies that $\Pr[X_i = 1] = 4/i$ for $i \geqslant 4$. For $i < 4$ we have $\Pr[X_i = 1] = 1 < 4/i$. Moreover, we note that the X_i are independent (unlike the random variables X_i defined on page 132).

Let $Y := \sum_{i=1}^n X_i$. The number of nodes on the search path for q is at most $3Y$, and we will bound the probability that Y exceeds $\lambda \ln(n+1)$. We will use *Markov's inequality*, which states that for any nonnegative random variable Z and any $\alpha > 0$ we have

$$\Pr[Z \geqslant \alpha] \leqslant \frac{E[Z]}{\alpha}.$$

So for any $t > 0$ we have

$$\Pr[Y \geqslant \lambda \ln(n+1)] = \Pr[e^{tY} \geqslant e^{t\lambda \ln(n+1)}] \leqslant e^{-t\lambda \ln(n+1)} E[e^{tY}].$$

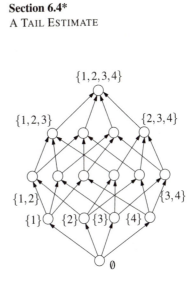

$\{1,2,3,4\}$

$\{1,2,3\}$ $\{2,3,4\}$

$\{1,2\}$ $\{3,4\}$

$\{1\}$ $\{2\}$ $\{3\}$ $\{4\}$

$\emptyset$

Recall that the expected value of the sum of random variables is the sum of the expected values. In general it is not true that the expected value of a product is the product of the expected values. But for random variables that are independent it is true. Our X_i are independent, so we have

$$E[e^{tY}] = E[e^{\sum_i t X_i}] = E[\prod_i e^{t X_i}] = \prod_i E[e^{t X_i}].$$

If we choose $t = \ln 1.25$, we get

$$E[e^{t X_i}] \leqslant e^t \frac{4}{i} + e^0 (1 - \frac{4}{i}) = (1 + 1/4)\frac{4}{i} + 1 - \frac{4}{i} = 1 + \frac{1}{i} = \frac{1+i}{i},$$

and we have

$$\prod_{i=1}^n E[e^{t X_i}] \leqslant \frac{2}{1} \frac{3}{2} \cdots \frac{n+1}{n} = n + 1.$$

Putting everything together, we get the bound we want to prove:

$$\Pr[Y \geqslant \lambda \ln(n+1)] \leqslant e^{-\lambda t \ln(n+1)}(n+1) = \frac{n+1}{(n+1)^{\lambda t}} = 1/(n+1)^{\lambda t - 1}.$$

We use this lemma to prove a bound on the expected maximum query time.

Lemma 6.7 *Let S be a set of n non-crossing line segments, and let λ be a parameter with $\lambda > 0$. Then the probability that the depth of the search structure computed by Algorithm* TRAPEZOIDALMAP *for S is more than $3\lambda \ln(n+1)$ is at most $2/(n+1)^{\lambda \ln 1.25 - 3}$.*

Proof. We will call two query points q and q' *equivalent* if they follow the same path through the search structure $\mathcal{D}$. Partition the plane into vertical slabs by passing a vertical line through every endpoint of S. Partition every slab into trapezoids by intersecting it with all possible segments in S. This defines a decomposition of the plane into at most $2(n+1)^2$ trapezoids. Any two points lying in the same trapezoid of this decomposition must be equivalent in every possible search structure for S. After all, the only comparisons made during a search are the test whether the query point lies to the left or to the right of the vertical line through a segment endpoint, and the test whether the query point lies above or below a segment.

This implies that the depth of $\mathcal{D}$ is the maximum of the length of the search path for at most $2(n+1)^2$ query points, one in each of these trapezoids. By Lemma 6.6, the probability that the length of the search path for a fixed point q exceeds $3\lambda \ln(n+1)$ is at most $1/(n+1)^{\lambda \ln 1.25 - 1}$. In the worst case, the probability that the length of the search path for any one of the $2(n+1)^2$ test points exceeds the bound is therefore at most $2(n+1)^2/(n+1)^{\lambda \ln 1.25 - 1}$.

This lemma implies that the expected maximum query time is $O(\log n)$. Take for instance $\lambda = 20$. Then the probability that the depth of $\mathcal{D}$ is more than $3\lambda \ln(n+1)$ is at most $2/(n+1)^{1.4}$, which is less than $1/4$ for $n > 4$. In other words, the probability that $\mathcal{D}$ has a good query time is at least $3/4$. Similarly one can show that the probability that the size of $\mathcal{D}$ is $O(n)$ is at least $3/4$, and that the probability that the running time of the construction algorithm is $O(n \log n)$ is at least $3/4$. Hence, the probability that the query time, the size, and the construction time are good is at least $1/4$.

Now we can obtain an algorithm to construct a search structure that has $O(\log n)$ *worst-case* query time and uses $O(n)$ space *in the worst case*. Only the construction has an expected time bound. What we do is the following. We run Algorithm TRAPEZOIDALMAP on the set S, keeping track of the size and depth of the search structure that is being created. As soon as the size exceeds $c_1 n$, or the depth exceeds $c_2 \log n$, for suitably chosen constants c_1 and c_2, we stop the algorithm, and start it again from the beginning, with a fresh random permutation. Since the probability that a permutation leads to a data structure with the desired size and depth is at least $1/4$, so we expect to be finished in four trials. (In fact, for large n the probability is almost one, so the expected number of trials is only slightly larger than one.) This leads to the following result.

Theorem 6.8 *Let S be a planar subdivision with n edges. There exists a point location data structure for S that uses $O(n)$ storage and has $O(\log n)$ query time in the worst case. The structure can be built in $O(n \log n)$ expected time.*

The constants in this example are not very convincing—a query time of $60 \log n$ isn't really so attractive. However, much better constants can be proven with the same technique—Exercise 6.16 ask you to do that.

6.5 Notes and Comments

The point location problem has a long history in computational geometry. Early results are surveyed by Preparata and Shamos [286]. Of all the methods suggested for the problem, four basically different approaches lead to optimal $O(\log n)$ search time, $O(n)$ storage solutions. These are the *chain method* by Edelsbrunner et al. [132], which is based on segment trees and *fractional cascading* (see also Chapter 10), the *triangulation refinement method* by Kirkpatrick [200], the use of *persistency* by Sarnak and Tarjan [298] and Cole [113], and the randomized incremental method by Mulmuley [255]. Our presentation here follows Seidel's [308] presentation and analysis of Mulmuley's algorithm.

Quite a lot of recent research has gone into dynamic point location, where the subdivision can be modified by adding and deleting edges [29, 94, 98]. A survey on dynamic point location is given by Chiang and Tamassia [99].

In more than two dimensions, the point location problem is still essentially open. A general structure for convex subdivisions in three dimensions is given by Preparata and Tamassia [287]. For the rest, point location structures are only known for special subdivisions, such as *arrangements* of hyperplanes [74, 83, 109]. Considering the subdivision induced by a set H of n hyperplanes in d-dimensional space, it is well known that the combinatorial complexity of this subdivision (the number of vertices, edges, and so on) is $\Theta(n^d)$ in the worst case [129]—see also the notes and comments of Chapter 8. Chazelle and Friedman [83] have shown that such subdivisions can be stored using $O(n^d)$ space, such that point location queries take $O(\log n)$ time. Other special subdivisions that allow for efficient point location are convex polytopes [109, 231], arrangements of triangles [43], and arrangements of algebraic varieties [81].

Other point location problems in 3- and higher-dimensional space that can be solved efficiently include those where assumptions on the shape of the cells are made. Two examples are rectangular subdivisions [41, 133], and so-called *fat* subdivisions [266, 273].

Usually, a point location query aks for the label of the cell of a subdivision that contains a given query point. For point location in a convex polytope in d-dimensional space, that means that there are only two possible answers: inside the polytope, or on the outside. Therefore, one could hope for point location structures that require considerably less storage than the combinatorial complexity of the subdivision. This can be as much as $\Theta(n^{\lfloor d/2 \rfloor})$ for convex polytopes defined by the intersection of n half-spaces [129]—see also the notes and comments of Chapter 11. Indeed, an $O(n)$ size data structure exists in which queries take $O(n^{1-1/\lfloor d/2 \rfloor} \log^{O(1)} n)$ time [229]. For arrangements of line segments in the plane, other *implicit point location* structures have been given that require less than quadratic space, even if the arrangement of line segments has quadratic complexity [131, 2].

6.6 Exercises

6.1 Draw the graph of the search structure $\mathcal{D}$ for the set of segments depicted in the margin, for some insertion order of the segments.

6.2 Give an example of a set of n line segments with an order on them that makes the algorithm create a search structure of size $\Theta(n^2)$ and query time $\Theta(n)$.

6.3 In this chapter we have looked at the point location problem with pre-processing. We have not looked at the *single shot* problem, where the subdivision and the query point are given at the same time, and we have no special preprocessing to speed up the searches. In this and the following exercises, we have a look at such problems.

Given a simple polygon $\mathcal{P}$ with n vertices and a query point q, here is an algorithm to determine whether q lies in $\mathcal{P}$. Consider the ray $\rho := \{(q_x + \lambda, q_y) : \lambda > 0\}$ (this is the horizontal ray starting in q and going rightwards). Determine for every edge e of $\mathcal{P}$ whether it intersects ρ. If the number of intersecting edges is odd, then $q \in \mathcal{P}$, otherwise $q \notin \mathcal{P}$.

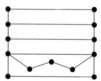

Prove that this algorithm is correct, and explain how to deal with degenerate cases. (One degenerate case is when ρ intersects an endpoint of an edge. Are there other special cases?) What is the running time of the algorithm?

6.4 Show that, given a planar subdivision S with n vertices and edges and a query point q, the face of S containing q can be computed in time $O(n)$.

6.5 Given a convex polygon P as an array of its n vertices in sorted order along the boundary. Show that, given a query point q, it can be tested in time $O(\log n)$ whether q lies inside P.

6.6 A polygon P is called *star-shaped* if there exists a point p in the interior of P such that, for any other point q in P, the line segment $\overline{pq}$ lies in P. Can you generalize the solution to the previous exercise to star-shaped polygons? What about y-monotone polygons?

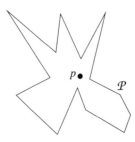

6.7 Design a *deterministic* algorithm, that is, one that doesn't make random choices, to compute the trapezoidal map of a set of non-crossing line segments. Use the plane sweep paradigm from Chapter 2. The worst-case running time of the algorithm should be $O(n \log n)$.

6.8* Give a randomized algorithm to compute all pairs of intersecting segments in a set of line segments in expected time $O(n \log n + A)$, where A is the number of intersecting pairs.

6.9 Design an algorithm with running time $O(n \log n)$ for the following problem: Given a set P of n points, determine a value of $\varepsilon > 0$ such that the shear transformation $\varphi : (x, y) \mapsto (x + \varepsilon y, y)$ does not change the order of the points in x-direction.

6.10 Let S be a set of non-crossing segments in the plane, and let s be a new segment not crossing any of the segments in S. Prove that a trapezoid Δ of $T(S)$ is also a trapezoid of $T(S \cup \{s\})$ if and only if s does not intersect the interior of Δ.

6.11 Prove that the number of inner nodes of the search structure D of algorithm TRAPEZOIDALMAP increases by $k_i - 1$ in iteration i, where k_i is the number of new trapezoids in $T(S_i)$ (and hence the number of new leaves of D).

6.12 Use a *plane sweep argument* to prove that the trapezoidal map of n line segments in general position has at most $3n + 1$ trapezoids. (Imagine a vertical line sweeping over the plane from left to right, stopping at all endpoints of segments. Count the number of trapezoids that are encountered by the sweep line.)

6.13 We have defined the trapezoidal map of a set S of n line segments only for the case that S is in general position. Give a definition for the trapezoidal map $T(S)$ of an arbitrary set of segments. Prove that the upper bound of $3n + 1$ for the number of trapezoids still holds.

6.14 Although we have started with the point location problem on the surface of the earth, we have only treated *planar* point location. But the earth is a globe. How would you define a *spherical subdivision*—a subdivision of the surface of a sphere? Give a point location structure for such a subdivision.

6.15 The *ray shooting problem* occurs in computer graphics (see Chapter 8). A 2-dimensional version can be given as follows: Store a set S of n non-crossing line segments such that one can quickly answer queries of the type: "Given a query ray ρ (a ray is a half-line), find the first segment in S intersected by ρ." (We leave it to you to define the behavior for degenerate cases.)

In this exercise, we look at *vertical* ray shooting, where the query ray must be a vertical ray pointing upwards.

Give a data structure for the vertical ray shooting problem for a set S of n non-crossing line segments in general position. Bound the query time and storage requirement of your data structure. What is the preprocessing time?

Can you do the same for a general set of line segments?

6.16 Modify the proof of Theorem 6.6 so that it gives you better constants. Prove a version of Theorem 6.8 that gives non-asymptotic bounds for the size and depth of the search structure.

7 Voronoi Diagrams

The Post Office Problem

Suppose you are on the advisory board for the planning of a supermarket chain, and there are plans to open a new branch at a certain location. To predict whether the new branch will be profitable, you must estimate the number of customers it will attract. For this you have to model the behavior of your potential customers: how do people decide where to do their shopping? A similar question arises in social geography, when studying the economic activities in a country: what is the trading area of certain cities? In a more abstract set-

Figure 7.1
The trading areas of the capitals of the twelve provinces in the Netherlands, as predicted by the Voronoi assignment model

ting we have a set of central places—called *sites*—that provide certain goods or services, and we want to know for each site where the people live who obtain their goods or services from that site. (In computational geometry the sites are traditionally viewed as post offices where customers want to post their letters—hence the subtitle of this chapter.) To study this question we make the following simplifying assumptions:

- the price of a particular good or service is the same at every site;

- the cost of acquiring the good or service is equal to the price plus the cost of transportation to the site;

- the cost of transportation to a site equals the Euclidean distance to the site times a fixed price per unit distance;

- consumers try to minimize the cost of acquiring the good or service.

Usually these assumptions are not completely satisfied: goods may be cheaper at some sites than at others, and inside a city the transportation cost from one point to another is probably not linear in the Euclidean distance between the points. But the model above can give a rough approximation of the trading areas of the sites. Areas where the behavior of the people differs from that predicted by the model can be subjected to further research, to see what caused the different behavior.

Our interest lies in the geometric interpretation of the model above. The assumptions in the model induce a subdivision of the total area under consideration into regions—the trading areas of the sites—such that the people who live in the same region all go to the same site. Our assumptions imply that people simply get their goods at the nearest site—a fairly realistic situation. This means that the trading area for a given site consists of all those points for which that site is closer than any other site. Figure 7.1 gives an example. The sites in this figure are the capitals of the twelve provinces in the Netherlands.

The model where every point is assigned to the nearest site is called the *Voronoi assignment model*. The subdivision induced by this model is called the *Voronoi diagram* of the set of sites. From the Voronoi diagram we can derive all kinds of information about the trading areas of the sites and their relations. For example, if the regions of two sites have a common boundary then these two sites are likely to be in direct competition for customers that live in the boundary region.

The Voronoi diagram is a versatile geometric structure. We have described an application to social geography, but the Voronoi diagram has applications in physics, astronomy, robotics, and many more fields. It is also closely linked to another important geometric structure, the so-called Delaunay triangulation, which we shall encounter in Chapter 9. In the current chapter we shall confine ourselves to the basic properties and the construction of the Voronoi diagram of a set of point sites in the plane.

7.1 Definition and Basic Properties

Denote the Euclidean distance between two points p and q by $\text{dist}(p,q)$. In the plane we have

$$\text{dist}(p,q) := \sqrt{(p_x - q_x)^2 + (p_y - q_y)^2}.$$

Let $P := \{p_1, p_2, \ldots, p_n\}$ be a set of n distinct points in the plane; these points are the sites. We define the Voronoi diagram of P as the subdivision of the plane into n cells, one for each site in P, with the property that a point q lies in the cell corresponding to a site p_i if and only if $\text{dist}(q, p_i) < \text{dist}(q, p_j)$ for each $p_j \in P$ with $j \neq i$. We denote the Voronoi diagram of P by $\text{Vor}(P)$. The cell that corresponds to a site p_i is denoted $\mathcal{V}(p_i)$; we call it the Voronoi cell of p_i. (In the terminology of the introduction to this chapter: $\mathcal{V}(p_i)$ is the trading area of site p_i.)

We now take a closer look at the Voronoi diagram. First we study the structure of a single Voronoi cell. For two points p and q in the plane we define the *bisector of p and q* as the perpendicular bisector of the line segment $\overline{pq}$. This bisector splits the plane into two half-planes. We denote the open half-plane that contains p by $h(p, q)$ and the open half-plane that contains q by $h(q, p)$. Notice that $r \in h(p, q)$ if and only if $\text{dist}(r, p) < \text{dist}(r, q)$. From this we obtain the following observation.

Observation 7.1 $\mathcal{V}(p_i) = \bigcap_{1 \leqslant j \leqslant n, j \neq i} h(p_i, p_j)$.

Thus $\mathcal{V}(p_i)$ is the intersection of $n - 1$ half-planes and, hence, a (possibly unbounded) open convex polygonal region bounded by at most $n - 1$ vertices and at most $n - 1$ edges.

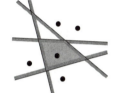

What does the complete Voronoi diagram look like? We just saw that each cell of the diagram is the intersection of a number of half-planes, so the Voronoi diagram is a planar subdivision whose edges are straight line segments. Some edges are line segments and others are half-lines. Unless all sites are collinear there will be no edges that are full lines:

Theorem 7.2 *Let P be a set of n point sites in the plane. If all the sites are collinear then $\text{Vor}(P)$ consists of $n - 1$ parallel lines and n cells. Otherwise, $\text{Vor}(P)$ is connected and its edges are either segments or half-lines.*

Proof. The first part of the theorem is easy to prove, so assume that not all sites in P are collinear.

We first show that the edges of $\text{Vor}(P)$ are either segments or half-lines. We already know that the edges of $\text{Vor}(P)$ are parts of straight lines, namely parts of the bisectors between pairs of sites. Now suppose for a contradiction that there is an edge e of $\text{Vor}(P)$ that is a full line. Let e be on the boundary of the Voronoi cells $\mathcal{V}(p_i)$ and $\mathcal{V}(p_j)$. Let $p_k \in P$ be a point that is not collinear with p_i and p_j. The bisector of p_j and p_k is not parallel to e and, hence, it intersects e. But then the part of e that lies in the interior of $h(p_k, p_j)$ cannot be on the boundary of $\mathcal{V}(p_j)$, because it is closer to p_k than to p_j, a contradiction.

It remains to prove that $\text{Vor}(P)$ is connected. If this were not the case then there would be a Voronoi cell $\mathcal{V}(p_i)$ splitting the plane into two. Because Voronoi cells are convex, $\mathcal{V}(p_i)$ would consist of a strip bounded by two parallel full lines. But we just proved that the edges of the Voronoi diagram cannot be full lines, a contradiction. $\square$

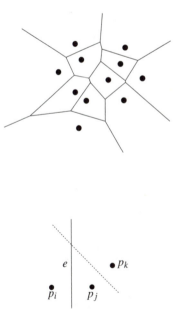

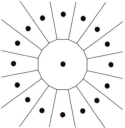

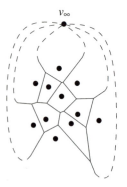

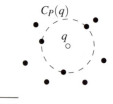

Now that we understand the structure of the Voronoi diagram we investigate its complexity, that is, the total number of its vertices and edges. Since there are n sites and each Voronoi cell has at most $n-1$ vertices and edges, the complexity of $\mathrm{Vor}(P)$ is at most quadratic. It is not clear, however, whether $\mathrm{Vor}(P)$ can actually have quadratic complexity: it is easy to construct an example where a single Voronoi cell has linear complexity, but can it happen that many cells have linear complexity? The following theorem shows that this is not the case and that the average number of vertices of the Voronoi cells is less than six.

Theorem 7.3 *The number of vertices in the Voronoi diagram of a set of n point sites in the plane is at most $2n-5$ and the number of edges is at most $3n-6$.*

Proof. If the sites are all collinear then the theorem immediately follows from Theorem 7.2, so assume this is not the case. We prove the theorem using *Euler's formula*, which states that for any connected planar embedded graph with m_v nodes, m_e arcs, and m_f faces the following relation holds:

$$m_v - m_e + m_f = 2.$$

We cannot apply Euler's formula directly to $\mathrm{Vor}(P)$, because $\mathrm{Vor}(P)$ has half-infinite edges and is therefore not a proper graph. To remedy the situation we add one extra vertex v_∞ "at infinity" to the set of vertices and we consider all half-infinite edges of $\mathrm{Vor}(P)$ to be connected to this vertex. We now have a connected planar graph to which we can apply Euler's formula. We obtain the following relation between n_v, the number of vertices of $\mathrm{Vor}(P)$, n_e, the number of edges of $\mathrm{Vor}(P)$, and n, the number of sites:

$$(n_v + 1) - n_e + n = 2. \tag{7.1}$$

Moreover, every edge in the augmented graph has exactly two vertices, so if we sum the degrees of all vertices we get twice the number of edges. Because every vertex, including v_∞, has degree at least three we get

$$2n_e \geqslant 3(n_v + 1). \tag{7.2}$$

Together with equation (7.1) this implies the theorem. □

We close this section with a characterization of the edges and vertices of the Voronoi diagram. We know that the edges are parts of bisectors of pairs of sites and that the vertices are intersection points between these bisectors. There is a quadratic number of bisectors, whereas the complexity of the $\mathrm{Vor}(P)$ is only linear. Hence, not all bisectors define edges of $\mathrm{Vor}(P)$ and not all intersections are vertices of $\mathrm{Vor}(P)$. To characterize which bisectors and intersections define features of the Voronoi diagram we make the following definition. For a point q we define the *largest empty circle of q with respect to P*, denoted by $C_P(q)$, as the largest circle with q as its center that does not contain any site of P in its interior. The following theorem characterizes the vertices and edges of the Voronoi diagram.

Theorem 7.4 *For the Voronoi diagram* $\text{Vor}(P)$ *of a set of points* P *the following holds:*

(i) *A point* q *is a vertex of* $\text{Vor}(P)$ *if and only if its largest empty circle* $C_P(q)$ *contains three or more sites on its boundary.*

(ii) *The bisector between sites* p_i *and* p_j *defines an edge of* $\text{Vor}(P)$ *if and only if there is a point* $q \in \mathbb{E}^2$ *such that* $C_P(q)$ *contains both* p_i *and* p_j *on its boundary but no other site.*

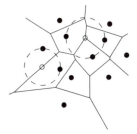

Proof. (i) Suppose there is a point q such that $C_P(q)$ contains three or more sites on its boundary. Let p_i, p_j, and p_k be three of those sites. Since the interior of $C_P(q)$ is empty q must be on the boundary of each of $\mathcal{V}(p_i)$, $\mathcal{V}(p_j)$, and $\mathcal{V}(p_k)$, and q must be a vertex of $\text{Vor}(P)$.

On the other hand, every vertex q of $\text{Vor}(P)$ is incident to at least three edges and, hence, to at least three Voronoi cells $\mathcal{V}(p_i)$, $\mathcal{V}(p_j)$, and $\mathcal{V}(p_k)$. Vertex q must be equidistant to p_i, p_j, and p_k and there cannot be another site closer to q, since otherwise $\mathcal{V}(p_i)$, $\mathcal{V}(p_j)$, and $\mathcal{V}(p_k)$ would not meet at q. Hence, the interior of the circle with p_i, p_j, and p_k on its boundary does not contain any site.

(ii) Suppose there is a point q with the property stated in the theorem. Since $C_P(q)$ does not contain any sites in its interior and p_i and p_j are on its boundary, we have $\text{dist}(q, p_i) = \text{dist}(q, p_j) \leqslant \text{dist}(q, p_k)$ for all $1 \leqslant k \leqslant n$. It follows that q lies on an edge or vertex of $\text{Vor}(P)$. The first part of the theorem implies that q cannot be a vertex of $\text{Vor}(P)$. Hence, q lies on an edge of $\text{Vor}(P)$, which is defined by the bisector of p_i and p_j.

Conversely, let the bisector of p_i and p_j define a Voronoi edge. The largest empty circle of any point q in the interior of this edge must contain p_i and p_j on its boundary and no other sites. ☐

7.2 Computing the Voronoi Diagram

In the previous section we studied the structure of the Voronoi diagram. We now set out to compute it. Observation 7.1 suggests a simple way to do this: for each site p_i, compute the common intersection of the half-planes $h(p_i, p_j)$, with $j \neq i$, using the algorithm presented in Chapter 4. This way we spend $O(n \log n)$ time per Voronoi cell, leading to an $O(n^2 \log n)$ algorithm to compute the whole Voronoi diagram, assuming we can assemble the cells to get the diagram. Can't we do better? After all, the total complexity of the Voronoi diagram is only linear. The answer is yes: the plane sweep algorithm described below—commonly known as *Fortune's algorithm* after its inventor—computes the Voronoi diagram in $O(n \log n)$ time. You may be tempted to look for an even faster algorithm, for example one that runs in linear time. This turns out to be too much to ask: the problem of sorting n real numbers is reducible to the problem of computing Voronoi diagrams, so any algorithm for computing

Voronoi diagrams must take $\Omega(n \log n)$ time in the worst case. Hence, Fortune's algorithm is optimal.

The strategy in a plane sweep algorithm is to sweep a horizontal line—the *sweep line*—from top to bottom over the plane. While the sweep is performed information is maintained regarding the structure that one wants to compute. More precisely, information is maintained about the intersection of the structure with the sweep line. While the sweep line moves downwards the information does not change, except at certain special points—the *event points*.

Let's try to apply this general strategy to the computation of the Voronoi diagram of a set $P = \{p_1, p_2, \dots, p_n\}$ of point sites in the plane. According to the plane sweep paradigm we move a horizontal sweep line ℓ from top to bottom over the plane. The paradigm involves maintaining the intersection of the Voronoi diagram with the sweep line. Unfortunately this is not so easy, because the part of Vor(P) above ℓ depends not only on the sites that lie above ℓ but also on sites below ℓ. Stated differently, when the sweep line reaches the topmost vertex of the Voronoi cell $\mathcal{V}(p_i)$ it has not yet encountered the corresponding site p_i. Hence, we do not have all the information needed to compute the vertex. We are forced to apply the plane sweep paradigm in a slightly different fashion: instead of maintaining the intersection of the Voronoi diagram with the sweep line, we maintain information about the part of the Voronoi diagram of the sites above ℓ that cannot be changed by sites below ℓ.

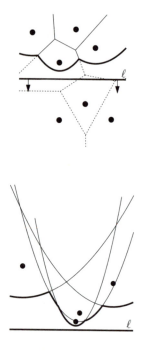

Denote the closed half-plane above ℓ by ℓ^+. What is the part of the Voronoi diagram above ℓ that cannot be changed anymore? In other words, for which points $q \in \ell^+$ do we know for sure what their nearest site is? The distance of a point $q \in \ell^+$ to any site below ℓ is greater than the distance of q to ℓ itself. Hence, the nearest site of q cannot lie below ℓ if q is at least as near to some site $p_i \in \ell^+$ as q is to ℓ. The locus of points that are closer to some site $p_i \in \ell^+$ than to ℓ is bounded by a parabola. Hence, the locus of points that are closer to any site above ℓ than to ℓ itself is bounded by parabolic arcs. We call this sequence of parabolic arcs the *beach line*. Another way to visualize the beach line is the following. Every site p_i above the sweep line defines a complete parabola β_i. The beach line is the function that—for each x-coordinate—passes through the lowest point of all parabolas.

Observation 7.5 *The beach line is x-monotone, that is, every vertical line intersects it in exactly one point.*

It is easy to see that one parabola can contribute more than once to the beach line. We'll worry later about how many pieces there can be. Notice that the *breakpoints* between the different parabolic arcs forming the beach line lie on edges of the Voronoi diagram. This is not a coincidence: the breakpoints exactly trace out the Voronoi diagram while the sweep line moves from top to bottom. These properties of the beach line can be proved using elementary geometric arguments.

So, instead of maintaining the intersection of Vor(P) with ℓ we maintain the beach line as we move our sweep line ℓ. We do not maintain the beach line

explicitly, since it changes continuously as ℓ moves. For the moment let's ignore the issue of how to represent the beach line until we understand where and
how its combinatorial structure changes. This happens when a new parabolic
arc appears on it, and when a parabolic arc shrinks to a point and disappears.

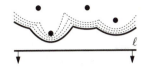

First we consider the events where a new arc appears on the beach line. One
occasion where this happens is when the sweep line ℓ reaches a new site. The
parabola defined by this site is at first a degenerate parabola with zero width: a
vertical line segment connecting the new site to the beach line. As the sweep
line continues to move downward the new parabola gets wider and wider. The
part of the new parabola below the old beach line is now a part of the new
beach line. Figure 7.2 illustrates this process. We call the event where a new
site is encountered a *site event*.

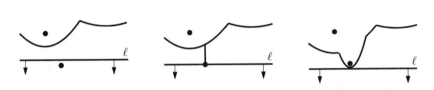

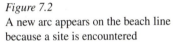

Figure 7.2
A new arc appears on the beach line
because a site is encountered

What happens to the Voronoi diagram at a site event? Recall that the breakpoints on the beach line trace out the edges of the Voronoi diagram. At a site
event two new breakpoints appear, which start tracing out edges. In fact, the
new breakpoints coincide at first, and then move in opposite directions to trace
out the same edge. Initially, this edge is not connected to the rest of the Voronoi
diagram above the sweep line. Later on—we will see shortly exactly when this
will happen—the growing edge will run into another edge, and it becomes connected to the rest of the diagram.

So now we understand what happens at a site event: a new arc appears on
the beach line, and a new edge of the Voronoi diagram starts to be traced out.
Is it possible that a new arc appears on the beach line in any other way? The
answer is no:

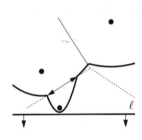

Lemma 7.6 *The only way in which a new arc can appear on the beach line is
through a site event.*

Proof. Suppose for a contradiction that an already existing parabola β_j defined
by a site p_j breaks through the beach line. There are two ways in which this
could happen.

The first possibility is that β_j breaks through in the middle of an arc of a
parabola β_i. The moment this is about to happen, β_i and β_j are tangent, that
is, they have exactly one point of intersection. Let ℓ_y denote the y-coordinate
of the sweep line at the moment of tangency. If $p_j := (p_{j,x}, p_{j,y})$, then the
parabola β_j is given by

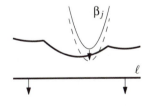

$$\beta_j := \quad y = \frac{1}{2(p_{j,y} - \ell_y)} \left(x^2 - 2p_{j,x}x + p_{j,x}^2 + p_{j,y}^2 - \ell_y^2 \right).$$

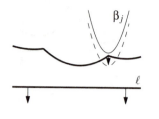

The formula for β_i is similar, of course. Using that both $p_{j,y}$ and $p_{i,y}$ are larger than ℓ_y, it is easy to show that it is impossible that β_i and β_j have only one point of intersection. Hence, a parabola β_j never breaks through in the middle of an arc of another parabola β_i.

The second possibility is that β_j appears in between two arcs. Let these arcs be part of parabolas β_i and β_k. Let q be the intersection point of β_i and β_k at which β_j is about to appear on the beach line, and assume that β_i is on the beach line left of q and β_k is on the beach line right of q, as in Figure 7.3. Then there is a circle C that passes through p_i, p_j, and p_k, the sites defining the parabolas. This circle is also tangent to the sweep line ℓ at some point p_ℓ. The cyclic clockwise order on C is p_ℓ, p_i, p_j, p_k, because β_j is assumed to appear in between the arcs of β_i and β_k. Consider an infinitesimal motion of the sweep line downward while keeping the circle C tangent to ℓ; see Figure 7.3. Then C cannot have empty interior and still pass through p_j: either p_i or p_k

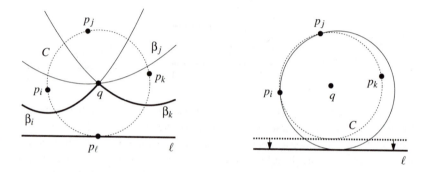

Figure 7.3
The situation when β_j would appear on the beach line, and the circle when the sweep line has proceeded

will penetrate the interior. Therefore, in a sufficiently small neighborhood of q the parabola β_j cannot appear on the beach line when the sweep line moves downward, because either p_i or p_k will be closer to ℓ than p_j. □

An immediate consequence of the lemma is that the beach line consists of at most $2n - 1$ parabolic arcs: each site encountered gives rise to one new arc and the splitting of at most one existing arc into two, and there is no other way an arc can appear on the beach line.

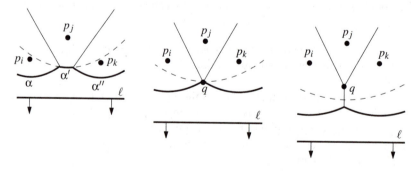

Figure 7.4
An arc disappears from the beach line

The second type of event in the plane sweep algorithm is where an existing arc of the beach line shrinks to a point and disappears, as in Figure 7.4. Let

α' be the disappearing arc, and let α and α'' be the two neighboring arcs of α' before it disappears. The arcs α and α'' cannot be part of the same parabola; this possibility can be excluded in the same way as the first possibility in the proof of Lemma 7.6 was excluded. Hence, the three arcs α, α', and α'' are defined by three distinct sites p_i, p_j, and p_k. At the moment α' disappears, the parabolas defined by these three sites pass through a common point q. Point q is equidistant from ℓ and each of the three sites, and there is a circle passing through p_i, p_j, and p_k with q as its center whose lowest point lies on ℓ. There cannot be a site in the interior of this circle: such a site would be closer to q than q is to ℓ, contradicting the fact that q is on the beach line. It follows that the point q is a vertex of the Voronoi diagram. This is not very surprising, since we observed earlier that the breakpoints on the beach line trace out the Voronoi diagram. So when an arc disappears from the beach line and two breakpoints meet, two edges of the Voronoi diagram meet as well. We call the event where the sweep line reaches the lowest point of a circle through three sites defining consecutive arcs on the beach line a *circle event*. From the above we can conclude the following lemma.

Lemma 7.7 *The only way in which an existing arc can disappear from the beach line is through a circle event.*

Now we know where and how the combinatorial structure of the beach line changes: at a site event a new arc appears, and at a circle event an existing arc drops out. We also know how this relates to the Voronoi diagram under construction: at a site event a new edge starts to grow, and at a circle event two growing edges meet to form a vertex. It remains to find the right data structures to maintain the necessary information during the sweep. Our goal is to compute the Voronoi diagram, so we need a data structure that stores the part of the Voronoi diagram computed thus far. We also need the two 'standard' data structures for any sweep line algorithm: an event queue and a structure that represents the status of the sweep line. Here the latter structure is a representation of the beach line. These data structures are implemented in the following way.

- We store the Voronoi diagram under construction in our usual data structure for subdivisions, the doubly-connected edge list. A Voronoi diagram, however, is not a true subdivision as defined in Chapter 2: it has edges that are half-lines or full lines, and these cannot be represented in a doubly-connected edge list. During the construction this is not a problem, because the representation of the beach line—described next—will make it possible to access the relevant parts of the doubly-connected edge list efficiently during its construction. But after the computation is finished we want to have a valid doubly-connected edge list. To this end we add a big bounding box to our scene, which is large enough so that it contains all vertices of the Voronoi diagram. The final subdivision we compute will then be the bounding box plus the part of the Voronoi diagram inside it.

- The beach line is represented by a balanced binary search tree $\mathcal{T}$; it is

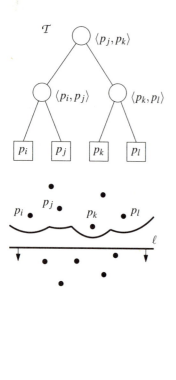

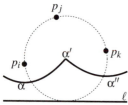

the status structure. Its leaves correspond to the arcs of the beach line—which is x-monotone—in an ordered manner: the leftmost leaf represents the leftmost arc, the next leaf represents the second leftmost arc, and so on. Each leaf μ stores the site that defines the arc it represents. The internal nodes of $\mathcal{T}$ represent the breakpoints on the beach line. A breakpoint is stored at an internal node by an ordered tuple of sites $\langle p_i, p_j \rangle$, where p_i defines the parabola left of the breakpoint and p_j defines the parabola to the right. Using this representation of the beach line, we can find in $O(\log n)$ time the arc of the beach line lying above a new site. At an internal node, we simply compare the x-coordinate of the new site with the x-coordinate of the breakpoint, which can be computed from the tuple of sites and the position of the sweep line in constant time. (Note that we do not explicitly store the parabolas.)

In $\mathcal{T}$ we also store pointers to the other two data structures used during the sweep. Each leaf of $\mathcal{T}$, representing an arc α, stores one pointer to a node in the event queue, namely, the node that represents the circle event in which α will disappear. Finally, every internal node has a pointer to a half-edge in the Voronoi diagram that is being traced out by the breakpoint it represents.

- The event queue Q is implemented as a priority queue, where the priority of an event is its y-coordinate. It stores the upcoming events that are already known. For a site event we simply store the site itself. For a circle event the event point that we store is the lowest point of the circle. For a circle event we also store a pointer to the leaf in $\mathcal{T}$ that represents the arc that will disappear in the event.

All the site events are known in advance, unlike the circle events. This brings us to one final issue that we must discuss, namely, the detection of the circle events. Recall that a circle event is caused by the disappearance of an arc from the beach line. Let α, α' and α'' be three consecutive arcs on the beach line, and assume that the circle event causes α' to disappear. This happens exactly when the three sites p_i, p_j, and p_k defining these arcs, define a circle $C(p_i, p_j, p_k)$ whose lowest point lies on the sweep line.

Our algorithm will make sure that for every three consecutive arcs on the beach line, the corresponding circle event is in the event queue Q if the circle intersects the sweep line. If the circle lies completely above the sweep line, then the event has already been dealt with. Also if the circle contains some other site in its interior the event should not be handled. In both cases the circle event isn't and shouldn't be in the event queue.

Lemma 7.8 *Every Voronoi vertex is detected by means of a circle event.*

Proof. For a Voronoi vertex q, let p_i, p_j, and p_k be the three sites through which a circle $C(p_i, p_j, p_k)$ passes with no sites in the interior. By Theorem 7.4, such a circle and three sites indeed exist. For simplicity we only prove the case where no other sites lie on $C(p_i, p_j, p_k)$, and the lowest point of $C(p_i, p_j, p_k)$ is not one of the defining sites. Assume without loss of generality that from the lowest point of $C(p_i, p_j, p_k)$, the clockwise traversal of $C(p_i, p_j, p_k)$ encounters the sites p_i, p_j, p_k in this order.

We must show that just before the sweep line reaches the lowest point of $C(p_i, p_j, p_k)$, there are three consecutive arcs α, α' and α'' on the beach line defined by the sites p_i, p_j, and p_k. Only then will the circle event take place. Consider the sweep line an infinitesimal amount before it reaches the lowest point of $C(p_i, p_j, p_k)$. Since $C(p_i, p_j, p_k)$ doesn't contain any other sites inside or on it, there exists a circle through p_i and p_j that is tangent to the sweep line, and doesn't contain sites in the interior. So there are adjacent arcs on the beach line defined by p_i and p_j. Similarly, there are adjacent arcs on the beach line defined by p_j and p_k. It is easy to see that the two arcs defined by p_j are actually the same arc, and it follows that there are three consecutive arcs on the beach line defined by p_i, p_j, and p_k. Therefore, the corresponding circle event is in Q just before the event takes place, and the Voronoi vertex is detected. $\square$

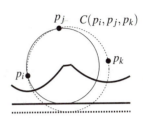

During the sweep the beach line changes its topological structure at every event. This means that new triples of consecutive arcs can show up at all events. At a site event, a new arc appears on the beach line and there will be new consecutive triples including that arc. At a circle event, an arc disappears from the beach line, and the neighboring arcs define new triples. Furthermore, the events can destroy triples of consecutive arcs. When an arc disappears at a circle event, all other triples involving that arc don't correspond to consecutive triples on the beach line anymore. When a new arc appears at a site event, another arc is split and triples involving the split arc may be destroyed.

Note that it is not true that circle events in Q will actually take place later on. A circle event will only take place if the triple of consecutive arcs lives on until the sweep line reaches the lowest point of the circle. Circle events that are deleted before they take place are also called false alarms.

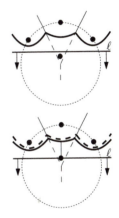

Let's summarize. The event queue Q contains a circle event if and only if there are three consecutive arcs on the beach line such that the circle through the sites defining the three arcs intersects the sweep line, and hasn't been deleted yet. Maintaining this invariant involves both insertions and deletions of circle events when the beach line changes its structure, that is, when other events are handled.

When there is a new triple of consecutive arcs, we first test whether these arcs are defined by three different sites. If only two sites are involved, we ignore the new triple: it cannot define a circle event. If three sites are involved, we test whether the circle event isn't in Q yet and the circle intersects the sweep line. If so, we insert the circle event; otherwise, we ignore it.

When a triple of consecutive arcs is destroyed, we delete the corresponding circle event from Q, if it occurs in Q. Suppose that an arc α disappears from the beach line, or it is split, and we wish to delete all circle events α participates in—see Figure 7.5. We first search in $\mathcal{T}$ for the leaf μ storing the site defining α. At that leaf there is a pointer to a circle event in Q, which is removed from Q. The other two circle events α may participate in can be found through the

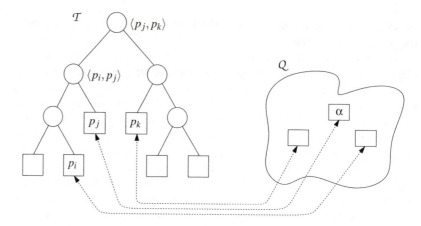

Figure 7.5
When an arc α defined by p_j
disappears, three circle events in Q
may be destroyed.

previous leaf and the next leaf in $\mathcal{T}$. These leaves also have a pointer to a node in Q, which must be deleted. Finally, the leaf μ is deleted from $\mathcal{T}$.

We can now describe the plane sweep algorithm in detail. Notice that after all events have been handled and the event queue Q is empty, the beach line hasn't disappeared yet. The breakpoints that are still present correspond to the half-infinite edges of the Voronoi diagram. As stated earlier, a doubly connected edge list cannot represent half-infinite edges, so we must add a bounding box to the scene to which these edges can be attached. The overall structure of the algorithm is as follows.

Algorithm VORONOIDIAGRAM(P)
Input. A set $P := \{p_1, \ldots, p_n\}$ of point sites in the plane.
Output. The Voronoi diagram Vor(P) given inside a bounding box in a doubly-connected edge list structure.
1. Initialize the event queue Q with all site events.
2. **while** Q is not empty
3.   **do** Consider the event with largest y-coordinate in Q.
4.   **if** the event is a site event, occurring at site p_i
5.   **then** HANDLESITEEVENT(p_i)
6.   **else** HANDLECIRCLEEVENT(p_ℓ), where p_ℓ is the lowest point of the circle causing the event
7.   Remove the event from Q.
8. The internal nodes still present in $\mathcal{T}$ correspond to the half-infinite edges of the Voronoi diagram. Compute a bounding box that contains all vertices of the Voronoi diagram in its interior, and attach the half-infinite edges to the bounding box by updating the doubly-connected edge list appropriately.
9. Traverse the half-edges of the doubly-connected edge list to add the cell records and the pointers to and from them.

The procedures to handle the events are defined as follows.

HANDLESITEEVENT(p_i)
1. Search in T for the arc α vertically above p_i, and delete all circle events involving α from Q.
2. Replace the leaf of T that represents α with a subtree having three leaves. The middle leaf stores the new site p_i and the other two leaves store the site p_j that was originally stored with α. Store the tuples $\langle p_j, p_i \rangle$ and $\langle p_i, p_j \rangle$ representing the new breakpoints at the two new internal nodes. Perform rebalancing operations on T if necessary.
3. Create new records in the Voronoi diagram structure for the two half-edges separating $\mathcal{V}(p_i)$ and $\mathcal{V}(p_j)$, which will be traced out by the two new breakpoints.
4. Check the triples of consecutive arcs involving one of the three new arcs. Insert the corresponding circle event only if the circle intersects the sweep line and the circle event isn't present yet in Q.

HANDLECIRCLEEVENT(p_ℓ)
1. Search in T for the arc α vertically above p_ℓ that is about to disappear, and delete all circle events that involve α from Q.
2. Delete the leaf that represents α from T. Update the tuples representing the breakpoints at the internal nodes. Perform rebalancing operations on T if necessary.
3. Add the center of the circle causing the event as a vertex record in the Voronoi diagram structure and create two half-edge records corresponding to the new breakpoint of the Voronoi diagram. Set the pointers between them appropriately.
4. Check the new triples of consecutive arcs that arise because of the disappearance of α. Insert the corresponding circle event into Q only if the circle intersects the sweep line and the circle event isn't present yet in Q.

Lemma 7.9 *The algorithm runs in $O(n \log n)$ time and it uses $O(n)$ storage.*

Proof. The primitive operations on the tree T and the event queue Q, such as inserting or deleting an element, take $O(\log n)$ time each. The primitive operations on the doubly-connected edge list take constant time. To handle an event we do a constant number of such primitive operations, so we spend $O(\log n)$ time to process an event. Obviously, there are n site events. As for the number of circle events, we observe that every such event that is processed defines a vertex of Vor(P). Note that false alarms are deleted from Q before they are processed. They are created and deleted while processing another, real event, and the time we spend on them is subsumed under the time we spend to process this event. Hence, the number of circle events that we process is at most $2n - 5$. The time and storage bounds follow. ⌑

Before we state the final result of this section we should say a few words about degenerate cases.

The algorithm handles the events from top to bottom, so there is a degeneracy when two or more events lie on a common horizontal line. This happens,

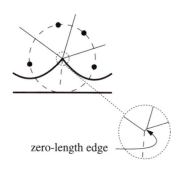

zero-length edge

for example, when there are two sites with the same y-coordinate. It is not difficult to see that these events can be handled in any order when their x-coordinates are distinct. So we can break ties between events with the same y-coordinate but with different x-coordinates arbitrarily. Now suppose there are event points that coincide. For instance, there will be several coincident circle events when there are four or more co-circular sites, such that the interior of the circle through them is empty. The center of this circle is a vertex of the Voronoi diagram. The degree of this vertex is at least four. We could write special code to handle such degenerate cases, but there is no need to do so. What will happen if we let the algorithm handle these events in arbitrary order? Instead of producing a vertex with degree four, it will just produce two vertices with degree three at the same location, with a zero length edge between them. These degenerate edges can be removed in a post-processing step, if required.

Besides these degeneracies in choosing the order of the events we may also encounter degeneracies while handling an event. This occurs when a site p_i that we process happens to be located exactly below the breakpoint between two arcs on the beach line. In this case the algorithm splits either of these two arcs and inserts the arc for p_i in between the two pieces, one of which has zero length. This piece of zero length now is the middle arc of a triple that defines a circle event. The lowest point of this circle coincides with p_i. The algorithm inserts this circle event into the event queue Q, because there are three consecutive arcs on the beach line that define it. When this circle event is handled, a vertex of the Voronoi diagram is correctly created and the zero length arc can be deleted later.

We conclude that the above algorithm handles degenerate cases correctly.

Theorem 7.10 *The Voronoi diagram of a set of n point sites in the plane can be computed with a sweep line algorithm in $O(n \log n)$ time using $O(n)$ storage.*

7.3 Notes and Comments

Although it is beyond the scope of this book to give an extensive survey of the history of Voronoi diagrams it is appropriate to make a few historical remarks. Voronoi diagrams are often attributed to Dirichlet [122]—hence the name *Dirichlet tessellations* that is sometimes used—and Voronoi [328, 329]. They can already be found in Descartes's treatment of cosmic fragmentation in Part III of his *Principia Philosophiae*, published in 1644. Also in this century the Voronoi diagram has been re-discovered several times. In biology this even happened twice in a very short period. In 1965 Brown [54] studied the intensity of trees in a forest. He defined the *area potentially available* to a tree, which was in fact the Voronoi cell of that tree. One year later Mead [238] used the same concept for plants, calling the Voronoi cells *plant polygons*. By now there is an impressive amount of literature concerning Voronoi diagrams and their applications in all kinds of research areas. The book by Okabe et al. [261] contains an ample treatment of Voronoi diagrams and their applications. We confine ourselves in this section to a discussion of the various aspects of Voronoi diagrams encountered in the computational geometry literature.

In this chapter we have proved some properties of the Voronoi diagram, but it has many more. For example, if one connects all the pairs of sites whose Voronoi cells are adjacent then the resulting set of segments forms a triangulation of the point set, called the Delaunay triangulation. This triangulation, which has some very nice properties, is the topic of Chapter 9.

There is a beautiful connection between Voronoi diagrams and convex polyhedra. Consider the transformation that maps a point $p = (p_x, p_y)$ in $\mathbb{E}^2$ to the non-vertical plane $h(p): z = 2p_x x + 2p_y y - (p_x^2 + p_y^2)$ in $\mathbb{E}^3$. Geometrically, $h(p)$ is the plane that is tangent to the unit paraboloid $\mathcal{U}: z = x^2 + y^2$ at the point vertically above $(p_x, p_y, 0)$. For a set P of point sites in the plane, let $H(P)$ be the set of planes that are the images of the sites in P. Now consider the convex polyhedron $\mathcal{P}$ that is the intersection of all positive half-spaces defined by the planes in $H(P)$, that is, $\mathcal{P} := \bigcap_{h \in H(P)} h^+$, where h^+ denotes the half-space above h. Surprisingly, if we project the edges and vertices of the polyhedron vertically downwards onto the xy-plane, we get the Voronoi diagram of P [138]. See Chapter 11 for a more extensive description of this transformation.

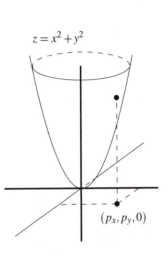

We have studied Voronoi diagrams in their most basic setting, namely for a set of point sites in the Euclidean plane. The first optimal $O(n \log n)$ time algorithm for this case was a divide-and-conquer algorithm presented by Shamos and Hoey [311]; since then many other optimal algorithms have been developed. The plane sweep algorithm that we described is due to Fortune [152]. Fortune's original description of the algorithm is a little different from ours, which follows the interpretation of the algorithm given by Guibas and Stolfi [172].

Voronoi diagrams can be generalized in many ways [16, 261]. One generalization is to point sets in higher-dimensional spaces. In $\mathbb{E}^d$ the maximum combinatorial complexity of the Voronoi diagram of a set of n point sites (the maximum number of vertices, edges, and so on, of the diagram) is $\Theta(n^{\lceil d/2 \rceil})$ [203] and it can be computed in $O(n \log n + n^{\lceil d/2 \rceil})$ optimal time [72, 111, 309]. The fact that the dual of the Voronoi diagram is a triangulation of the set of sites, and the connection between Voronoi diagrams and convex polyhedra as discussed above, still hold in higher dimensions.

Another generalization concerns the metric that is used. In the L_1-metric or Manhattan metric, the distance between two points p and q is defined as

$$\text{dist}_1(p, q) := |p_x - q_x| + |p_y - q_y|;$$

the sum of the absolute differences in the x- and y-coordinates. In a Voronoi diagram in the L_1-metric, all edges are horizontal, vertical, or diagonal (an angle of 45° with the coordinate axes). In the more general L_p-metric, the distance between two points p and q is defined as

$$\text{dist}_p(p, q) := \sqrt[p]{|p_x - q_x|^p + |p_y - q_y|^p}.$$

Note that the L_2-metric is simply the Euclidean metric. There are several papers dealing with Voronoi diagrams in these metrics [96, 213, 217]. One can also define a distance function by assigning weights to the sites: a multiplicative weight and an additive weight. Now the distance of a site to a point is its multiplicative weight times the Euclidean distance to the point, plus its additive weight. The resulting diagrams are called weighted Voronoi diagrams [17]. Power diagrams [13, 14, 15, 18] are another generalization of Voronoi diagrams where a different distance function is used. It is even possible to drop the distance function altogether and define the Voronoi diagram in terms of bisectors between any two sites only. Such diagrams are called abstract Voronoi diagrams [204, 205, 206, 240].

Other generalizations concern the shape of the sites. Such generalizations occur when the Voronoi diagram is used for motion planning purposes. An important special case is the Voronoi diagram of the edges of a simple polygon, interior to the polygon itself. This Voronoi diagram is also known as the medial axis or skeleton, and it has applications in shape analysis. The medial axis can be computed in time linear in the number of edges of the polygon [101].

Instead of partitioning the space into regions according to the closest sites, one can also partition it according to the k closest sites, for some $1 \leqslant k \leqslant n - 1$. The diagrams obtained in this way are called higher-order Voronoi diagrams, and for given k the diagram is called the order-k Voronoi diagram [1, 19, 50, 77]. Note that the order-1 Voronoi diagram is nothing more than the standard Voronoi diagram. The order-$(n - 1)$ Voronoi diagram is also called the farthest-point Voronoi diagram, because the Voronoi cell of a point p_i now is the region of points for which p_i is the farthest site. The maximum complexity of the order-k Voronoi diagram of a set of n point sites in the plane is $\Theta(k(n - k))$ [214]. Currently the best known algorithm for computing the order-k Voronoi diagram runs in $O(n \log^3 n + k(n - k))$ time [1].

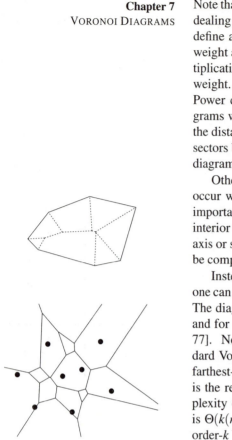

7.4 Exercises

7.1 Prove that for any $n > 3$ there is a set of n point sites in the plane such that one of the cells of Vor(P) has $n - 1$ vertices.

7.2 Show that Theorem 7.3 implies that the average number of vertices of a Voronoi cell is less than six.

7.3 Show that $\Omega(n \log n)$ is a lower bound for computing Voronoi diagrams by reducing the sorting problem to the problem of computing Voronoi diagrams. You can assume that the Voronoi diagram algorithm should be able to compute for every vertex of the Voronoi diagram its incident edges in cyclic order around the vertex.

7.4 Prove that the breakpoints of the beach line, as defined in Section 7.2, trace out the Voronoi diagram while the sweep line moves from top to bottom.

7.5 Give an example where the parabola defined by some site p_i contributes more than one arc to the beach line. Can you give an example where it contributes a linear number of arcs?

7.6 Give an example of six sites such that the plane sweep algorithm encounters the six site events before any of the circle events.

7.7 Do the breakpoints of the beach line always move downwards? Prove this or give a counterexample.

7.8 Write a procedure to compute a big enough bounding box from the incomplete doubly-connected edge list and the tree $\mathcal{T}$ after the sweep is completed. The box should contain all sites and all Voronoi vertices.

7.9 Write a procedure to add all cell records and the corresponding pointers to the incomplete doubly-connected edge list after the bounding box has been added.

7.10 Let P be a set of n points in the plane. Give an $O(n \log n)$ time algorithm to find two points in P that are closest together. Show that your algorithm is correct.

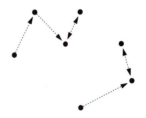

7.11 Let P be a set of n points in the plane. Give an $O(n \log n)$ time algorithm to find for each point p in P another point in P that is closest to it.

7.12 Let the Voronoi diagram of a point set P be stored in a doubly-connected edge list as it is computed in this chapter, inside a bounding box. Give an algorithm to compute the convex hull of P in time linear in the output size.

7.13* In the Voronoi assignment model the goods or services that the consumers want to acquire have the same market price at every site. Suppose this is not the case, and that the price of the good at site p_i is w_i. The trading areas of the sites now correspond to the cells in the weighted Voronoi diagram of the sites (see Section 7.3), where site p_i has an additive weight w_i. Generalize the sweep line algorithm of Section 7.2 to this case.

7.14* Suppose that we are given a subdivision of the plane into n convex regions. We suspect that this subdivision is a Voronoi diagram, but we do not know the sites. Develop an algorithm that finds a set of n point sites whose Voronoi diagram is exactly the given subdivision, if such a set exists.

8 Arrangements and Duality

Supersampling in Ray Tracing

Computer generated images of 3-dimensional scenes are becoming more and more realistic. Nowadays, they can hardly be distinguished from photographs. A technique that has played an important role in this development is *ray tracing*. It works as follows.

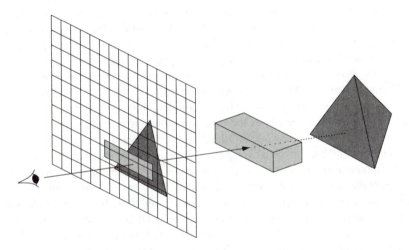

Figure 8.1
Determining visible objects using ray tracing

The screen of a computer monitor is composed of many small dots, called pixels. A nice screen consists of, say, 1280 by 1024 pixels. Suppose that we are given a 3-dimensional scene consisting of several objects, a light source, and a view point. Generating an image of this scene—also called *rendering* the scene—amounts to determining for every pixel on the screen which object is visible at that pixel, and determining the intensity of light that the object emits in the direction of the view point at that particular point. Let's first look at the first task, determining the visible object at each pixel. Ray tracing performs this task by shooting a ray through each pixel, as in Figure 8.1. The first object that is hit is the one that is visible at the pixel. Once the visible object is determined, one has to compute the intensity of light emitted by the object at the visible point. Here we have to take into account how much light that point

163

receives from the light source, either directly or indirectly via reflections on other objects. The strength of ray tracing is that it can perform this second task—which is crucial to getting a realistic image—quite well. In this chapter, however, we are mainly interested in the first part.

There is one issue in determining the visible object for each pixel that we have swept under the rug: a pixel is not a point, but a small square area. In general this is not a problem. Most pixels are covered completely by one of the objects, and shooting a ray through the pixel center will tell us which object this is. But problems arise near the edges of the objects. When an edge of an object crosses a pixel, the object may cover 49% of the pixel area, but be missed by the ray through the pixel center. If, on the other hand, the object covered 51% of the pixel area, it would be hit by the ray and we would falsely assume that the whole pixel was covered. This results in the well-known jaggies in the image. This artifact is diminished if we had not just two categories, "hit" and "miss", but also categories like "49% hit". Then we could set the intensity of the pixel to, say, 0.49 times the object intensity. Or if there is more than one object visible inside the pixel, we could make the pixel intensity a mixture of the object intensities.

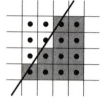

How can we incorporate such different pixel intensities in our ray tracing scheme? The solution is to shoot more than one ray per pixel. If, for instance, we shoot 100 rays per pixel and 35 of them hit an object then we expect that object to be visible in roughly 35% of the pixel area. This is called supersampling: instead of taking one sample point per pixel, we take many.

How should we distribute the rays over the pixel to make this work? An obvious choice would be to distribute them regularly; for 100 rays this would mean having a regular grid of 10 by 10 sample points in the pixel. Indeed, if 35 of these rays hit the object then the area where it is visible cannot be far from 35%. There is, however, a disadvantage to choosing a regular sample pattern: although the error at each particular pixel will be small, there will be a certain regularity across rows (and columns) of pixels. Regularity in the errors triggers the human visual system, and as a result we see annoying artifacts. So choosing a regular sample pattern isn't such a good idea. It's better to choose the sample points in a somewhat random fashion. Of course not every random pattern works equally well; we still want the sample points to be distributed in such a way that the number of hits is close to the percentage of covered area.

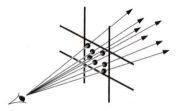

Suppose that we have generated a random set of sample points. We'd like to have a way of deciding whether this set is good. What we want is that the difference between the percentage of hits for an object and the percentage of the pixel area where that object is visible is small. This difference is called the *discrepancy of the sample set with respect to the object*. Of course we do not know in advance which objects will be visible in the pixel, so we have to prepare for the worst-case scenario: we want the maximum discrepancy over all possible ways that an object can be visible inside the pixel to be small. This is called the *discrepancy of the sample set*, and it depends on the type of objects that are present in the scene. So formally the discrepancy of a sample set is defined with respect to a given class of objects. Based on the discrepancy of the given set of sample points we can decide if it is good enough: if the discrepancy is low enough we decide to keep it, and otherwise we generate a new random set. For this we need an algorithm that computes the discrepancy of a given point set.

8.1 Computing the Discrepancy

We mentioned above that the discrepancy of a point set is defined with respect to a class of objects. The objects that we must consider are the projections of the 3-dimensional objects that make up our scene. As is common in computer graphics, we assume that curved objects are approximated using polygonal meshes. So the 2-dimensional objects that we must consider are the projections of the facets of polyhedra. In other words, we are interested in the discrepancy with respect to the class of polygons. In general scenes, most pixels will be crossed by at most one edge of a given polygon, unless the scene consists of many polygons that are extremely thin or tiny. If a pixel is intersected by one polygon edge, the polygon behaves inside the pixel like a half-plane. The situation that a pixel is intersected by more polygon edges is much less common. Also, it doesn't cause regularity in the error, which was the source of the annoying artifacts. Therefore we restrict our attention to half-plane discrepancy.

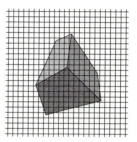

Let $U := [0:1] \times [0:1]$ be the unit square—U models a pixel—and let S be a set of n sample points in U. Let $\mathcal{H}$ denote the (infinite) set of all possible closed half-planes. The *continuous measure* of a half-plane $h \in \mathcal{H}$ is defined as the area of $h \cap U$. We denote it by $\mu(h)$. For a half-plane h that completely covers U we have $\mu(h) = 1$, for instance. The *discrete measure* of h, which we denote as $\mu_S(h)$, is defined as the fraction of the sample points that is contained in h, so $\mu_S(h) := \operatorname{card}(S \cap h) / \operatorname{card}(S)$, where $\operatorname{card}(\cdot)$ denotes the cardinality of a set. The *discrepancy* of h with respect to the sample set S, denoted as $\Delta_S(h)$, is the absolute difference between the continuous and the discrete measure:

$$\Delta_S(h) := |\mu(h) - \mu_S(h)|.$$

The discrepancy of the half-plane in the margin figure, for example, is $|0.25 - 0.3| = 0.05$. Finally, the *half-plane discrepancy* of S is the supremum of the discrepancies over all possible half-planes:

$$\Delta_{\mathcal{H}}(S) := \sup_{h \in \mathcal{H}} \Delta_S(h).$$

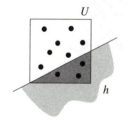

We have defined what it is that we want to compute. Now let's see how to compute it.

The supremum of the discrepancy over all closed half-planes equals the maximum discrepancy over all open or closed half-planes. To get the search for the half-plane with maximum discrepancy started, we first identify a finite set of candidate half-planes. It's always a good idea to replace an infinite set of candidates by a finite set, provided the latter contains the ones we're interested in. So the finite set we identify must contain the half-plane that has maximum discrepancy. We'll select each half-plane that *locally* has maximal discrepancy. This means that if we translated or rotated such a half-plane slightly, its discrepancy would decrease. One of these half-planes with locally maximal discrepancy will be the one that gives the real maximum.

Any half-plane that does not contain a point of S on its boundary can be translated slightly such that its continuous measure increases while its discrete measure remains the same. A slight translation in the opposite direction will decrease the continuous measure while the discrete measure remains the same. So one of the two translations results in an increase in the discrepancy. Hence, the half-plane we are looking for has a point of S on its boundary. Now consider a half-plane h having only one point $p \in S$ on its boundary. Can we always rotate h around p such that the discrepancy increases? In other words, does the half-plane with the maximum discrepancy always have two points on its boundary? The answer is no: when we rotate h around p we can reach a local extremum in the continuous measure function. Suppose this local extremum is a local maximum. Then any slight rotation will decrease the continuous measure. If the discrete measure is smaller than the continuous measure at the local maximum, rotating will decrease the discrepancy. Similarly, any slight rotation at a local minimum of the continuous measure where the discrete measure is greater than the continuous measure will decrease the discrepancy. Hence, the maximum discrepancy could very well be achieved at such an extremum.

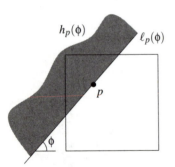

Let's investigate the extrema more closely. Let $p := (p_x, p_y)$ be a point in S. For $0 \leqslant \phi < 2\pi$, let $\ell_p(\phi)$ be the line through p that makes an angle ϕ with the positive x-axis. Consider the continuous measure function of the half-plane initially lying above $\ell_p(\phi)$; we denote this half-plane by $h_p(\phi)$. We are interested in the local extrema of the function $\phi \mapsto \mu(h_p(\phi))$. When ϕ increases from 0 to 2π, the line $\ell_p(\phi)$ rotates around p. First of all, an extremum may occur when $\ell_p(\phi)$ sweeps over one of the vertices of U. This happens at most eight times. In between two such events $\ell_p(\phi)$ intersects two fixed edges of U. A little calculation yields the continuous measure functions for the various cases that occur. For example, when $\ell_p(\phi)$ intersects the top and left boundary of U we have

$$\mu(h_p(\phi)) = \frac{1}{2}(1 - p_y + p_x \tan \phi)(p_x + \frac{1 - p_y}{\tan \phi}).$$

In this case there are at most two local extrema. The continuous measure function is similar when $\ell_p(\phi)$ intersects two other edges of U, so we may conclude that there is a constant number of local extrema per point $p \in S$. Thus the total number of candidate half-planes with one point on their boundary is $O(n)$. Moreover, we can find the extrema and the corresponding half-planes in $O(1)$ time per point. We have proved the following lemma.

Lemma 8.1 *Let S be a set of n points in the unit square U. A half-plane h that achieves the maximum discrepancy with respect to S is of one of the following types:*

(i) h contains one point $p \in S$ on its boundary,
(ii) h contains two or more points of S on its boundary.
The number of type (i) candidates is $O(n)$, and they can be found in $O(n)$ time.

The number of type (ii) candidates is clearly quadratic. Because the number of type (i) candidates is much smaller than this, we treat them in a brute-force way: for each of the $O(n)$ half-planes we compute their continuous measure in constant time, and their discrete measure in $O(n)$ time. This way the maximum of the discrepancies of these half-planes can be computed in $O(n^2)$ time. For the type (ii) candidates we must be more careful when computing the discrete measures. For this we need some new techniques. In the remainder of this chapter we introduce these techniques and we show how to use them to compute all discrete measures in $O(n^2)$ time. We can then compute the discrepancy of these half-planes in constant time per half-plane, and take the maximum. Finally, by comparing this maximum to the maximum discrepancy of the type (i) candidates we find the discrepancy of S. This leads to the following theorem.

Theorem 8.2 *The half-plane discrepancy of a set S of n points in the unit square can be computed in $O(n^2)$ time.*

8.2 Duality

A point in the plane has two parameters: its x-coordinate and its y-coordinate. A (non-vertical) line in the plane also has two parameters: its slope and its intersection with the y-axis. Therefore we can map a set of points to a set of lines, and vice versa, in a one-to-one manner. We can even do this in such a way that certain properties of the set of points translate to certain other properties for the set of lines. For instance, three points on a line become three lines through a point. Several different mappings that achieve this are possible; they

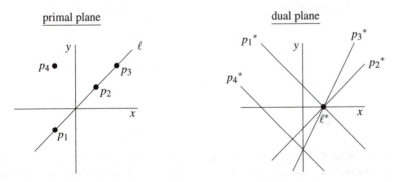

Figure 8.2
An example of duality

are called *duality transforms*. The image of an object under a duality transform is called the *dual* of the object. A simple duality transform is the following. Let $p := (p_x, p_y)$ be a point in the plane. The dual of p, denoted p^*, is the line defined as

$$p^* := (y = p_x x - p_y).$$

The dual of a line $\ell : y = mx + b$ is the point p such that $p^* = \ell$. In other words,

$$\ell^* := (m, -b).$$

The duality transform is not defined for vertical lines. In most cases vertical lines can be handled separately, so this is not a problem. Another solution is to rotate the scene so that there are no vertical lines.

We say that the duality transform maps objects from the *primal plane* to the *dual plane*. Certain properties that hold in the primal plane also hold in the dual plane:

Observation 8.3 *Let p be a point in the plane and let ℓ be a non-vertical line in the plane. The duality transform $o \mapsto o^*$ has the following properties.*

■ *It is incidence preserving: $p \in \ell$ if and only if $\ell^* \in p^*$.*

■ *It is order preserving: p lies above ℓ if and only if ℓ^* lies above p^*.*

Figure 8.2 illustrates these properties. The three points p_1, p_2, and p_3 lie on the line ℓ in the primal plane; the three lines p_1^*, p_2^*, and p_3^* go through the point ℓ^* in the dual plane. The point p_4 lies above the line ℓ in the primal plane; the point ℓ^* lies above the line p_4^* in the dual plane.

The duality transform can also be applied to other objects than points and lines. What would be the dual of a line segment $s := \overline{pq}$, for example? A logical choice for s^* is the union of the duals of all points on s. What we get is an infinite set of lines. All the points on s are collinear, so all the dual lines pass through one point. Their union forms a *double wedge*, which is bounded by the duals of the endpoints of s. The lines dual to the endpoints of s define two double wedges, a left-right wedge and a top-bottom wedge; s^* is the left-right wedge. Figure 8.3 shows the dual of a segment s. It also shows a line

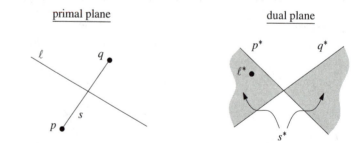

primal plane   dual plane

Figure 8.3
The dual transform applied to a line segment

ℓ intersecting s, whose dual ℓ^* lies in s^*. This is not a coincidence: any line that intersects s must have either p or q above it and the other point below it, so the dual of such a line lies in s^* by the order preserving property of the dual transform.

The dual transform defined above has a nice geometric interpretation. Let $\mathcal{U}$ denote the parabola $\mathcal{U} : y = x^2/2$. Let's first look at the dual of a point p that lies on $\mathcal{U}$. The derivative of $\mathcal{U}$ at p is p_x, so p^* has the same slope as the tangent line of $\mathcal{U}$ at p. As a matter of fact, the dual of a point $p \in \mathcal{U}$ is the tangent line at p, because the intersection of the tangent with the y-axis is $(0, -p_x^2/2)$. Now suppose that a point q does not lie on $\mathcal{U}$. What is the slope of q^*? Well, any two points on the same vertical line have duals with equal slope. In particular, q^* is parallel to p^*, where p is the point that lies on $\mathcal{U}$ and has the same x-coordinate as q. Let q' be the point with the same x-coordinate as q (and as p) such that $q'_y - p_y = p_y - q_y$. The vertical distance between the duals of points with the same x-coordinate is equal to the difference in y-coordinates of these points. Hence, q^* is the line through q' that is parallel to the tangent of $\mathcal{U}$ at p.

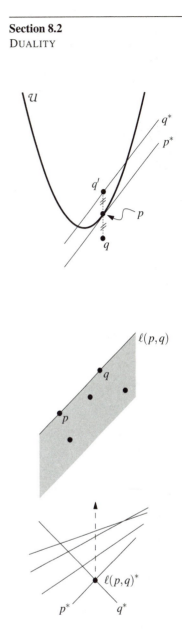

When you think about duality for a few minutes you may wonder how duality can be useful. If you can solve a problem in the dual plane, you could have solved it in the primal plane as well by mimicking the solution to the dual problem in the primal plane. After all, the primal and dual problems are essentially the same. Still, transforming a problem to the dual plane has one important advantage: it provides a new perspective. Looking at a problem from a different angle can give the insight needed to solve it.

Let's see what happens when we consider the discrepancy problem in the dual plane. In the previous section we were left with the following problem: Given a set S of n points, compute the discrete measure of every half-plane bounded by a line through two of the points. When we dualize the set S of points we get a set $S^* := \{p^* : p \in S\}$ of lines. Let $\ell(p,q)$ denote the line through two points $p, q \in S$. The dual of this line is the intersection point of the two lines $p^*, q^* \in S^*$. Consider the open half-plane bounded by and below $\ell(p,q)$. The discrete measure of this half-plane is the number of points strictly below $\ell(p,q)$. This means that in dual plane we are interested in the number of lines strictly above $\ell(p,q)^*$. For the closed half-plane below $\ell(p,q)$ we must also take the lines through $\ell(p,q)^*$ into account. Similarly, for the half-plane bounded by and above $\ell(p,q)$ we are interested in the number of lines below $\ell(p,q)^*$. In the next section we study sets of lines, and we give an efficient algorithm to compute the number of lines above every intersection point, through every intersection point, and below every intersection point. When we apply this algorithm to S^* we get all the information we need to compute the discrete measure of all half-planes bounded by lines through two points in S.

There is one thing that we should be careful about: two points in S with the same x-coordinate dualize to lines with the same slope. So the line through these points does not show up as an intersection in the dual plane. This makes sense, because the dual transform is undefined for vertical lines. In our application this calls for an additional step. For every vertical line through at least two points, we must determine the discrete measures of the corresponding half-planes. Since there is only a linear number of vertical lines through two (or more) points in S, the discrete measures for these lines can be computed in a brute-force manner in $O(n^2)$ time in total.

8.3 Arrangements of Lines

Let L be a set of n lines in the plane. The set L induces a subdivision of the plane that consists of vertices, edges, and faces. Some of the edges and faces are unbounded. This subdivision is usually referred to as the *arrangement* induced by L, and it is denoted by $\mathcal{A}(L)$. An arrangement is called *simple* if no three lines pass through the same point and no two lines are parallel. The *(combinatorial) complexity* of an arrangement is the total number of vertices, edges, and faces of the arrangement. Arrangements of lines and their higher-dimensional counterparts occur frequently in computational geometry. Often a problem that is defined on a set of points is dualized and turned into a problem on arrangements. This is done because the structure of a line arrangement is more apparent than the structure of a point set. A line through a pair of points in the primal plane, for instance, becomes a vertex in the dual arrangement—a much more explicit feature. The extra structure in an arrangement does not come for free: constructing a full arrangement is a time- and storage-consuming task, because the combinatorial complexity of an arrangement is high.

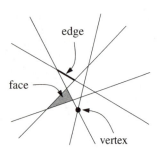

Theorem 8.4 *Let L be a set of n lines in the plane, and let $\mathcal{A}(L)$ be the arrangement induced by L.*
(i) The number of vertices of $\mathcal{A}(L)$ is at most $n(n-1)/2$.
(ii) The number of edges of $\mathcal{A}(L)$ is at most n^2.
(iii) The number of faces of $\mathcal{A}(L)$ is at most $n^2/2 + n/2 + 1$.
Equality holds in these three statements if and only if $\mathcal{A}(L)$ is simple.

Proof. First we show that the number of vertices, edges, and faces in a planar arrangement is maximum when the arrangement is simple. Let ℓ be a line that is parallel to one or more other lines. If we rotate ℓ slightly then it will intersect the lines it was parallel to. If the rotation is small enough then these are the only changes in the combinatorial structure of the arrangement. Hence, the complexity of the arrangement increases by the rotation. Similarly, let ℓ be a line that passes through a vertex v that lies on at least two more lines of L. If we translate ℓ slightly, at least one tiny triangle appears at the vertex v, at least two new vertices appear and at least three new edges. If the translation is small enough then these are the only changes in the combinatorial structure of the arrangement. Hence, the complexity of the arrangement increases by the translation. We conclude that an arrangement of maximum complexity must be simple.

We now analyze the complexity of the arrangement $\mathcal{A}(L)$ under the assumption that it be simple. Let n_v denote the number of vertices of $\mathcal{A}(L)$, let n_e denote the number of edges, and let n_f denote the number of faces. Bounding the number of vertices of $\mathcal{A}(L)$ is easy: any pair of lines gives rise to exactly one vertex, so $n_v = n(n-1)/2$. Bounding the number of edges is not difficult either: the number of edges lying on a fixed line is one plus the number of intersections on that line, which adds up to n. So $n_e = n^2$. Bounding the number of faces is more challenging. We use Euler's formula together with the bounds on the number of vertices and edges to do this. Recall that Euler's formula states that for any connected planar embedded graph with m_v nodes, m_e arcs, and m_f faces the following relation holds:

$$m_v - m_e + m_f = 2.$$

Because $\mathcal{A}(L)$ has edges that have only one incident vertex, it is not a proper graph. Therefore we augment it with an extra vertex v_∞ "at infinity"; all edges that are not proper are connected to this vertex. (The same trick was used in Chapter 7 to bound the complexity of the Voronoi diagram.) We get a connected planar embedded graph with $n_v + 1$ nodes, n_e arcs, and n_f faces. Hence,

$$
\begin{aligned}
n_f &= 2 - (n_v + 1) + n_e \\
&= 2 - (n(n-1)/2 + 1) + n^2 \\
&= n^2/2 + n/2 + 1.
\end{aligned}
$$

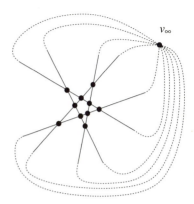

So the arrangement $\mathcal{A}(L)$ induced by a set L of lines is a planar subdivision of at most quadratic complexity. The doubly-connected edge list seems a suitable way to store an arrangement; with this representation we can efficiently list the edges of a given face, step from one face to a neighboring one, and so on. A doubly-connected edge list, however, can only store bounded edges, and an arrangement also has a number of unbounded edges. Therefore we place a large bounding box that encloses the interesting part of the arrangement, that is, a bounding box that contains all vertices of the arrangement in its interior. The subdivision defined by the bounding box plus the part of the arrangement inside it has bounded edges only and can be stored in a doubly-connected edge list.

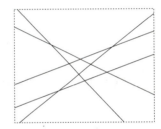

How can we construct this doubly-connected edge list? The approach that immediately comes to mind is plane sweep. In Chapter 2 all intersection points of a set of line segments were computed by plane sweep, and this algorithm was subsequently used to compute the doubly-connected edge list for the overlay of two planar subdivisions. Indeed, it's not so difficult to adapt the algorithms of Chapter 2 to compute the arrangement $\mathcal{A}(L)$. Since the number of intersection points is quadratic, the algorithm would run in $O(n^2 \log n)$ time. Not bad, but not optimal. So let's try another approach that may come to mind: an incremental construction algorithm.

A bounding box $\mathcal{B}(L)$ that contains all vertices of $\mathcal{A}(L)$ in its interior can easily be computed in quadratic time: compute all intersection points of pairs of lines, and choose the leftmost one, the rightmost one, the bottom one, and the top one. An axis-parallel rectangle that contains these four points contains all vertices of the arrangement.

The incremental algorithm adds the lines $\ell_1, \ell_2, \ldots, \ell_n$ one after the other and updates the doubly-connected edge list after each addition. Let $\mathcal{A}_i$ denote the subdivision of the plane induced by the bounding box $\mathcal{B}(L)$ and the part of $\mathcal{A}(\{\ell_1, \ldots, \ell_i\})$ inside $\mathcal{B}(L)$. To add the line ℓ_i, we must split the faces in $\mathcal{A}_{i-1}$

that are intersected by ℓ_i. We can find these faces by walking along ℓ_i from left to right, as follows. Suppose we enter a face f through an edge e. We walk along the boundary of f following *Next*()-pointers in the doubly-connected edge list until we find the half-edge of the edge e' where ℓ_i leaves f. We then step to the next face using the *Twin*()-pointer of that half-edge to reach the other half-edge for e' in the doubly-connected edge list. This way we find the next face in time proportional to the complexity of f. It can also happen that we leave f through a vertex v. In this case we walk around v, visiting its incident edges, until we find the next face intersected by ℓ_i. The doubly-connected edge list allows us to do this in time proportional to the degree of v. Figure 8.4 illustrates how we traverse the arrangement.

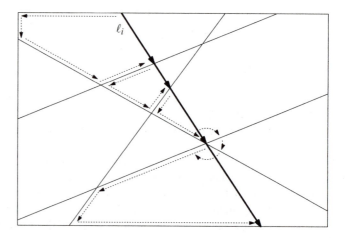

Figure 8.4
Traversing the arrangement

Two things are left: how do we find the leftmost edge intersected by ℓ_i—this is the edge where we start the walk through $\mathcal{A}_{i-1}$—and how do we actually split the faces we encounter?

The first issue is easy. The leftmost intersection point of ℓ_i and $\mathcal{A}_{i-1}$ is an edge on $\mathcal{B}(L)$. We simply test all of them to locate the one where the traversal can be started. The face incident to this edge and inside $\mathcal{B}(L)$ is the first face that is split by ℓ_i. In case ℓ_i intersects $\mathcal{A}_{i-1}$ first in a corner of $\mathcal{B}(L)$, the first face split by ℓ_i is the unique face incident to this corner and inside $\mathcal{B}(L)$. If ℓ_i is a vertical line we can locate the bottom intersection point of ℓ_i and $\mathcal{A}_i$ to start off the traversal. Since $\mathcal{A}_{i-1}$ contains at most $2i+4$ edges on $\mathcal{B}(L)$, the time needed for this step is linear for each line.

Suppose we have to split a face f, and assume that the face intersected by ℓ_i to the left of f has already been split. In particular, we assume that the edge e where we enter f has already been split. Splitting f is done as follows—see Figure 8.5. First of all we create two new face records, one for the part of f above ℓ_i and one for the part of f below ℓ_i. Next we split e', the edge where ℓ_i leaves f, and create a new vertex for $\ell_i \cap e'$. Thus we create one new vertex record, and two new half-edge records for both new edges. (If ℓ_i leaves f through a vertex, then this step is omitted.) Furthermore, we create half-edge records for the edge $\ell_i \cap f$. It remains to correctly initialize the various pointers

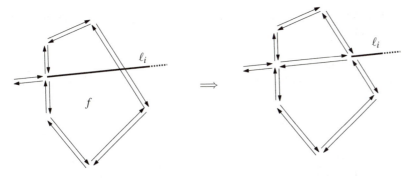

Figure 8.5
Splitting a face

in the new face, vertex, and half-edge records, set some existing pointers to the new vertex record, half-edge records, and face records, and destroy the face record for f and the half-edge records for e'. This is done in the same way as in Section 2.3, where the overlay of two subdivisions was constructed. The total time for the split is linear in the complexity of f.

The algorithm for constructing an arrangement can be summarized as follows:

Algorithm CONSTRUCTARRANGEMENT(L)
Input. A set L of n lines in the plane.
Output. The doubly-connected edge list for the subdivision induced by $\mathcal{B}(L)$
    and the part of $\mathcal{A}(L)$ inside $\mathcal{B}(L)$, where $\mathcal{B}(L)$ is a bounding box containing
    all vertices of $\mathcal{A}(L)$ in its interior.
1.   Compute a bounding box $\mathcal{B}(L)$ that contains all vertices of $\mathcal{A}(L)$ in its
    interior.
2.   Construct the doubly-connected edge list for the subdivision induced by
    $\mathcal{B}(L)$.
3.   **for** $i \leftarrow 1$ **to** n
4.   **do** Find the edge e on $\mathcal{B}(L)$ that contains the leftmost intersection
            point of ℓ_i and $\mathcal{A}_i$.
5. $f \leftarrow$ the bounded face incident to e
6.   **while** f is not the unbounded face, that is, the face outside $\mathcal{B}(L)$
7.   **do** Split f, and set f to be the next intersected face.

We have given a simple incremental algorithm for computing an arrangement. Next we analyze its running time. Step 1 of the algorithm, computing $\mathcal{B}(L)$, can be done in $O(n^2)$ time. Step 2 takes only constant time. Finding the first face split by ℓ_i takes $O(n)$ time, as we noted before. We now bound the time it takes to split the faces intersected by ℓ_i.

    First, assume that $\mathcal{A}(L)$ is simple. In this case the time we spend to split a face f and to find the next intersected face is linear in the complexity of f. Hence, the total time we need to insert line ℓ_i is linear in the sum of the complexities of the faces of $\mathcal{A}_{i-1}$ intersected by ℓ_i. When $\mathcal{A}(L)$ is not simple, we may leave f through a vertex v. In that case we have to walk around v to find the next face to split, and we encounter edges that are not on the boundary of an intersected face. But notice that the edges we encounter in this case are on the boundary of faces whose *closure* is intersected by ℓ_i. This leads us to the concept of zones.

The *zone* of a line ℓ in the arrangement $\mathcal{A}(L)$ induced by a set L of lines in the plane is the set of faces of $\mathcal{A}(L)$ whose closure intersects ℓ. Figure 8.6 gives

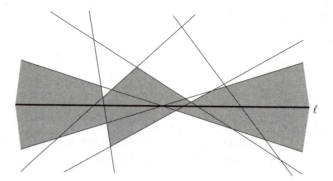

Figure 8.6
The zone of a line in an arrangement of lines

an example of a zone consisting of nine faces. The complexity of a zone is defined as the total complexity of all faces it consists of, that is, the sum of the number of edges and vertices of these faces. In Figure 8.6 you can see that some vertices are counted once in the zone complexity, others are counted twice, three times, or even four times. The time we need to insert line ℓ_i is linear in the complexity of the zone of ℓ_i in $\mathcal{A}(\{\ell_1, \ldots, \ell_i\})$. The Zone Theorem tells us that this quantity is linear:

Theorem 8.5 (Zone Theorem) *The complexity of the zone of a line in an arrangement of m lines in the plane is $O(m)$.*

Proof. Let L be a set of m lines in the plane, and let ℓ be another line. Without loss of generality we assume that ℓ coincides with the x-axis; we can change the coordinate system for this. We assume that no line of L is horizontal. This assumption is removed at the end of the proof.

Each edge in $\mathcal{A}(L)$ bounds two faces. We say that an edge is a *left bounding edge* for the face lying to the right of it and a *right bounding edge* for the face lying to the left of it. We shall prove that the number of left bounding edges of the faces in the zone of ℓ is at most $4m$. By symmetry, the number of right bounding edges is then bounded by $4m$ as well, and the theorem follows.

The proof is by induction on m. The base case, $m = 1$, is trivially true. Now let $m > 1$. Of the lines in L, let ℓ_1 be the one that has the rightmost intersection with ℓ. We first assume that this line is uniquely defined. By induction, the zone of ℓ in $\mathcal{A}(L \setminus \{\ell_1\})$ has at most $4(m-1)$ left bounding edges. When we add the line ℓ_1, the number of left bounding edges increases in two ways: there are new left bounding edges on ℓ_1 and there are old left bounding edges that are split by ℓ_1. Let v be the first intersection point of ℓ_1 with another line in L above ℓ, and let w be the first intersection point of ℓ_1 below ℓ. The edge connecting v and w is a new left bounding edge on ℓ_1. Furthermore, ℓ_1 splits a left bounding edge at the points v and w. This adds up to an increase of three in

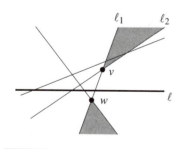

the number of left bounding edges. If v or w doesn't exist, the increase is only two. We claim that this is the only increase.

Consider the part of ℓ_1 above v. Let ℓ_2 be a line that intersects ℓ_1 at v. The region above v enclosed by ℓ_1 and ℓ_2 is not in the zone of ℓ. Because ℓ_2 crosses ℓ_1 from left to right at v, the region lies to the right of ℓ_1. Hence, the part of ℓ_1 above v cannot contribute any left bounding edges to the zone. Moreover, if a left bounding edge e that was in the zone is intersected by ℓ_1 somewhere above v, then the part of e to the right of ℓ_1 is no longer in the zone. Hence, there is no increase in the number of left bounding edges due to such an intersection.

In the same way it can be shown that the part of ℓ_1 below w does not increase the number of left bounding edges in the zone of ℓ. Therefore the total increase is three, as claimed.

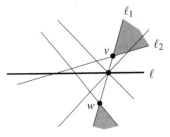

Up to now we assumed that the line ℓ_1 through the rightmost intersection point on ℓ is unique. If there is more than one line passing through the rightmost intersection point, then we take an arbitrary one to be ℓ_1. Following the same arguments as above, the increase in the number of left bounding edges can be shown to be four. This time there are two new ones on ℓ_1. Hence, the total number of left bounding edges is at most $4(m-1)+4=m$.

Finally, we remove the assumption that no line of L is horizontal. For a horizontal line that doesn't coincide with ℓ, a slight rotation only increases the complexity of the zone of ℓ in $\mathcal{A}(L)$. Since we are proving an upper bound on the zone complexity, we can safely assume such lines don't exist. If L contains a line ℓ_i that coincides with ℓ, then the proof above shows that the zone of ℓ in $\mathcal{A}(L \setminus \{\ell_i\})$ has at most $8m-8$ edges, and the addition of ℓ_i increases this quantity by at most $4m-2$: at most m edges on ℓ_i for the faces above ℓ_i, at most m edges on ℓ_i for the faces below ℓ_i, and at most $m-1$ edges are split into two, each of which is counted as a left bounding edge and a right bounding edge. This concludes the proof of the Zone Theorem. ◻

We can now bound the running time of the incremental algorithm for constructing an arrangement. We have seen that the time needed to insert ℓ_i is linear in the complexity of the zone of ℓ_i in $\mathcal{A}(\{\ell_1,\ldots,\ell_{i-1}\})$. By the Zone Theorem this is $O(i)$, so the time required to insert all lines is

$$\sum_{i=1}^{n} O(i) = O(n^2).$$

Steps 1–2 of the algorithm together take $O(n^2)$ time, so the total running time of the algorithm is $O(n^2)$. Because the complexity of $\mathcal{A}(L)$ is $\Theta(n^2)$ when $\mathcal{A}(L)$ is simple, our algorithm is optimal.

Theorem 8.6 *A doubly-connected edge list for the arrangement induced by a set of n lines in the plane can be constructed in $O(n^2)$ time.*

8.4 Levels and Discrepancy

It's time to go back to the discrepancy problem. We had dualized the set S of n sample points into a set S^* of n lines, and we needed to compute for every vertex of $\mathcal{A}(S^*)$ how many lines lie above it, pass through it, and lie below it. For each vertex, these three numbers add up to exactly n, so it is sufficient to compute two of the numbers. After we have constructed a doubly-connected edge list for $\mathcal{A}(S^*)$ we know how many lines pass through each vertex. We define the *level* of a point in an arrangement of lines to be the number of lines strictly above it. We next show how to compute the level of each vertex in $\mathcal{A}(S^*)$.

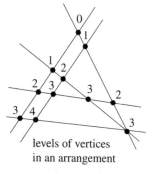

levels of vertices
in an arrangement

To compute the levels of the vertices of $\mathcal{A}(S^*)$ we do the following for each line $\ell \in S^*$. First, we compute the level of the leftmost vertex on ℓ in $O(n)$ time, by checking for each of the remaining lines whether it lies strictly above that vertex. Next we walk along ℓ from left to right to visit the other vertices on ℓ, using the doubly-connected edge list. It is easy to maintain the

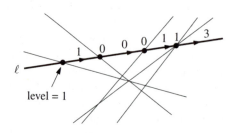

Figure 8.7
Maintaining the level while walking
along a line

level while we walk: the level only changes at a vertex, and the change can be computed by inspecting the edges incident to the vertex that is encountered. In Figure 8.7, for instance, the leftmost vertex on ℓ has level one. The points on the edge incident to that vertex and going to the right also have level one. At the second vertex a line crosses ℓ coming from above; the level decreases by one and becomes zero. Since the level is defined as the number of lines *strictly* above a point, the level of the second vertex itself is also zero. At the third vertex a line crosses ℓ coming from below. Hence, the level increases to one after the vertex is passed; the vertex itself still has level zero. And so on. Note that we needn't worry about vertical lines; our set is obtained by dualizing a set of points. This way the time to compute the levels of the vertices on ℓ is $O(n)$. Hence, the levels of all vertices of $\mathcal{A}(S^*)$ can be computed in $O(n^2)$ time.

The number of lines above, through, and below each vertex of $\mathcal{A}(S^*)$ gives us all the information we need to compute the discrete measure of the half-planes bounded by lines containing two points in S. Hence, these discrete measures can be computed in $O(n^2)$ time. This finally finishes the proof of Theorem 8.2.

8.5 Notes and Comments

In this chapter some important non-algorithmic concepts were introduced: geometric duality and arrangements. Duality is a transform that can shed a different light on a geometric problem, and is a standard tool for the computational geometer. The duality transform of Section 8.2 is not defined for vertical lines. Usually, vertical lines can be treated as a special case or by a perturbation of the setting. There exist different duality transforms that can handle vertical lines, but these have other drawbacks—see Edelsbrunner's book [129]. Duality applies to higher-dimensional point sets as well. For a point $p = (p_1, p_2, \ldots, p_d)$, its dual p^* is the hyperplane $x_d = p_1 x_1 + p_2 x_2 + \cdots + p_{d-1} x_{d-1} - p_d$. For a hyperplane $x_d = a_1 x_1 + a_2 x_2 + \cdots + a_{d-1} x_{d-1} + a_d$, its dual is the point $(a_1, a_2, \ldots, a_{d-1}, -a_d)$. The transform is incidence and order preserving.

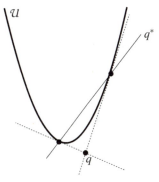

Recall that with the geometric interpretation of the duality transform using the parabola $y = x^2/2$, the dual of any point can be constructed. Interestingly, the dual of a point q can also be constructed without measuring distances. Assume that q lies below $\mathcal{U}$. Draw the two lines that pass through q and are tangent to $\mathcal{U}$. The line q^* is the line through the two points where these tangents touch $\mathcal{U}$. By construction, point q is the intersection of the two tangent lines. Hence, the dual of q must go through the duals of these two tangents, which are the points where the tangents touch $\mathcal{U}$. The dual of a point above $\mathcal{U}$ can be constructed without measuring distances as well. We won't show how to do this here. (One hint: you will need to be able to draw the line through a given point that is parallel to a given line.)

Another geometric transform that has been applied successfully in computational geometry is *inversion*. It can change the point-inside-circle relation in the plane to a point-below-plane relation in 3-dimensional space. More specifically, a point $p := (p_x, p_y)$ is lifted to the unit paraboloid $z = x^2 + y^2$ in 3-dimensional space, so

$$p^\circ := (p_x, p_y, p_x^2 + p_y^2).$$

A circle $C := (x-a)^2 + (y-b)^2 = r^2$ in the plane is transformed to a plane in 3-dimensional space by lifting the circle to the unit paraboloid and taking the plane through the lifted circle. In particular,

$$C^\circ := (z = a(x-a) + b(y-b) + r^2).$$

Now p lies inside C if and only if p° is below C°. This transform can be extended to higher dimensions, where a hypersphere in d-dimensional space becomes a hyperplane in $(d+1)$-dimensional space.

Arrangements have been studied extensively in computational and combinatorial geometry. Arrangements are not restricted to the plane. A 3-dimensional arrangement is induced by a set of planes, and a higher-dimensional arrangement is induced by a set of hyperplanes. The book of Edelsbrunner [129] is an excellent exposition of the research on arrangements up to 1987. It also contains the references to earlier textbooks on combinatorial—but not computational—geometry. For a more recent survey, see the paper by Guibas and Sharir [169]. We list a selection of results on arrangements in the plane and in higher dimensions.

The complexity of an arrangement of n hyperplanes in d-dimensional space is $\Theta(n^d)$ in the worst case. Any simple arrangement—one where any d hyperplanes but no $d+1$ hyperplanes intersect in one point—achieves this bound. Edelsbrunner et al. [136] presented the first optimal algorithm for constructing arrangements. The optimality of this incremental construction algorithm depends on a higher-dimensional version of the Zone Theorem, which states that the zone of a hyperplane in an arrangement of n hyperplanes in d-dimensional space has complexity $O(n^{d-1})$. A proof of this theorem is given by Edelsbrunner et al. [139].

The concept of levels in arrangements extends to higher dimensions as well—see Edelsbrunner's book [129]. The k-level in an arrangement $\mathcal{A}(H)$ induced by a set H of n hyperplanes is defined as the set of points with at most $k-1$ hyperplanes strictly above it, and at most $n-k$ hyperplanes strictly below. Tight bounds on the maximum complexity of k-levels are still unknown, even in the planar case. In the dual setting, the problem is closely related to the following question: given an set of n points, how many subsets of k points can be separated from the other $n-k$ points by a hyperplane? Such subsets are called k-sets, and the maximum number of k-sets in a set of n points is again unknown. The best known bounds in the planar case—both for k-sets and k-levels—are $\Omega(n \log(k+1))$ [129] and $O(n\sqrt{k}/\log^*(k+1))$ [277].

Given a set of n points in the plane, how many subsets of *at most k* points can be separated from the other $n-k$ points by a line? Such subsets are called $(\leqslant k)$-sets, and the maximum number of them is $O(nk)$, which is tight in the worst case. The same bounds hold for $(\leqslant k)$-levels in arrangements. The higher-dimensional bound is $O(n^{\lfloor d/2 \rfloor} k^{\lceil d/2 \rceil})$, which was shown by Clarkson and Shor [111].

In the notes and comments of Chapter 7, a connection between Voronoi diagrams and convex polyhedra in one dimension higher was explained: the Voronoi diagram of a set of points in the plane is the same as the projection of the boundary of the common intersection of a set of half-spaces in 3-dimensional space. This boundary is in fact the 0-level of the arrangement of planes bounding these half-spaces. This connection extends to order-k Voronoi diagrams and k-levels in arrangements: the k-level in the same arrangement of planes projects down to the order-k Voronoi diagram of the points.

Arrangements can be defined for objects other than lines and hyperplanes. A set of line segments in the plane, for instance, also forms an arrangement. For such arrangements, even bounds on the maximum complexity of a single face aren't simple to prove. Since faces can be non-convex, line segments can appear several times on the boundary. Indeed, the maximum complexity of a single face can be superlinear: it is $\Theta(n\alpha(n))$ in the worst case, where $\alpha(n)$ is the extremely slowly growing functional inverse of Ackermann's function. The upper bound can be proved using *Davenport-Schinzel sequences*; the interested reader is referred to the book by Sharir and Agarwal [313].

The main motivation for studying combinatorial structures like arrangements, single cells in arrangements, and envelopes, lies in motion planning. Several motion planning problems can be formulated as problems on arrangements and their substructures [170, 174, 175, 195, 305, 306].

Our original motivation for studying arrangements arose from computer graphics and the quality of random samples. The use of discrepancy was introduced to computer graphics by Shirley [316], and developed algorithmically by Dobkin and Mitchell [124], Dobkin and Eppstein [123], Chazelle [75], and de Berg [38].

8.6 Exercises

8.1 Prove that the duality transform introduced in this chapter is indeed incidence and order preserving, as claimed in Observation 8.3.

8.2 The dual of a line segment is a left-right double wedge, as was shown in Section 8.2.
   a. What is the dual of the collection of points inside a given triangle with vertices p, q, and r?
   b. What thing in the primal plane would dualize to a top-bottom double wedge?

8.3 Let L be a set of n lines in the plane. Give an $O(n \log n)$ time algorithm to compute an axis-parallel rectangle that contains all the vertices of $\mathcal{A}(L)$ in its interior.

8.4 Let S be a set of n points in the plane. In this chapter an algorithm was given to determine for every line ℓ through two points of S how many points of S lie strictly above ℓ. This was done by dualizing the problem first. Transform the algorithm for the dual problem back to the primal plane, and give the corresponding $O(n^2)$ time algorithm for the given problem.

8.5 Let S be a set of n points in the plane and let L be a set of m lines in the plane. Suppose we wish to determine whether there is a point in S that lies on a line of L. What is the dual of this problem?

8.6 You may have noticed that the dual transform of Section 8.2 has minus signs. Suppose we change them to plus signs, so the dual of a point (p_x, p_y) is the line $y = p_x x + p_y$, and the dual of the line $y = mx + b$ is the point (m, b). Is this dual transform incidence and order preserving?

8.7 Let L be a set of n non-vertical lines in the plane. Suppose the arrangement $\mathcal{A}(L)$ only has vertices with level 0. What can you say about this arrangement? Next suppose that lines of L can be vertical. What can you say now about the arrangement?

8.8 When constructing the arrangement of a set L of lines, we traversed every line of L from left to right to include it. When computing the discrepancy, we needed the level of each vertex in the arrangement. To determine these levels, we traversed every line of L from left to right again. Is it possible to combine these two traversals, that is, can you add the lines to the arrangement and compute the levels of the intersection points immediately?

8.9 Given a set L of n lines in the plane, give an $O(n \log n)$ time algorithm to compute the maximum level of any vertex in the arrangement $\mathcal{A}(L)$.

8.10 Let S be a set of n points in the plane. Give an $O(n^2)$ time algorithm to find the line containing the maximum number of points in S.

8.11 Let S be a set of n segments in the plane. We want to preprocess S into a data structure that can answer the following query: Given a query line ℓ, how many segments in S does it intersect?

    a.   Formulate the problem in the dual plane.
    b.   Describe a data structure for this problem that uses $O(n^2)$ expected storage and has $O(\log n)$ expected query time.
    c.   Describe how the data structure can be built in $O(n^2 \log n)$ expected time.

8.12 Let S be a set of n segments in the plane. Give an algorithm to decide if there exists a line ℓ that intersects all segments of S. Such a line is called a *transversal* or *stabber*. The algorithm should run in quadratic time.

9 Delaunay Triangulations

Height Interpolation

When we talked about maps of a piece of the earth's surface in previous chapters, we implicitly assumed there is no relief. This may be reasonable for a country like the Netherlands, but it is a bad assumption for Switzerland. In this chapter we set out to remedy this situation.

We can model a piece of the earth's surface as a *terrain*. A terrain is a 2-dimensional surface in 3-dimensional space with a special property: every vertical line intersects it in a point, if it intersects it at all. In other words, it is the graph of a function $f : A \subset \mathbb{R}^2 \to \mathbb{R}$ that assigns a height $f(p)$ to every point p in the *domain*, A, of the terrain. (The earth is round, so on a global scale terrains defined in this manner are not a good model of the earth. But on a more local scale terrains provide a fairly good model.) A terrain can be visualized with a perspective drawing like the one in Figure 9.1, or with contour lines—lines of equal height—like on a topographic map.

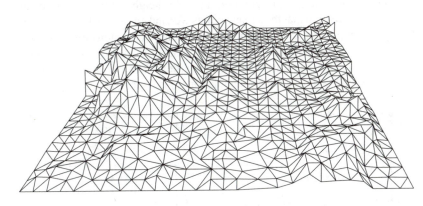

Figure 9.1
A perspective view of a terrain

Of course, we don't know the height of every point on earth; we only know it where we've measured it. This means that when we talk about some terrain, we only know the value of the function f at a finite set $P \subset A$ of sample points. From the height of the sample points we somehow have to approximate the height at the other points in the domain. A naive approach assigns to every $p \in A$ the height of the nearest sample point. However, this gives a

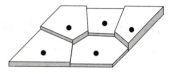

discrete terrain, which doesn't look very natural. Therefore our approach for approximating a terrain is as follows. We first determine a *triangulation* of P: a planar subdivision whose bounded faces are triangles and whose vertices are the points of P. (We assume that the sample points are such that we can make the triangles cover the domain of the terrain.) We then lift each sample point to its correct height, thereby mapping every triangle in the triangulation to a triangle in 3-space. Figure 9.2 illustrates this. What we get is a *polyhedral terrain*, the graph of a continuous function that is piecewise linear. We can use the polyhedral terrain as an approximation of the original terrain.

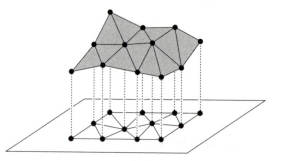

Figure 9.2
Obtaining a polyhedral terrain from a
set of sample points

The question remains: how do we triangulate the set of sample points? In general, this can be done in many different ways. But which triangulation is the most appropriate one for our purpose, namely to approximate a terrain? There is no definitive answer to this question. We *do not know* the original terrain, we only know its height at the sample points. Since we have no other information, and the height at the sample points is the correct height for any triangulation, all triangulations of P seem equally good. Nevertheless, some triangulations look more natural than others. For example, have a look at Figure 9.3, which shows two triangulations of the same point set. From the heights of the samples points we get the impression that the sample points were taken from a mountain ridge. Triangulation (a) reflects this intuition. Triangulation (b), however, where one single edge has been "flipped," has introduced a narrow valley cutting through the mountain ridge. Intuitively, this looks wrong. Can we turn this intuition into a criterion that tells us that triangulation (a) is better triangulation (b)?

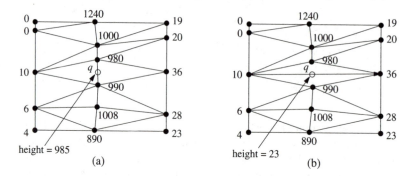

Figure 9.3
Flipping one edge can make a big
difference

The problem with triangulation (b) is that the height of the point q is determined by two points that are relatively far away. This happens because q lies in the middle of an edge of two long and sharp triangles. The skinniness of these triangles causes the trouble. So it seems that a triangulation that contains small angles is bad. Therefore we will rank triangulations by comparing their smallest angle. If the minimum angles of two triangulations are identical, then we can look at the second smallest angle, and so on. Since there is only a finite number of different triangulations of a given point set P, this implies that there must be an optimal triangulation, one that maximizes the minimum angle. This will be the triangulation we are looking for.

9.1 Triangulations of Planar Point Sets

Let $P := \{p_1, p_2, \ldots, p_n\}$ be a set of points in the plane. To be able to formally define a triangulation of P, we first define a *maximal planar subdivision* as a subdivision S such that no edge connecting two vertices can be added to S without destroying its planarity. In other words, any edge that is not in S intersects one of the existing edges. A *triangulation* of P is now defined as a maximal planar subdivision whose vertex set is P.

With this definition it is obvious that a triangulation exists. But does it consist of triangles? Yes, every face except the unbounded one must be a triangle: a bounded face is a polygon, and we have seen in Chapter 3 that any polygon can be triangulated. What about the unbounded face? It is not difficult to see that any segment connecting two consecutive points on the boundary of the convex hull of P is an edge in any triangulation $\mathcal{T}$. This implies that the union of the bounded faces of $\mathcal{T}$ is always the convex hull of P, and that the unbounded face is always the complement of the convex hull. (In our application this means that if the domain is a rectangular area, say, we have to make sure that the corners of the domain are included in the set of sample points, so that the triangles in the triangulation cover the domain of the terrain.) The number of triangles is the same in any triangulation of P. This also holds for the number of edges. The exact numbers depend on the number of points in P that are on the boundary of the convex hull of P. (Here we also count points in the interior of convex hull edges. Hence, the number of points on the convex hull boundary is not necessarily the same as the number of convex hull vertices.) This is made precise in the following theorem.

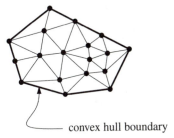

— convex hull boundary

Theorem 9.1 *Let P be a set of n points in the plane, not all collinear, and let k denote the number of points in P that lie on the boundary of the convex hull of P. Then any triangulation of P has $2n - 2 - k$ triangles and $3n - 3 - k$ edges.*

Proof. Let $\mathcal{T}$ be a triangulation of P, and let m denote the number of triangles of $\mathcal{T}$. Note that the number of faces of the triangulation, which we denote by n_f, is $m + 1$. Every triangle has three edges, and the unbounded face has k edges. Furthermore, every edge is incident to exactly two faces. Hence, the total number of edges of $\mathcal{T}$ is $n_e := (3m + k)/2$. Euler's formula tells us that

$$n - n_e + n_f = 2.$$

Plugging the values for n_e and n_f into the formula, we get $m = 2n - 2 - k$, which in turn implies $n_e = 3n - 3 - k$. □

Let $\mathcal{T}$ be a triangulation of P, and suppose it has m triangles. Consider the $3m$ angles of the triangles of $\mathcal{T}$, sorted by increasing value. Let $\alpha_1, \alpha_2, \ldots, \alpha_{3m}$ be the resulting sequence of angles; hence, $\alpha_i \leqslant \alpha_j$, for $i < j$. We call $A(\mathcal{T}) := (\alpha_1, \alpha_2, \ldots, \alpha_{3m})$ the *angle-vector* of $\mathcal{T}$. Let $\mathcal{T}'$ be another triangulation of the same point set P, and let $A(\mathcal{T}') := (\alpha_1', \alpha_2', \ldots, \alpha_{3m}')$ be its angle-vector. We say that the angle-vector of $\mathcal{T}$ is larger than the angle-vector of $\mathcal{T}'$ if $A(\mathcal{T})$ is lexicographically larger than $A(\mathcal{T}')$, or, in other words, if there exists an index i with $1 \leqslant i \leqslant 3m$ such that

$$\alpha_j = \alpha_j' \text{ for all } j < i, \quad \text{and} \quad \alpha_i > \alpha_i'.$$

We denote this as $A(\mathcal{T}) > A(\mathcal{T}')$. A triangulation $\mathcal{T}$ is called *angle-optimal* if $A(\mathcal{T}) \geqslant A(\mathcal{T}')$ for all triangulations $\mathcal{T}'$ of P. Angle-optimal triangulations are interesting because, as we have seen in the introduction to this chapter, they are good triangulations if we want to construct a polyhedral terrain from a set of sample points.

Below we will study when a triangulation is angle-optimal. To do this it is useful to know the following theorem, often called Thales's Theorem. Denote the smaller angle defined by three points p, q, r by $\angle pqr$.

Theorem 9.2 *Let C be a circle, ℓ a line intersecting C in points a and b, and p, q, r, and s points lying on the same side of ℓ. Suppose that p and q lie on C, that r lies inside C, and that s lies outside C. Then*

$$\angle arb > \angle apb = \angle aqb > \angle asb.$$

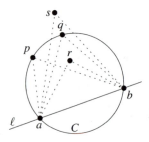

Now consider an edge $e = \overline{p_i p_j}$ of a triangulation $\mathcal{T}$ of P. If e is not an edge of the unbounded face, it is incident to two triangles $p_i p_j p_k$ and $p_i p_j p_l$. If these two triangles form a convex quadrilateral, we can obtain a new triangulation $\mathcal{T}'$ by removing $\overline{p_i p_j}$ from $\mathcal{T}$ and inserting $\overline{p_k p_l}$ instead. We call this operation an *edge flip*. The only difference in the angle-vector of $\mathcal{T}$ and $\mathcal{T}'$ are the six angles

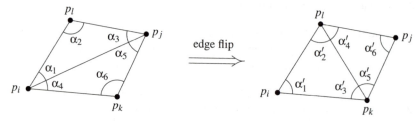

Figure 9.4
Flipping an edge

$\alpha_1, \ldots, \alpha_6$ in $A(\mathcal{T})$, which are replaced by $\alpha_1', \ldots, \alpha_6'$ in $A(\mathcal{T}')$. Figure 9.4 illustrates this. We call the edge $e = \overline{p_i p_j}$ an *illegal edge* if

$$\min_{1 \leqslant i \leqslant 6} \alpha_i < \min_{1 \leqslant i \leqslant 6} \alpha_i'.$$

In other words, an edge is illegal if we can locally increase the smallest angle by flipping that edge. The following observation immediately follows from the definition of an illegal edge.

Observation 9.3 *Let T be a triangulation with an illegal edge e. Let T' be the triangulation obtained from T by flipping e. Then $A(T') > A(T)$.*

It turns out that it is not necessary to compute the angles $\alpha_1, \ldots, \alpha_6, \alpha_1', \ldots, \alpha_6'$ to check whether a given edge is legal. Instead, we can use the simple criterion stated in the next lemma. The correctness of this criterion follows from Thales's Theorem.

Lemma 9.4 *Let edge $\overline{p_i p_j}$ be incident to triangles $p_i p_j p_k$ and $p_i p_j p_l$, and let C be the circle through p_i, p_j, and p_k. The edge $\overline{p_i p_j}$ is illegal if and only if the point p_l lies in the interior of C. Furthermore, if the points p_i, p_j, p_k, p_l form a convex quadrilateral and do not lie on a common circle, then exactly one of $\overline{p_i p_j}$ and $\overline{p_k p_l}$ is an illegal edge.*

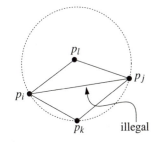

Observe that the criterion is symmetric in p_k and p_l: p_l lies inside the circle through p_i, p_j, p_k if and only if p_k lies inside the circle through p_i, p_j, p_l. When all four points lie on a circle, both $\overline{p_i p_j}$ and $\overline{p_k p_l}$ are legal. Note that the two triangles incident to an illegal edge must form a convex quadrilateral, so that it is always possible to flip an illegal edge.

We define a *legal triangulation* to be a triangulation that does not contain any illegal edge. From the observation above it follows that any angle-optimal triangulation is legal. Computing a legal triangulation is quite simple, once we are given an initial triangulation. We simply flip illegal edges until all edges are legal.

Algorithm LEGALTRIANGULATION(T)
Input. Some triangulation T of a point set P.
Output. A legal triangulation of P.
1. **while** T contains an illegal edge $\overline{p_i p_j}$
2.   **do** (* Flip $\overline{p_i p_j}$ *)
3.   Let $p_i p_j p_k$ and $p_i p_j p_l$ be the two triangles adjacent to $\overline{p_i p_j}$.
4.   Remove $\overline{p_i p_j}$ from T, and add $\overline{p_k p_l}$ instead.
5. **return** T

Why does this algorithm terminate? It follows from Observation 9.3 that the angle-vector of T increases in every iteration of the loop. Since there is only a finite number of different triangulations of P, this proves termination of the algorithm. Once it terminates, the result is a legal triangulation. Although the algorithm is guaranteed to terminate, it is too slow to be interesting. We have given the algorithm anyway, because later we shall need a similar procedure. But first we will look at something completely different—or so it seems.

9.2 The Delaunay Triangulation

Let P be a set of n points—or *sites*, as we shall sometimes call them—in the plane. Recall from Chapter 7 that the Voronoi diagram of P is the subdivision of the plane into n regions, one for each site in P, such that the region of a site $p \in P$ contains all points in the plane for which p is the closest site. The Voronoi diagram of P is denoted by Vor(P). The region of a site p is called

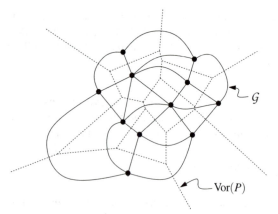

Figure 9.5
The dual graph of Vor(P)

the Voronoi cell of p; it is denoted by $\mathcal{V}(p)$. In this section we will study the dual graph of the Voronoi diagram. This graph G has a node for every Voronoi cell—equivalently, for every site—and it has an arc between two nodes if the corresponding cells share an edge. Note that this means that G has an arc for every edge of Vor(P). As you can see in Figure 9.5, there is a one-to-one correspondence between the bounded faces of G and the vertices of Vor(P).

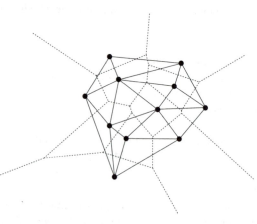

Figure 9.6
The Delaunay graph $\mathcal{D}G(P)$

   Consider the straight-line embedding of G, where the node corresponding to the Voronoi cell $\mathcal{V}(p)$ is the point p, and the arc connecting the nodes of $\mathcal{V}(p)$ and $\mathcal{V}(q)$ is the segment $\overline{pq}$—see Figure 9.6. We call this embedding the *Delaunay graph* of P, and we denote it by $\mathcal{D}G(P)$. (Although the name sounds French, Delaunay graphs have nothing to do with the French painter. They are called after the mathematician Boris Nikolaevich Delone, who probably wrote his own name as "Делоне," which would be transliterated into English as "Delone." However, since his work was published in French—at his time, the languages of science were French and German—his name is now better known in the French transliteration.) The Delaunay graph of a point set turns out to have a number of surprising properties. The first is that it is always a plane graph: no two edges in the embedding cross.

Theorem 9.5 *The Delaunay graph of a planar point set is a plane graph.*

Proof. To prove this, we need a property of the edges in the Voronoi diagram stated in Theorem 7.4(ii). For completeness we repeat the property, phrased here in terms of Delaunay graphs.

The edge $\overline{p_i p_j}$ is in the Delaunay graph $\mathcal{DG}(P)$ if and only if there is a closed disc C_{ij} with p_i and p_j on its boundary and no other site of P contained in it. (The center of such a disc lies on the common edge of $\mathcal{V}(p_i)$ and $\mathcal{V}(p_j)$.)

Define t_{ij} to be the triangle whose vertices are p_i, p_j, and the center of C_{ij}. Now let $\overline{p_k p_l}$ be another edge of the $\mathcal{DG}(P)$, and define the circle C_{kl} and the triangle t_{kl} similar to the way C_{ij} and t_{ij} were defined. Since the discs C_{ij} and C_{kl} are empty, the triangle t_{ij} doesn't contain a vertex of t_{kl} and vice versa. Hence, if t_{ij} and t_{kl} intersect, then one of the segments of t_{ij} incident to the center of C_{ij} must intersect one of the segments of t_{kl} incident to the center of C_{kl}. But this is impossible, since these segments are contained in disjoint Voronoi cells. We conclude that t_{ij} and t_{kl}—and, hence, $\overline{p_i p_j}$ and $\overline{p_k p_l}$—cannot intersect. $\square$

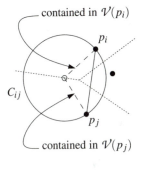

The Delaunay graph of P is an embedding of the dual graph of the Voronoi diagram. As observed earlier, it has a face for every vertex of Vor(P). The edges around a face correspond to the Voronoi edges incident to the corresponding Voronoi vertex. In particular, if a vertex v of Vor(P) is a vertex of the Voronoi cells for the sites $p_1, p_2, p_3, \ldots, p_k$, then the corresponding face f in $\mathcal{DG}(P)$ has $p_1, p_2, p_3, \ldots, p_k$ as its vertices. Theorem 7.4(i) tells us that in this situation the points $p_1, p_2, p_3, \ldots, p_k$ lie on a circle around v, so we not only know that f is a k-gon, but even that it is convex.

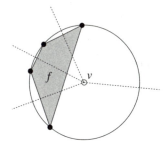

If the points of P are distributed at random, the chance that four points happen to lie on a circle is very small. We will—in this chapter—say that a set of points is in *general position* if it contains no four points on a circle. If P is in general position, then all vertices of the Voronoi diagram have degree three, and consequently all bounded faces of $\mathcal{DG}(P)$ are triangles. This explains why $\mathcal{DG}(P)$ is often called the *Delaunay triangulation* of P. We shall be a bit more careful, and will call $\mathcal{DG}(P)$ the *Delaunay graph* of P. We define a *Delaunay triangulation* to be any triangulation obtained by adding edges to the Delaunay graph. Since all faces of $\mathcal{DG}(P)$ are convex, obtaining such a triangulation is easy. Observe that the Delaunay triangulation of P is unique if and only if $\mathcal{DG}(P)$ is a triangulation, which is the case if P is in general position.

We now rephrase Theorem 7.4 about Voronoi diagrams in terms of Delaunay graphs.

Theorem 9.6 *Let P be a set of points in the plane.*

(i) *Three points $p_i, p_j, p_k \in P$ are vertices of the same face of the Delaunay graph of P if and only if the circle through p_i, p_j, p_r contains no point of P in its interior.*

(ii) *Two points $p_i, p_j \in P$ form an edge of the Delaunay graph of P if and only if there is a closed disc C that contains p_i and p_j on its boundary and does not contain any other point of P.*

Theorem 9.6 readily implies the following characterization of Delaunay triangulations.

Theorem 9.7 *Let P be a set of points in the plane, and let $\mathcal{T}$ be a triangulation of P. Then $\mathcal{T}$ is a Delaunay triangulation of P if and only if the circumcircle of any triangle of $\mathcal{T}$ does not contain a point of P in its interior.*

Since we argued before that a triangulation is good for the purpose of height interpolation if its angle-vector is as large as possible, our next step should be to look at the angle-vector of Delaunay triangulations. We do this by a slight detour through legal triangulations.

Theorem 9.8 *Let P be a set of points in the plane. A triangulation $\mathcal{T}$ of P is legal if and only if $\mathcal{T}$ is a Delaunay triangulation of P.*

Proof. It follows immediately from the definitions that any Delaunay triangulation is legal.

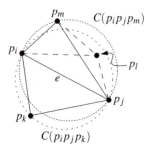

We shall prove that any legal triangulation is a Delaunay triangulation by contradiction. So assume $\mathcal{T}$ is a legal triangulation of P that is not a Delaunay triangulation. By Theorem 9.6, this means that there is a triangle $p_i p_j p_k$ such that the circumcircle $C(p_i p_j p_k)$ contains a point $p_l \in P$ in its interior. Let $e := \overline{p_i p_j}$ be the edge of $p_i p_j p_l$ such that the triangle $p_i p_j p_l$ does not intersect $p_i p_j p_k$. Of all such pairs $(p_i p_j p_k, p_l)$ in $\mathcal{T}$, choose the one that maximizes the angle $\angle p_i p_l p_j$. Now look at the triangle $p_i p_j p_m$ adjacent to $p_i p_j p_k$ along e. Since $\mathcal{T}$ is legal, e is legal. By Lemma 9.4 this implies that p_m does not lie in the interior of $C(p_i p_j p_k)$. The circumcircle $C(p_i p_j p_m)$ of $p_i p_j p_m$ contains the part of $C(p_i p_j p_k)$ that is separated from $p_i p_j p_k$ by e. Consequently, $p_l \in C(p_i p_j p_m)$. Assume that $\overline{p_j p_m}$ is the edge of $p_i p_j p_m$ such that $p_j p_m p_l$ does not intersect $p_i p_j p_m$. But now $\angle p_j p_l p_m > \angle p_i p_l p_j$ by Thales's Theorem, contradicting the definition of the pair $(p_i p_j p_k, p_l)$. $\qquad \square$

Since any angle-optimal triangulation must be legal, Theorem 9.8 implies that any angle-optimal triangulation of P is a Delaunay triangulation of P. When P is in general position, there is only one legal triangulation, which is then the only angle-optimal triangulation, namely the unique Delaunay triangulation that coincides with the Delaunay graph. When P is not in general position, then any triangulation of the Delaunay graph is legal. Not all these Delaunay triangulations need to be angle-optimal. However, their angle-vectors do not differ too much. Moreover, using Thales's Theorem one can show that the minimum angle in any triangulation of a set of co-circular points is the same, that is, the minimum angle is independent of the triangulation. This implies that any triangulation turning the Delaunay graph into a Delaunay triangulation has the same minimum angle. The following theorem summarizes this.

Theorem 9.9 *Let P be a set of points in the plane. Any angle-optimal triangulation of P is a Delaunay triangulation of P. Furthermore, any Delaunay triangulation of P maximizes the minimum angle over all triangulations of P.*

9.3 Computing the Delaunay Triangulation

We have seen that for our purpose—approximating a terrain by constructing a polyhedral terrain from a set P of sample points—a Delaunay triangulation of P is a suitable triangulation. This is because the Delaunay triangulation maximizes the minimum angle. So how do we compute such a Delaunay triangulation?

We already know from Chapter 7 how to compute the Voronoi diagram of P. From $\text{Vor}(P)$ we can easily obtain the Delaunay graph $\mathcal{DG}(P)$, and by triangulating the faces with more than three vertices we can obtain a Delaunay triangulation. In this section we describe a different approach: we will compute a Delaunay triangulation directly, using the randomized incremental approach we have so successfully applied to the linear programming problem in Chapter 4 and to the point location problem in Chapter 6.

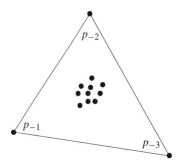

In Chapter 6 we found it convenient to start with a large rectangle containing the scene, to avoid problems caused by unbounded trapezoids. In the same spirit we now start with a large triangle $p_{-1}p_{-2}p_{-3}$ that contains the set P. This means we are now computing a Delaunay triangulation of $P \cup \Omega$ instead of P, where $\Omega := \{p_{-1}, p_{-2}, p_{-3}\}$. Later we want to obtain the Delaunay triangulation of P by discarding p_{-1}, p_{-2}, and p_{-3}, together with all incident edges. For this to work we have to choose p_{-1}, p_{-2}, and p_{-3} far enough away, so that they don't destroy any triangles in the Delaunay triangulation of P. In particular, we must ensure they do not lie in any circle defined by three points in P. We postpone the details of this to a later stage; first we have a look at the algorithm.

The algorithm is randomized incremental, so it adds the points in random order and it maintains a Delaunay triangulation of the current point set. Consider the addition of a point p_r. We first find the triangle of the current triangulation that contains p_r—how this is done will be explained later—and we add edges from p_r to the vertices of this triangle. If p_r happens to fall on an edge e of the triangulation, we have to add edges from p_r to the opposite vertices in the triangles sharing e. Figure 9.7 illustrates these two cases. We now have a triangulation again, but not necessarily a Delaunay triangulation. This is because the addition of p_r can make some of the existing edges illegal. To remedy this, we call a procedure LEGALIZEEDGE with each potentially illegal

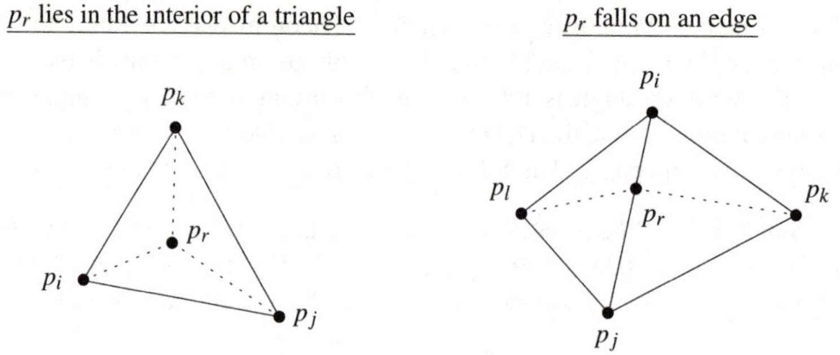

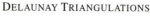

Figure 9.7
The two cases when adding a point p_r

edge. This procedure replaces illegal edges by legal ones through edge flips. Before we come to the details of this, we give a precise description of the main algorithm.

Algorithm DELAUNAYTRIANGULATION(P)
Input. A set P of n points in the plane.
Output. A Delaunay triangulation of P.

1. Let p_{-1}, p_{-2}, and p_{-3} be a suitable set of three points such that P is contained in the triangle $p_{-1}p_{-2}p_{-3}$.
2. Initialize $\mathcal{T}$ as the triangulation consisting of the single triangle $p_{-1}p_{-2}p_{-3}$.
3. Compute a random permutation $p_1, p_2, \ldots, p_n$ of P.
4. **for** $r \leftarrow 1$ **to** n
5.   **do** (∗ Insert p_r into $\mathcal{T}$: ∗)
6.   Find a triangle $p_ip_jp_k \in \mathcal{T}$ containing p_r.
7.   **if** p_r lies in the interior of the triangle $p_ip_jp_k$
8.   **then** Add edges from p_r to the three vertices of $p_ip_jp_k$, thereby splitting $p_ip_jp_k$ into three triangles.
9.   LEGALIZEEDGE($p_r, \overline{p_ip_j}, \mathcal{T}$)
10.   LEGALIZEEDGE($p_r, \overline{p_jp_k}, \mathcal{T}$)
11.   LEGALIZEEDGE($p_r, \overline{p_kp_i}, \mathcal{T}$)
12.   **else** (∗ p_r lies on an edge of $p_ip_jp_k$, say the edge $\overline{p_ip_j}$ ∗)
13.   Add edges from p_r to p_k and to the third vertex p_l of the other triangle that is incident to $\overline{p_ip_j}$, thereby splitting the two triangles incident to $\overline{p_ip_j}$ into four triangles.
14.   LEGALIZEEDGE($p_r, \overline{p_ip_l}, \mathcal{T}$)
15.   LEGALIZEEDGE($p_r, \overline{p_lp_j}, \mathcal{T}$)
16.   LEGALIZEEDGE($p_r, \overline{p_jp_k}, \mathcal{T}$)
17.   LEGALIZEEDGE($p_r, \overline{p_kp_i}, \mathcal{T}$)
18. Discard p_{-1}, p_{-2}, and p_{-3} with all their incident edges from $\mathcal{T}$.
19. **return** $\mathcal{T}$

Next we discuss the details of turning the triangulation we get after line 8 (or line 13) into a Delaunay triangulation. We know from Theorem 9.8 that a triangulation is a Delaunay triangulation if all its edges are legal. In the spirit of algorithm LEGALTRIANGULATION, we therefore flip illegal edges until the triangulation is legal again. The question that remains is which edges may become illegal due to the insertion of p_r. Observe that an edge $\overline{p_ip_j}$ that was

legal before can only become illegal if one of the triangles incident to it has changed. So only the edges of the new triangles need to be checked. This is done using the subroutine LEGALIZEEDGE, which tests and possibly flips an edge. If LEGALIZEEDGE flips an edge, other edges may become illegal. Therefore LEGALIZEEDGE calls itself recursively with such potentially illegal edges.

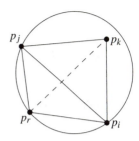

LEGALIZEEDGE($p_r, \overline{p_i p_j}, \mathcal{T}$)
1.   (* The point being inserted is p_r, and $\overline{p_i p_j}$ is the edge of $\mathcal{T}$ that may need to be flipped. *)
2.   **if** $\overline{p_i p_j}$ is illegal
3.   **then** Let $p_i p_j p_k$ be the triangle adjacent to $p_r p_i p_j$ along $\overline{p_i p_j}$.
4.   (* Flip $\overline{p_i p_j}$: *) Replace $\overline{p_i p_j}$ with $p_r p_k$.
5.   LEGALIZEEDGE($p_r, \overline{p_i p_k}, \mathcal{T}$)
6.   LEGALIZEEDGE($p_r, \overline{p_k p_j}, \mathcal{T}$)

The test in line 2 whether an edge is illegal can normally be done by applying Lemma 9.4. However, there are some complications because of the presence of the special points p_{-1}, p_{-2}, and p_{-3}. We shall come back to this later; first we prove that the algorithm is correct.

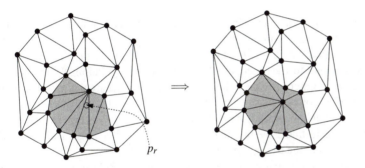

Figure 9.8
All edges created are incident to p_r.

To ensure the correctness of the algorithm, we need to prove that no illegal edges remain after all calls to LEGALIZEEDGE have been processed. From the code of LEGALIZEEDGE it is clear that every new edge created due to the insertion of p_r is incident to p_r. Figure 9.8 illustrates this; the triangles that are destroyed and the new triangles are shows in grey. The crucial observation (proved below) is that every new edge must be legal, so there is no need to test them. Together with the earlier observation that an edge can only become illegal if one of its incident triangles changes, this proves that the algorithm tests any edge that may become illegal. Hence, the algorithm is correct. Note that, as in Algorithm LEGALTRIANGULATION, the algorithm cannot get into an infinite loop, because every flip makes the angle-vector of the triangulation larger.

Lemma 9.10 *Every new edge created in* DELAUNAYTRIANGULATION *or in* LEGALIZEEDGE *during the insertion of p_r is an edge of the Delaunay graph of* $\Omega \cup \{p_1, \ldots, p_r\}$.

191

Proof. Consider first the edges $\overline{p_r p_i}$, $\overline{p_r p_j}$, $\overline{p_r p_k}$ (and perhaps $\overline{p_r p_l}$) created by splitting $p_i p_j p_k$ (and maybe $p_i p_j p_l$). Since $p_i p_j p_k$ is a triangle in the Delaunay triangulation before the addition of p_r, the circumcircle C of $p_i p_j p_k$ contains no point p_j with $j < r$ in its interior. By shrinking C we can find a circle C' through p_i and p_r contained in C. Because $C' \subset C$ we know that C' is empty. This implies that $\overline{p_r p_i}$ is an edge of the Delaunay graph after the addition of p_r. The same holds for $\overline{p_r p_j}$ and $\overline{p_r p_k}$ (and for $\overline{p_r p_l}$, if it exists).

Now consider an edge flipped by LEGALIZEEDGE. Such an edge flip always replaces an edge $\overline{p_i p_j}$ of a triangle $p_i p_j p_l$ by an edge $\overline{p_r p_l}$ incident to p_r. Since $p_i p_j p_l$ was a Delaunay triangle before the addition of p_r and because its circumcircle C contains p_r—otherwise $\overline{p_i p_j}$ would not be illegal—we can shrink the circumcircle to obtain an empty circle C' with only p_r and p_l on its boundary. Hence, $\overline{p_r p_l}$ is an edge of the Delaunay graph after the addition. $\square$

We have proved the correctness of the algorithm. What remains is to describe how to implement two important steps: how to find the triangle containing the point p_r in line 6 of DELAUNAYTRIANGULATION, and how to deal correctly with the points p_{-1}, p_{-2}, and p_{-3} in the test in line 2 in LEGALIZEEDGE. We start with the former issue.

To find the triangle containing p_r we use an approach quite similar to what we did in Chapter 6: while we build the Delaunay triangulation, we also build a point location structure $\mathcal{D}$, which is a directed acyclic graph. The leaves of $\mathcal{D}$ correspond to the triangles of the current triangulation $\mathcal{T}$, and we maintain cross-pointers between those leaves and the triangulation. The internal nodes of $\mathcal{D}$ correspond to triangles that were in the triangulation at some earlier stage, but have already been destroyed. The point location structure is built as follows. In line 2 we initialize $\mathcal{D}$ as a DAG with a single leaf node, which corresponds to the triangle $p_{-1}p_{-2}p_{-3}$.

Now suppose that at some point we split a triangle $p_i p_j p_k$ of the current triangulation into three (or two) new triangles. The corresponding change in $\mathcal{D}$ is to add three (or two) new leaves to $\mathcal{D}$, and to make the leaf for $p_i p_j p_k$ into an internal node with outgoing pointers to those three (or two) leaves. Similarly, when we replace two triangles $p_k p_i p_j$ and $p_i p_j p_l$ by triangles $p_k p_i p_l$ and $p_k p_l p_j$ by an edge flip, we create leaves for the two new triangles, and the nodes of $p_k p_i p_j$ and $p_i p_j p_l$ get pointers to the two new leaves. Figure 9.9 shows an example of the changes in $\mathcal{D}$ caused by the addition of a point. Observe that when we make a leaf into an internal node, it gets at most three outgoing pointers.

Using $\mathcal{D}$ we can locate the next point p_r to be added in the current triangulation. This is done as follows. We start at the root of $\mathcal{D}$, which corresponds to the initial triangle $p_{-1}p_{-2}p_{-3}$. We check the three children of the root to see in which triangle p_r lies, and we descend to the corresponding child. We then check the children of this node, descend to a child whose triangle contains p_r, and so on, until we reach a leaf of $\mathcal{D}$. This leaf corresponds to a triangle in the current triangulation that contains p_r. Since the out-degree of any node is at

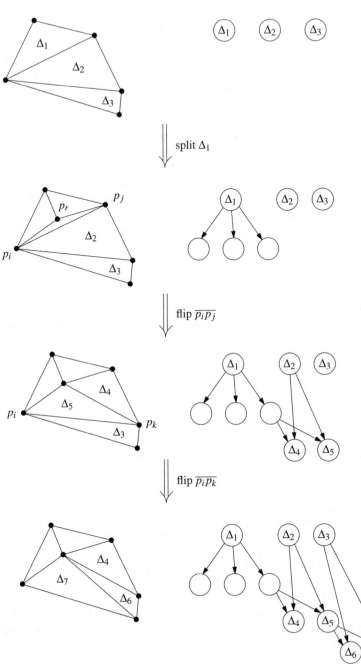

Figure 9.9
The effect of inserting point p_r into triangle Δ_1 on the data structure $\mathcal{D}$ (the part of $\mathcal{D}$ that does not change is omitted in the figure)

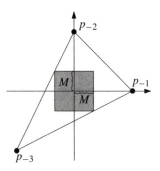

most three, this takes linear time in the number of nodes on the search path, or, in other words, in the number of triangles stored in $\mathcal{D}$ that contain p_r.

There is only one detail left, namely how to choose p_{-1}, p_{-2}, and p_{-3}, and how to implement the test whether an edge is legal. On the one hand, we have to choose p_{-1}, p_{-2}, and p_{-3} far away, because we don't want their presence to influence the Delaunay triangulation of P. One the other hand, we don't want to introduce the huge coordinates needed for that. So what we do is to choose these points such that the triangle $p_{-1}p_{-2}p_{-3}$ contains P, and to modify the test for illegal edges such that it works *as if* we had chosen them very far away.

More precisely, we take $p_{-1} := (3M, 0)$, we take $p_{-2} := (0, 3M)$, and we take $p_{-3} := (-3M, -3M)$, where M is the maximum absolute value of any coordinate of a point in P. This ensures that P is contained in the triangle $p_{-1}p_{-2}p_{-3}$. However, when we check whether an edge is illegal, we don't use these coordinates. Instead, we pretend that p_{-1} lies outside all circles defined by any three points in P, that p_{-2} lies outside all circles defined by any three points in $P \cup \{p_{-1}\}$, and that p_{-3} lies outside all circles defined by any three points in $P \cup \{p_{-1}, p_{-2}\}$. Hence, we implement the test as follows. Let $\overline{p_i p_j}$ be the edge to be tested.

case (i): both i and j are negative.

In this case we decide that $\overline{p_i p_j}$ is legal, because we must keep the edges of the large triangle $p_{-1}p_{-2}p_{-3}$.

For the other cases, let p_k and p_l be the two other vertices of the triangles incident to $\overline{p_i p_j}$.

case (ii): the indices i, j, k, l are all positive.

This is the normal case; none of the points involved in the test is a special point. Hence, $\overline{p_i p_j}$ is illegal if and only if p_l lies inside the circle defined by p_i, p_j, and p_k.

case (iii): exactly one of the indices i, j, k, l is negative.

We don't want a special point to destroy any Delaunay edges between points in P. Hence, if i or j is negative (that is, p_i or p_j is a special point) then we decide that $\overline{p_i p_j}$ is illegal so that it will be replaced by $\overline{p_k p_l}$; otherwise we decide that $\overline{p_i p_j}$ is legal.

case (iv): exactly two of the indices i, j, k, l are negative.

Now one of the indices i, j and one of the indices k, l must be negative; we cannot have that both k, l are negative and the case that both i, j are negative was case (i) above. If the negative index of i, j is smaller than the negative index of k, l then we decide that $\overline{p_i p_j}$ is legal, and otherwise we decide it is illegal.

case (v): exactly three of the indices i, j, k, l are negative.

This cannot occur: i and j cannot both be negative—that was case (i)—and k and l cannot both be negative because one of p_k, p_l must be the point p_r that we just inserted.

9.4 The Analysis

We first look at the *structural change* generated by the algorithm. This is the number of triangles created and deleted during the course of the algorithm. Before we start the analysis, we introduce some notation: $P_r := \{p_1, \ldots, p_r\}$ and $\mathcal{D}G_r := \mathcal{D}G(\Omega \cup P_r)$.

Lemma 9.11 *The expected number of triangles created by algorithm* DELAU-NAYTRIANGULATION *is at most* $9n + 1$.

Proof. In the beginning, we create a single triangle whose vertices are given by Ω. In iteration r of the algorithm, when we insert p_r, we first split one or two triangles, creating three or four new triangles. This splitting creates the same number of edges in $\mathcal{D}G_r$, namely $\overline{p_r p_i}$, $\overline{p_r p_j}$, $\overline{p_r p_k}$ (and maybe $\overline{p_r p_l}$). Furthermore, for every edge that we flip in procedure LEGALIZEEDGE, we create two new triangles. Again, the flipping creates an edge of $\mathcal{D}G_r$ incident to p_r. To summarize: if after the insertion of p_r there are k edges of $\mathcal{D}G_r$ incident to p_r, then we have created at most $2(k-3) + 3 = 2k - 3$ new triangles. The number k is the degree of p_r in $\mathcal{D}G_r$; we denote this degree by $\deg(p_r, \mathcal{D}G_r)$.

So what is the expected degree of p_r, over all possible permutations of the set P? As in Chapter 4 and 6 we use *backwards analysis* to bound this value. So, for the moment, we fix the set P_r. We want to bound the expected degree of the point p_r, which is a *random* element of the set P_r. By Theorem 7.3, the Delaunay graph $\mathcal{D}G_r$ has at most $3(r+3) - 6$ edges. Three of these are the edges of $p_{-1}p_{-2}p_{-3}$, and therefore the total degree of the vertices in P_r is less than $2[3(r+3) - 9] = 6r$. This means that the expected degree of a random point of P_r is at most 6. Summarizing the above, we can bound the number of triangles created in step r as follows.

$$
\begin{aligned}
\mathrm{E}\big[\text{number of triangles created in step } r\big] &\leqslant \mathrm{E}\big[2\deg(p_r, \mathcal{D}G_r) - 3\big] \\
&= 2\mathrm{E}\big[\deg(p_r, \mathcal{D}G_r)\big] - 3 \\
&\leqslant 2 \cdot 6 - 3 = 9
\end{aligned}
$$

The total number of created triangles is one for the triangle $p_{-1}p_{-2}p_{-3}$ that we start with, plus the number of triangles created in each of the insertion steps. Using linearity of expectation, we get that the expected total number of created triangles is bounded by $1 + 9n$. $\square$

We now state the main result.

Theorem 9.12 *The Delaunay triangulation of a set P of n points in the plane can be computed in $O(n \log n)$ expected time, using $O(n)$ expected storage.*

Proof. The correctness of the algorithm follows from the discussion above. As for the storage requirement, we note that only the search structure $\mathcal{D}$ could use more than linear storage. However, every node of $\mathcal{D}$ corresponds to a triangle created by the algorithm, and by the previous lemma the expected number of these is $O(n)$.

To bound the expected running time we first ignore the time spent in the point location step (line 6). Now the time spent by the algorithm is proportional to the number of created triangles. From the previous lemma we can therefore conclude that the expected running time, not counting the time for point location, is $O(n)$.

It remains to account for the point location steps. The time to locate the point p_r in the current triangulation is linear in the number of nodes of $\mathcal{D}$ that we visit. Any visited node corresponds to a triangle that was created at some earlier stage and that contains p_r. If we count the triangle of the current triangulation separately, then the time for locating p_r is $O(1)$ plus linear time in the number of triangles that were present at some earlier stage, but have been destroyed, and contain p_r.

A triangle $p_i p_j p_k$ can be destroyed from the triangulation for one of two reasons:

- A new point p_l has been inserted inside (or on the boundary of) $p_i p_j p_k$, and $p_i p_j p_k$ was split into three (or two) subtriangles.

- An edge flip has replaced $p_i p_j p_k$ and an adjacent triangle $p_i p_j p_l$ by the pair $p_k p_i p_l$ and $p_k p_j p_l$.

In the first case, the triangle $p_i p_j p_k$ was a Delaunay triangle before p_l was inserted. In the second case, either $p_i p_j p_k$ was a Delaunay triangle and p_l was inserted, or $p_i p_j p_l$ was a Delaunay triangle and p_k was inserted. If $p_i p_j p_l$ was the Delaunay triangle, then the fact that the edge $\overline{p_i p_j}$ was flipped means that both p_k and p_r lie inside the circumcircle of $p_i p_j p_l$.

In all cases we can charge the fact that triangle $p_i p_j p_k$ was visited to a Delaunay triangle Δ that has been destroyed in the same stage as $p_i p_j p_k$, and such that the circumcircle of Δ contains p_r. Denote the subset of points in P that lie in a given triangle Δ by $K(\Delta)$. In the argument above the visit to a triangle during the location of p_r is charged to a triangle Δ with $p_r \in K(\Delta)$. It is easy to see that a triangle Δ can be charged at most once for every one of the points in $K(\Delta)$. Therefore the total time for the point location steps is

$$O(n + \sum_\Delta \text{card}(K(\Delta))), \qquad (9.1)$$

where the summation is over all Delaunay triangles Δ created by the algorithm. We shall prove later that the expected value of this sum is $O(n \log n)$. This proves the theorem. ⬚

It remains to bound the expected size of the sets $K(\Delta)$. If Δ is a triangle of a Delaunay triangulation of $\Omega \cup P_r$, then what would we expect $\text{card}(K(\Delta))$ to be? For $r = 1$ we would expect it to be roughly n, and for $r = n$ we know that it is zero. What happens in between? The nice thing about randomization is that it "interpolates" between those two extremes. The right intuition would be that, since P_r is a random sample, the number of points lying inside the circumcircle of a triangle $\Delta \in \mathcal{D}G_r$ is about $O(n/r)$. But be warned: this is not really true for *all* triangles in $\mathcal{D}G_r$. Nevertheless, the sum in expression (9.1) behaves as if it were true.

In the remainder of this section we will give a quick proof of this fact for the case of a point set in general position. The result is true for the general case as well, but to see that we have to work a little bit harder, so we postpone that to the next section, where we treat the problem in more generality.

Lemma 9.13 *If P is a point set in general position, then*

$$\sum_{\Delta} \operatorname{card}(K(\Delta)) = O(n \log n),$$

where the summation is over all Delaunay triangles Δ created by the algorithm.

Proof. Since P is in general position, every subset P_r is in general position. This implies that the triangulation after adding the point p_r is the unique triangulation $\mathcal{D}G(\Omega \cup P_r)$. We denote the set of triangles of $\mathcal{D}G(\Omega \cup P_r)$ by $\mathcal{T}_r$. Now the set of Delaunay triangles created in stage r equals $\mathcal{T}_r \setminus \mathcal{T}_{r-1}$ by definition. Hence, we can rewrite the sum we want to bound as

$$\sum_{r=1}^{n} \left(\sum_{\Delta \in \mathcal{T}_r \setminus \mathcal{T}_{r-1}} \operatorname{card}(K(\Delta)) \right).$$

For a point q, let $k(P_r, q)$ denote the number of triangles $\Delta \in \mathcal{T}_r$ such that $q \in K(\Delta)$, and let $k(P_r, q, p_r)$ be the number of triangles $\Delta \in \mathcal{T}_r$ such that not only $q \in K(\Delta)$ but for which we also have that p_r is incident to Δ. Recall that any Delaunay triangle created in stage r is incident to p_r, so we have

$$\sum_{\Delta \in \mathcal{T}_r \setminus \mathcal{T}_{r-1}} \operatorname{card}(K(\Delta)) = \sum_{q \in P \setminus P_r} k(P_r, q, p_r). \qquad (9.2)$$

For the moment, we fix P_r. In other words, we consider all expectations to be over the set of permutations of the set P where P_r is equal to a fixed set P_r^*. The value of $k(P_r, q, p_r)$ then depends only on the choice of p_r. Since a triangle $\Delta \in \mathcal{T}_r$ is incident to a random point $p \in P_r^*$ with probability at most $3/r$, we get

$$\mathrm{E}\big[k(P_r, q, p_r)\big] \leqslant \frac{3k(P_r, q)}{r}.$$

If we sum this over all $q \in P \setminus P_r$ and use (9.2), we get

$$\mathrm{E}\Big[\sum_{\Delta \in \mathcal{T}_r \setminus \mathcal{T}_{r-1}} \operatorname{card}(K(\Delta)) \Big] \leqslant \frac{3}{r} \sum_{q \in P \setminus P_r} k(P_r, q). \qquad (9.3)$$

Every $q \in P \setminus P_r$ is equally likely to appear as p_{r+1}, and so we have

$$\mathrm{E}\big[k(P_r, p_{r+1})\big] = \frac{1}{n-r} \sum_{q \in P \setminus P_r} k(P_r, q).$$

We can substitute this into (9.3), and get

$$\mathrm{E}\Big[\sum_{\Delta \in \mathcal{T}_r \setminus \mathcal{T}_{r-1}} \mathrm{card}(K(\Delta))\Big] \leqslant 3\Big(\frac{n-r}{r}\Big) \mathrm{E}\big[k(P_r, p_{r+1})\big].$$

What is $k(P_r, p_{r+1})$? It is the number of triangles Δ of $\mathcal{T}_r$ that have $p_{r+1} \in K(\Delta)$. By the criterion from Theorem 9.6 (i), these triangles are exactly the triangles of $\mathcal{T}_r$ that will be destroyed by the insertion of p_{r+1}. Hence, we can rewrite the previous expression as

$$\mathrm{E}\Big[\sum_{\Delta \in \mathcal{T}_r \setminus \mathcal{T}_{r-1}} \mathrm{card}(K(\Delta))\Big] \leqslant 3\Big(\frac{n-r}{r}\Big) \mathrm{E}\big[\mathrm{card}(\mathcal{T}_r \setminus \mathcal{T}_{r+1})\big].$$

Theorem 9.1 shows that the number of triangles in $\mathcal{T}_m$ is precisely $2(m+3) - 2 - 3 = 2m - 1$. Therefore, the number of triangles *destroyed* by the insertion of point p_{r+1} is exactly two less than the number of triangles *created* by the insertion of p_{r+1}, and we can rewrite the sum as

$$\mathrm{E}\Big[\sum_{\Delta \in \mathcal{T}_r \setminus \mathcal{T}_{r-1}} \mathrm{card}(K(\Delta))\Big] \leqslant 3\Big(\frac{n-r}{r}\Big) \Big(\mathrm{E}\big[\mathrm{card}(\mathcal{T}_{r+1} \setminus \mathcal{T}_r)\big] - 2\Big).$$

Until now we considered P_r to be fixed. At this point, we can simply take the average over all choices of $P_r \subset P$ on both sides of the inequality above, and find that it also holds if we consider the expectation to be over all possible permutations of the set P.

We already know that the number of triangles created by the insertion of p_{r+1} is identical to the number of edges incident to p_{r+1} in $\mathcal{T}_{r+1}$, and that the expected number of these edges is at most 6. We conclude that

$$\mathrm{E}\Big[\sum_{\Delta \in \mathcal{T}_r \setminus \mathcal{T}_{r-1}} \mathrm{card}(K(\Delta))\Big] \leqslant 12\Big(\frac{n-r}{r}\Big).$$

Summing over r proves the lemma. $\qquad\qquad\qquad\qquad\qquad\qquad\square$

9.5* A Framework for Randomized Algorithms

Up to now we have seen three randomized incremental algorithms in this book: one for linear programming in Chapter 4, one for computing a trapezoidal map in Chapter 6, and one for computing a Delaunay triangulation in this chapter. (We will see one more in Chapter 11.) These algorithms, and most other randomized incremental algorithms in the computational geometry literature, all work according to the following principle.

Suppose the problem is to compute some geometric structure $\mathcal{T}(X)$, defined by a set X of geometric objects. (For instance, a Delaunay triangulation defined by a set of points in the plane.) A randomized incremental algorithm does this by adding the objects in X in random order, meanwhile maintaining the structure $\mathcal{T}$. To add the next object, the algorithm first finds out where the current structure has to be changed because there is a conflict with the object—the *location step*—and then it updates the structure locally—the *update step*. Because all randomized incremental algorithms are so much alike, their analyses are quite similar as well. To avoid having to prove the same bounds over and over again for different problems, an axiomatic framework has been developed that captures the essence of randomized incremental algorithms. This framework—called a *configuration space*—can be used to prove ready-to-use bounds for the expected running time of many randomized incremental algorithms. (Unfortunately, the term "configuration space" is also used in motion planning, where it means something completely different—see Chapter 13.) In this section we describe this framework, and we give a theorem that can be used to analyze any randomized incremental algorithm that fits into the framework. For instance, the theorem can immediately be applied to prove Lemma 9.13, this time without assuming that P has to be in general position.

A *configuration space* is defined to be a four-tuple (X, Π, D, K). Here X is the input to the problem, which is a finite set of (geometric) *objects*; we denote the cardinality of X by n. The set Π is a set whose elements are called *configurations*. Finally, D and K both assign to every configuration $\Delta \in \Pi$ a subset of X, denoted $D(\Delta)$ and $K(\Delta)$ repectively. Elements of the set $D(\Delta)$ are said to *define* the configuration Δ, and the elements of the set $K(\Delta)$ are said to be *in conflict with*, or to *kill*, Δ. The number of elements of $K(\Delta)$ is called the *conflict size* of the configuration Δ. We require that (X, Π, D, K) satisfies the following conditions.

- The number $d := \max\{\operatorname{card}(D(\Delta)) \mid \Delta \in \Pi\}$ is a constant. We call this number the *maximum degree* of the configuration space. Moreover, the number of configurations sharing the same defining set should be bounded by a constant.

- We have $D(\Delta) \cap K(\Delta) = \emptyset$ for all configurations $\Delta \in \Pi$.

A configuration Δ is called *active* over a subset $S \subseteq X$ if $D(\Delta)$ is contained in S and $K(\Delta)$ is disjoint from S. We denote the set of configurations active over S by $\mathcal{T}(S)$, so we have

$$\mathcal{T}(S) := \{\Delta \in \Pi \ : \ D(\Delta) \subseteq S \text{ and } K(\Delta) \cap S = \emptyset\}.$$

The active configurations form the structure we want to compute. More precisely, the goal is to compute $\mathcal{T}(X)$. Before we continue our discussion of this abstract framework, let's see how the geometric structures we have met so far fit in.

Half-plane intersection. In this case the input set X is a set of half-planes in the plane. We want to define Π, D, and K in such a way that $\mathcal{T}(X)$ is what we want to compute, namely the intersection of the half-planes in X. We can achieve this as follows. The set Π of configurations consists of all the intersection points of the lines bounding the half-planes in X. The defining set $D(\Delta)$ of a configuration $\Delta \in \Pi$ consists of the two lines defining the intersection, and the killing set $K(\Delta)$ consists of all half-planes that do not contain the intersection point. Hence, for any subset $S \subset X$, and in particular for X itself, $\mathcal{T}(S)$ is the set of vertices of the common intersection of the half-planes in S.

Trapezoidal maps. Here the input set X is a set of segments in the plane. The set Π of configurations contains all trapezoids appearing in the trapezoidal map of any $S \subseteq X$. The defining set $D(\Delta)$ of a configuration Δ is the set of segments that are necessary to define Δ. The killing set $K(\Delta)$ of a trapezoid Δ is the set of segments that intersect Δ. With these definitions, $\mathcal{T}(S)$ is exactly the set of trapezoids of the trapezoidal map of S.

Delaunay Triangulation. The input set X is a set of points in general position in the plane. The set Π of configurations consists of triangles formed by three (non-collinear) points in X. The defining set $D(\Delta)$ consists of the points that form the vertices of Δ, and the killing set $K(\Delta)$ is the set of points lying inside the circumcircle of Δ. By Theorem 9.6, $\mathcal{T}(S)$ is exactly the set of triangles of the unique Delaunay triangulation of S.

As stated earlier, the goal is to compute the structure $\mathcal{T}(X)$. Randomized incremental algorithms do this by computing a random permutation $x_1, x_2, \ldots, x_n$ of the objects in X and then adding the objects in this order, meanwhile maintaining $\mathcal{T}(X_r)$, where $X_r := \{x_1, x_2, \ldots, x_r\}$. The fundamental property of configuration spaces that makes this possible is that we can decide whether or not a configuration Δ appears in $\mathcal{T}(X_r)$ by looking at it *locally*—we only need to look for the defining and killing objects of Δ. In particular, $\mathcal{T}(X_r)$ does not depend on the order in which the objects in X_r were added. For instance, a triangle Δ is in the Delaunay triangulation of S if and only if the vertices of Δ are in S, and no point of S lies in the circumcircle of Δ.

The first thing we usually did when we analyzed a randomized incremental algorithm was to prove a bound on the expected structural change—see for instance Lemma 9.11. The next theorem does the same, but now in the abstract configuration-space framework.

Theorem 9.14 *Let (X, Π, D, K) be a configuration space, and let $\mathcal{T}$ and X_r be defined as above. Then the expected number of configurations in $\mathcal{T}(X_r) \setminus \mathcal{T}(X_{r-1})$ is at most*

$$\frac{d}{r} E[\mathrm{card}(\mathcal{T}(X_r))],$$

where d is the maximum degree of the configuration space.

Proof. As in previous occasions where we wanted to bound the structural change, we use backwards analysis: instead of trying to argue about the number of configurations that appear due to the addition of x_r into X_{r-1}, we shall argue about the number of configurations that disappear when we remove x_r

from X_r. To this end we temporarily let X_r be some fixed subset $X_r^* \subset X$ of cardinality r. We now want to bound the expected number of configurations $\Delta \in \mathcal{T}(X_r)$ that disappear when we remove a random object x_r from X_r. By definition of $\mathcal{T}$, such a configuration Δ must have $x_r \in D(\Delta)$. Since there are at most $d \cdot \text{card}(\mathcal{T}(X_r))$ pairs (x, Δ) with $\Delta \in \mathcal{T}(X_r)$ and $x \in D(\Delta)$, we have

$$\sum_{x \in X_r} \text{card}(\{\Delta \in \mathcal{T}(X_r) \mid x \in D(\Delta)\}) \leqslant d \cdot \text{card}(\mathcal{T}(X_r)).$$

Hence, the expected number of trapezoids disappearing due to the removal of a random object from X_r is $\frac{d}{r} \text{card}(\mathcal{T}(X_r))$. In this argument, the set X_r was a fixed subset $X_r^* \subset X$ of cardinality r. To obtain the general bound, we have to average over all possible subsets of size r, which gives a bound of $\frac{d}{r} \mathbb{E}\big[\text{card}(\mathcal{T}(X_r))\big]$. $\qquad\Box$

This theorem gives a generic bound for the expected size of the structural change during a randomized incremental algorithm. But what about the cost of the location steps? In many cases we will need a bound of the same form as in this chapter, namely we need to bound

$$\sum_{\Delta} \text{card}(K(\Delta)),$$

where the summation is over all configurations Δ that are created by the algorithm, that is, all configurations that appear in one of the $\mathcal{T}(X_r)$. This bound is given in the following theorem.

Theorem 9.15 *Let* (X, Π, D, K) *be a configuration space, and let* $\mathcal{T}$ *and* X_r *be defined as above. Then the expected value of*

$$\sum_{\Delta} \text{card}(K(\Delta)),$$

where the summation is over all configurations Δ *appearing in at least one* $\mathcal{T}(X_r)$ *with* $1 \leqslant r \leqslant n$, *is at most*

$$\sum_{r=1}^{n} d^2 \left(\frac{n-r}{r}\right) \left(\frac{\mathbb{E}\big[\text{card}(\mathcal{T}(X_r))\big]}{r}\right),$$

where d *is the maximum degree of the configuration space.*

Proof. We can follow the proof of Lemma 9.13 quite closely. We first rewrite the sum as

$$\sum_{r=1}^{n} \left(\sum_{\Delta \in \mathcal{T}_r \setminus \mathcal{T}_{r-1}} \text{card}(K(\Delta))\right).$$

Next, let $k(X_r, y)$ denote the number of configurations $\Delta \in \mathcal{T}(X_r)$ such that $y \in K(\Delta)$, and let $k(X_r, y, x_r)$ be the number of configurations $\Delta \in \mathcal{T}(X_r)$ such that not only $y \in K(\Delta)$ but for which we also have $x_r \in D(\Delta)$. Any new configuration appearing due to the addition of x_r must have $x_r \in D(\Delta)$. This implies that

$$\sum_{\Delta \in \mathcal{T}_r \setminus \mathcal{T}_{r-1}} \mathrm{card}(K(\Delta)) = \sum_{y \in X \setminus X_r} k(X_r, y, x_r).$$
(9.4)

We now fix the set X_r. The expected value of $k(X_r, y, x_r)$ then depends only on the choice of $x_r \in X_r$. Since the probability that $y \in D(\Delta)$ for a configuration $\Delta \in \mathcal{T}(X_r)$ is at most d/r, we have

$$\mathrm{E}\big[k(X_r, y, x_r)\big] \leqslant \frac{dk(X_r, y)}{r}.$$

If we sum this over all $y \in X \setminus X_r$ and use (9.4), we get

$$\mathrm{E}\Big[\sum_{\Delta \in \mathcal{T}_r \setminus \mathcal{T}_{r-1}} \mathrm{card}(K(\Delta))\Big] \leqslant \frac{d}{r} \sum_{y \in X \setminus X_r} k(X_r, y).$$
(9.5)

On the other hand, every $y \in X \setminus X_r$ is equally likely to appear as x_{r+1}, so

$$\mathrm{E}\big[k(X_r, x_{r+1})\big] = \frac{1}{n-r} \sum_{y \in X \setminus X_r} k(X_r, y).$$

Substituting this into (9.5) gives

$$\mathrm{E}\Big[\sum_{\Delta \in \mathcal{T}_r \setminus \mathcal{T}_{r-1}} \mathrm{card}(K(\Delta))\Big] \leqslant d\Big(\frac{n-r}{r}\Big) \mathrm{E}\big[k(X_r, x_{r+1})\big].$$

Now observe that $k(X_r, x_{r+1})$ is the number of configurations Δ of $\mathcal{T}(X_r)$ that will be destroyed in the next stage, when x_{r+1} is inserted. This means we can rewrite the last expression as

$$\mathrm{E}\Big[\sum_{\Delta \in \mathcal{T}_r \setminus \mathcal{T}_{r-1}} \mathrm{card}(K(\Delta))\Big] \leqslant d\Big(\frac{n-r}{r}\Big) \mathrm{E}\big[\mathrm{card}(\mathcal{T}(X_r) \setminus \mathcal{T}(X_{r+1}))\big].$$
(9.6)

Unlike in the proof of Lemma 9.13, however, we cannot simply bound the number of configurations destroyed in stage $r+1$ by the number of configurations created at that stage, because that need not be true in a general configuration space. Hence, we proceed somewhat differently.

First we observe that we can take the average over all choices of X_r on both sides of (9.6) and find that it also holds if the expectation is over all permutations of X. Next, we sum over all r, and rewrite the sum as follows:

$$\sum_{r=1}^{n} d\Big(\frac{n-r}{r}\Big) \mathrm{card}(\mathcal{T}(X_r) \setminus \mathcal{T}(X_{r+1})) = \sum_{\Delta} d\Big(\frac{n-[j(\Delta)-1]}{j(\Delta)-1}\Big),$$
(9.7)

where the summation on the right hand side is over all configurations Δ that are created and later destroyed by the algorithm, and where $j(\Delta)$ denotes the stage when configuration Δ is destroyed. Let $i(\Delta)$ denote the stage when the configuration Δ is created. Since $i(\Delta) \leqslant j(\Delta) - 1$, we have

$$\frac{n - [j(\Delta) - 1]}{j(\Delta) - 1} = \frac{n}{j(\Delta) - 1} - 1 \leqslant \frac{n}{i(\Delta)} - 1 = \frac{n - i(\Delta)}{i(\Delta)}.$$

If we substitute this into (9.7), we see that

$$\sum_{r=1}^{n} d\left(\frac{n-r}{r}\right) \operatorname{card}(\mathcal{T}(X_r) \setminus \mathcal{T}(X_{r+1})) \leqslant \sum_{\Delta} d\left(\frac{n - i(\Delta)}{i(\Delta)}\right).$$

The right hand side of this expression is at most

$$\sum_{r=1}^{n} d\left(\frac{n-r}{r}\right) \operatorname{card}(\mathcal{T}(X_r) \setminus \mathcal{T}(X_{r-1}))$$

(the difference being only those configurations that are created but never destroyed) and so we have

$$\mathrm{E}\Big[\sum_{r=1}^{n} \sum_{\Delta \in \mathcal{T}_r \setminus \mathcal{T}_{r-1}} \operatorname{card}(K(\Delta))\Big] \leqslant \sum_{r=1}^{n} d\left(\frac{n-r}{r}\right) \mathrm{E}\big[\operatorname{card}(\mathcal{T}(X_r) \setminus \mathcal{T}(X_{r-1}))\big].$$

By Theorem 9.14, we get the bound we wanted to prove:

$$\mathrm{E}\Big[\sum_{r=1}^{n} \sum_{\Delta \in \mathcal{T}_r \setminus \mathcal{T}_{r-1}} \operatorname{card}(K(\Delta))\Big] \leqslant \sum_{r=1}^{n} d\left(\frac{n-r}{r}\right) \frac{d}{r} \mathrm{E}\big[\operatorname{card}(\mathcal{T}(X_r))\big].$$

$\square$

This finishes the analysis in the abstract setting. As an example, we will show how to apply the results to our randomized incremental algorithm for computing the Delaunay triangulation. In particular, we will prove that

$$\sum_{\Delta} \operatorname{card}(K(\Delta)) = O(n \log n),$$

where the summation is over all triangles Δ created by the algorithm, and where $K(\Delta)$ is the set of points in the circumcircle of the triangle.

Unfortunately, it seems impossible to properly define a configuration space whose configurations are triangles when the points are not in general position. Therefore we shall choose the configurations slightly differently.

Let X be a set of points in the plane, not necessarily in general position. Recall that Ω denotes the set of three extra points that we added to start the construction. The points in Ω were chosen such that they do not destroy any Delaunay edges between points in P. Every triple $\Delta = (p_i, p_j, p_k)$ of points in $X \cup \Omega$ that do not lie on a line defines a configuration with $D(\Delta) := \{p_i, p_j, p_k\}$ and $K(\Delta)$ is the set of points that lie either in the interior of the circumcircle of the

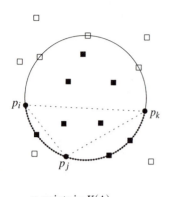

■ points in $K(\Delta)$

□ points not in $K(\Delta)$

triangle $p_i p_j p_k$ or on the circular arc on the circumcircle from p_i to p_k containing p_j. We call such a configuration Δ a *Delaunay corner* of X, because Δ is active over $S \subseteq X$ if and only if p_i, p_j, and p_k are consecutive points on the boundary of one face of the Delaunay graph $\mathcal{D}\mathcal{G}(\Omega \cup S)$. Note that any set of three non-collinear points defines three different configurations.

The important observation is that whenever DELAUNAYTRIANGULATION creates a new triangle, this triangle is of the form $p_i p_r p_j$, where p_r is the point inserted in this stage, and $\overline{p_r p_i}$ and $\overline{p_r p_j}$ are edges of the Delaunay graph $\mathcal{D}\mathcal{G}(\Omega \cup P_r)$—see Lemma 9.10. It follows that when the triangle $p_i p_r p_j$ is created, the triple (p_j, p_r, p_i) is a Delaunay corner of $\mathcal{D}\mathcal{G}(\Omega \cup P_r)$ and, hence, it is an active configuration over the set P_r. The set $K(\Delta)$ defined for this configuration contains all points contained in the circumcircle of the triangle $p_i p_r p_j$. We can therefore bound the original sum by

$$\sum_{\Delta} \text{card}(K(\Delta)),$$

where the sum is over all *Delaunay corners* Δ that appear in some intermediate Delaunay graph $\mathcal{D}\mathcal{G}(\Omega \cup P_r)$.

Now Theorem 9.15 applies. How many Delaunay corners are there in the Delaunay graph of $S \cup \Omega$ points? The worst case is when the Delaunay graph is a triangulation. If S contains r points, then the triangulation has $2(r+3) - 5$ triangles, and therefore $6(r+3) - 15 = 6r - 3$ Delaunay corners. It follows from Theorem 9.15 that

$$\sum_{\Delta} \text{card}(K(\Delta)) \leqslant \sum_{r=1}^{n} 9 \left(\frac{n-r}{r} \right) \left(\frac{6r-3}{r} \right) \leqslant 54n \sum_{r=1}^{n} \frac{1}{r} = 54n(\ln n + 1).$$

This finally completes the proof of Theorem 9.12.

9.6 Notes and Comments

The problem of triangulating a set of points is a topic in computational geometry that is well known outside this field. Triangulations of point sets in two and more dimensions are of paramount importance in numerical analysis, for instance for finite element methods, but also in computer graphics. In this chapter we looked at the case of triangulations that only use the given points as vertices. If additional points—so-called Steiner points—are allowed, the problem is also known as *meshing* and is treated in more detail in Chapter 14.

Lawson [209] proved that any two triangulations of a planar point set can be transformed into each other by flipping edges. He later suggested finding a good triangulation by iteratively flipping edges, where such an edge-flip improves some cost function of the triangulation [210].

It had been observed for some time that triangulations that lead to good interpolations avoid long and skinny triangles [26]. The result that there is—if we ignore degenerate cases—only one locally optimal triangulation with respect to the angle-vector, namely the Delaunay triangulation, is due to Sibson [317].

Looking only at the angle-vector completely ignores the height of the data points, and is therefore also called the *data-independent* approach. A good motivation for this approach is given by Rippa [290], who proves that the Delaunay triangulation is the triangulation that minimizes the *roughness* of the resulting terrain, no matter what the actual height data is. Here, roughness is defined as the integral of the square of the L_2-norm of the gradient of the terrain. More recent research tries to find improved triangulations by taking the height information into account. This *data-dependent* approach was first proposed by Dyn et al. [126], who suggest different cost criteria for triangulations, which depend on the height of the data points. Interestingly, they compute their improved triangulations by starting with the Delaunay triangulation and iteratively flipping edges. The same approach is taken by Quak and Schumaker [288], who consider piecewise cubic interpolation, and Brown [55]. Quak and Schumaker observe that their triangulations are small improvements compared to the Delaunay triangulation when they try to approximate smooth surfaces, but that they can be drastically different for non-smooth surfaces.

More references relevant to Delaunay triangulations as the dual of Voronoi diagrams can be found in Chapter 7.

The randomized incremental algorithm we have given here is due to Guibas et al. [164], but our analysis of $\sum_\Delta \mathrm{card}(K(\Delta))$ is from Mulmuley's book [256]. The argument that extends the analysis to the case of points in degenerate position is from Matoušek et al. [232]. Alternative randomized algorithms were given by Boissonnat et al. [49, 51], and by Clarkson and Shor [111].

Various *geometric graphs* defined on a set of points P have been found to be subgraphs of the Delaunay triangulation of P. The most important one is probably the *Euclidean minimum spanning tree* (EMST) of the set of points [310]; others are the *Gabriel graph* [155] and the *relative neighborhood graph* [324]. Another important triangulation is the minimum weight triangulation [120, 121, 5, 30]. We treat these geometric graphs in the exercises.

9.7 Exercises

9.1 In this exercise we look at the number of different triangulations that a set of n points may allow.

   a. Prove that no set of n points can be triangulated in more than $2^{\binom{n}{2}}$ ways.

   b. Prove that there are sets of n points that can be triangulated in at least $2^{n-2\sqrt{n}}$ different ways.

9.2 Prove that any two triangulations of a planar point set can be transformed into each other by edge flips. *Hint:* Show first that any two triangulations of a convex polygon can be transformed into each other by edge flips.

9.3 Prove that the smallest angle of any triangulation of a convex polygon whose vertices lie on a circle is the same. This implies that *any* completion of the Delaunay triangulation of a set of points maximizes the minimum angle.

9.4 Given four points x, y, z, t in the plane, prove that point t lies in the interior of the circle through x, y, and z if and only if the following condition holds.

$$\det \begin{pmatrix} x_1 & x_2 & x_1^2 + x_2^2 & 1 \\ y_1 & y_2 & y_1^2 + y_2^2 & 1 \\ z_1 & z_2 & z_1^2 + z_2^2 & 1 \\ t_1 & t_2 & t_1^2 + t_2^2 & 1 \end{pmatrix} > 0.$$

Why is this a good way to determine whether an edge of a triangulation is illegal?

9.5 We have described algorithm DELAUNAYTRIANGULATION by calling a recursive procedure LEGALIZEEDGE. Give an iterative version of the algorithm. Once p_r has been inserted, use a counterclockwise traversal of the triangulation to flip the illegal edges.

Is this a better way to implement the algorithm?

9.6 Prove that all edges of $\mathcal{DG}(P_r)$ that are not in $\mathcal{DG}(P_{r-1})$ are incident to p_r. In other words, the new edges of $\mathcal{DG}(P_r)$ form a star as in Figure 9.8. Give a direct proof, without referring to algorithm DELAUNAYTRIANGULATION.

9.7 A *Euclidean minimum spanning tree* (EMST) of a set of points in the plane is a tree of minimum total edge length connecting all the points. EMST's are interesting in applications where we want to connect sites in a planar environment by communication lines (local area networks), roads, railroads, or the like. (Another important application of the EMST is for an approximative solution to the traveling salesman problem.)

    a.   Prove that the set of edges of a Delaunay triangulation of a set of points P in the plane contains an EMST for P.

    b.   Use this result to give an $O(n \log n)$ algorithm to compute an EMST for a given set of n points.

9.8 The *Gabriel graph* of a set P of points in the plane is defined as follows: Two points p and q are connected by an edge of the Gabriel graph if and only if the circle with diameter pq does not contain any other point of P in its interior.

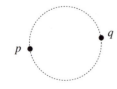

    a.   Prove that $\mathcal{DG}(P)$ contains the Gabriel graph of P.

    b.   Prove that p and q are adjacent in the Gabriel graph of p if and only if the Delaunay edge between p and q intersects its dual Voronoi edge.

    c.   Give an $O(n \log n)$ time algorithm to compute the Gabriel graph of a set of n points.

9.9 Given a point set P, a point $q \in P$ is called a *nearest neighbor* of a point $p \in P$ if no other point $r \in P$ is closer to p than q, that is, $d(p,q) \leqslant d(p,r)$ for all $r \in P, r \neq p$. Two points p and q are called *mutual nearest neighbors* if q is a nearest neighbor of p and p is a nearest neighbor of q.

The *relative neighborhood graph* of a set P of points in the plane is defined as follows: Two points p and q are connected by an edge of the relative neighborhood graph if and only if p and q are mutual nearest neighbors.

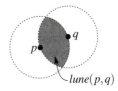

a. Given two points p and q, let $lune(p,q)$ be the moon-shaped region formed as the intersection of the two circles around p and q whose radius is $d(p,q)$. Prove that p and q are mutual nearest neighbors in a set P if and only $lune(p,q)$ does not contain any point of P in its interior.

b. Prove that $\mathcal{D}\mathcal{G}(P)$ contains the relative neighborhood graph of P.

c. Design an algorithm to compute the relative neighborhood graph of a given point set.

9.10 Prove the following relationship between the edge sets of an EMST, of the relative neighborhood graph (RNG), the Gabriel graph (GG), and the Delaunay graph ($\mathcal{D}\mathcal{G}$) of a point set P.

$$EMST \subseteq RNG \subseteq GG \subseteq \mathcal{D}\mathcal{G}.$$

(See the previous exercises for the definition of these graphs.)

9.11* Give an example of a geometric configuration space (X, Π, D, K) where $\mathcal{T}(X_r) \setminus \mathcal{T}(X_{r+1})$ can be arbitrarily large compared to $\mathcal{T}(X_{r+1}) \setminus \mathcal{T}(X_r)$.

9.12* Apply configuration spaces to analyze the randomized incremental algorithm of Chapter 6.

9.13 The *weight* of a triangulation is the sum of the length of all edges of the triangulation. Determining a *minimum weight triangulation*, that is, a triangulation whose weight is minimal among all triangulations of a given point set, is one of the major open problems in computational geometry, as it is not known whether a minimum weight triangulation can be computed in polynomial time, nor has the problem be proven to be NP-complete.

Disprove the conjecture that the Delaunay triangulation is a minimum weight triangulation.

10   More Geometric Data Structures
Windowing

In the future most cars will be equipped with a vehicle navigation system to help you determine your position and to guide you to your destination. Such a system stores a roadmap of, say, the whole of the U.S. It also keeps track of where you are, so that it can show the appropriate part of the roadmap at any time on a little computer screen; this will usually be a rectangular region around your present position. Sometimes the system will do even more for you. For example, it might warn you when a turn is coming up that you should take to get to your destination.

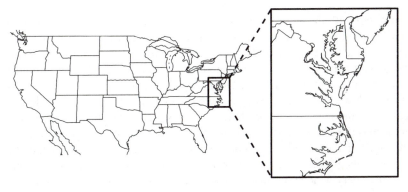

Figure 10.1
A windowing query in a map of the U.S.

To be of any use, the map should contain sufficient detail. A detailed map of the whole of Europe contains an enormous amount of data. Fortunately, only a small part of the map has to be displayed. Nevertheless, the system still has to find that part of the map: given a rectangular region, or a *window*, the system must determine the part of the map (roads, cities, and so on) that lie in the window, and display them. This is called a *windowing query*.

Checking every single feature of the map to see if it lies inside the window is not a workable method with the amount of data that we are dealing with. What we should do is to store the map in some kind of data structure that allows us to retrieve the part inside a window quickly.

Windowing queries are not only useful operations on geographic maps. They also play an important role in several applications in computer graphics and

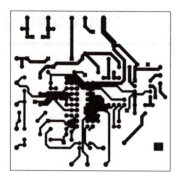

CAD/CAM. One example is flight simulation. A model of a landscape can consist of a huge number of triangles, but only a small part of the landscape will be within sight of the pilot. So we have to select the part of the landscape that lies within a given region. Here the region is 3-dimensional, and it is called the *viewing volume*. Another example comes from the design of printed circuit boards. Such a design typically consists of (a number of layers of) a planar drawing showing the location of traces and components. (See also Chapter 14.) In the design process one often wants to zoom in onto a certain portion of the board to inspect it more closely. Again, what we need is to determine the traces and components on the board that lie inside the window. In fact, windowing is required whenever one wants to inspect a small portion of a large, complex object.

Windowing queries are similar to the range queries studied in Chapter 5. The difference is the type of data that is dealt with: the data for range queries are points, whereas the data for windowing queries are typically line segments, polygons, curves, and so on. Also, for range queries we often deal with higher-dimensional search spaces, while for windowing queries the search space usually is 2- or 3-dimensional.

10.1 Interval Trees

Let's start with the easiest of the examples that we gave above, namely windowing for a printed circuit board. The reason that this example is easier than the others is that the type of data is restricted: the objects on a printed circuit board normally have boundaries that consist of line segments with a limited number of possible orientations. Often they are parallel to one of the sides of the board or make 45 degree angles with the sides. Here we only consider the case were the segments are parallel to the sides. In other words, if we consider the *x*-axis to be aligned with the bottom side of the board, and the *y*-axis to be aligned with the left side of the board, then any segment is parallel to either the *x*-axis or the *y*-axis: the segments are *axis-parallel*, or *orthogonal*. We assume that the query window is an axis-parallel rectangle, that is, a rectangle whose edges are axis-parallel.

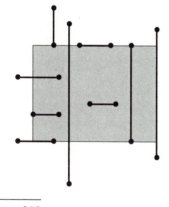

Let S be a set of n axis-parallel line segments. To solve windowing queries we need a data structure that stores S in such a way that the segments intersecting a query window $W := [x : x'] \times [y : y']$ can be reported efficiently. Let's first see in what ways a segment can intersect the window. There are a number of different cases: the segment can lie entirely inside W, it can intersect the boundary of W once, it can intersect the boundary twice, or it can (partially) overlap the boundary of W. In most cases the segment has at least one endpoint inside W. We can find such segments by performing a range query with W in the set of $2n$ endpoints of the segments in S. In Chapter 5 we have seen a data structure for this: the range tree. A 2-dimensional range tree uses $O(n \log n)$ storage, and a range query can be answered in $O(\log^2 n + k)$ time, where k is the number of reported points. We have also shown that we can apply fractional cascading to reduce the query time to $O(\log n + k)$. There is one little problem. If we do a range query with W in the set of segment endpoints, we report the segments that have both endpoints inside W twice. This can be avoided by marking a segment when we report it for the first time, and only reporting segments that are not yet marked. Alternatively, when we find an endpoint of a segment to lie inside W, we can check whether the other endpoint lies inside W as well. If not we report the segment. If the other endpoint does lie inside W, we only report the segment when the current endpoint is the leftmost or bottom endpoint. This leads to the following lemma.

Lemma 10.1 *Let S be a set of n axis-parallel line segments in the plane. The segments that have at least one endpoint inside an axis-parallel query window W can be reported in $O(\log n + k)$ time with a data structure that uses $O(n \log n)$ storage and preprocessing time, where k is the number of reported segments.*

It remains to find the segments that do not have an endpoint inside the query window. Such segments either cross the boundary of W twice or contain one edge of the boundary. When the segment is vertical it will cross both horizontal edges of the boundary. When it is horizontal it will cross both vertical edges. Hence, we can find such segments by reporting all segments that intersect the left edge of the boundary and all segments that intersect the bottom edge of the boundary. (Note that there is no need to query with the other two edges of the boundary.) To be precise, we should only report those segments that do not have an endpoint inside W, because the others were already reported before. Let's consider the problem of finding the horizontal segments intersected by the left edge of W; to deal with the top edge we just have to reverse the roles of the x- and y-coordinates.

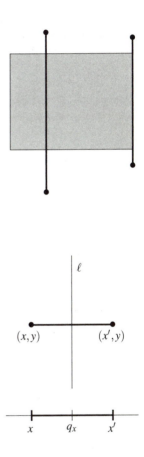

We have arrived at the following problem: preprocess a set S of horizontal line segments in the plane such that the segments intersecting a vertical query segment can be reported efficiently. To gain some insight into the problem we first look at a simpler version, namely where the query segment is a full line. Let $\ell := (x = q_x)$ denote the query line. A horizontal segment $s := (x, y)(x', y)$ is intersected by ℓ if and only if $x \leq q_x \leq x'$. So only the x-coordinates of the segments play a role here. In other words, the problem becomes 1-dimensional: given a set of intervals on the real line, report the ones that contain the query point q_x.

Let $I := \{[x_1 : x_1'], [x_2 : x_2'], \ldots, [x_n : x_n']\}$ be a set of closed intervals on the real line. To keep the connection with the 2-dimensional problem alive we image the real line to be horizontal, and we say "to the left (right) of" instead of "less (greater) than". Let x_{mid} be the median of the $2n$ interval endpoints, So at most half of the interval endpoints lies to the left of x_{mid} and at most half of the endpoints lies to the right of x_{mid}. If the query value q_x lies to the left of x_{mid} then the intervals that lie completely to the right of x_{mid} obviously do not contain q_x. We construct a binary tree based on this idea. The right subtree of the tree stores the set I_{right} of the intervals lying completely to the right of x_{mid}, and the left subtree stores the set I_{left} of intervals completely to the left of x_{mid}. These subtrees are constructed recursively in the same way. There is one problem that we still have to deal with: what to do with the intervals that contain x_{mid}? One possibility would be to store such intervals in both subtrees.

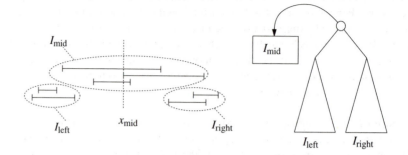

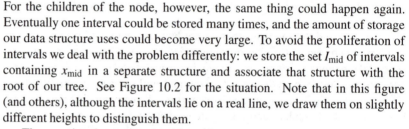

Figure 10.2
Classification of the segments with
respect to x_{mid}

For the children of the node, however, the same thing could happen again. Eventually one interval could be stored many times, and the amount of storage our data structure uses could become very large. To avoid the proliferation of intervals we deal with the problem differently: we store the set I_{mid} of intervals containing x_{mid} in a separate structure and associate that structure with the root of our tree. See Figure 10.2 for the situation. Note that in this figure (and others), although the intervals lie on a real line, we draw them on slightly different heights to distinguish them.

The associated structure should enable us to report the intervals in I_{mid} that contain q_x. So we ended up with the same problem that we started with: given a set I_{mid} of intervals, find those that contain q_x. But if we have bad luck I_{mid} could be the same as I. It seems we are back to exactly the same problem, but there is a difference. We know that all the intervals in I_{mid} contain x_{mid}, and this helps a great deal. Suppose, for example, that q_x lies to the left of x_{mid}. In that case we already know that the right endpoint of all intervals in I_{mid} lies right of q_x. So only the left endpoints of the intervals are important: q_x is contained in an interval $[x_j : x_j'] \in I_{mid}$ if and only if $x_j \leq q_x$. If we store the intervals in a list ordered on increasing left endpoint, then q_x can only be contained in an interval if q_x is also contained in all its predecessors in the sorted list. In other words, we can simply walk along the sorted list reporting intervals, until we come to an interval that does not contain q_x. At that point we can stop: none of the remaining intervals can contain q_x. Similarly, when q_x lies right of x_{mid} we can walk along a list that stores the intervals sorted on right endpoint. This list must be sorted on decreasing right endpoint, because it is traversed if the query point q_x lies to the right of x_{mid}. Finally, when $q_x = x_{mid}$ we can report all intervals in I_{mid}. (We do not need to treat this as a separate case. We can simply walk along one of the sorted lists.)

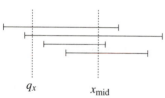

We now give a succinct description of the whole data structure that stores the intervals in I. The data structure is called an *interval tree*. Figure 10.3 shows an interval tree; the dotted vertical segments indicate the values x_{mid} for each node.

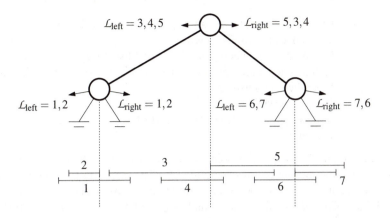

$\mathcal{L}_{\text{left}} = 3, 4, 5$ $\mathcal{L}_{\text{right}} = 5, 3, 4$

$\mathcal{L}_{\text{left}} = 1, 2$ $\mathcal{L}_{\text{right}} = 1, 2$ $\mathcal{L}_{\text{left}} = 6, 7$ $\mathcal{L}_{\text{right}} = 7, 6$

Figure 10.3
An interval tree

- If $I = \emptyset$ then the interval tree is a leaf.

- Otherwise, let x_{mid} be the median of the endpoints of the intervals. Let

$$
\begin{aligned}
I_{\text{left}} &:= \{[x_j : x'_j] \in I : x'_j < x_{\text{mid}}\}, \\
I_{\text{mid}} &:= \{[x_j : x'_j] \in I : x_j \leqslant x_{\text{mid}} \leqslant x'_j\}, \\
I_{\text{right}} &:= \{[x_j : x'_j] \in I : x_{\text{mid}} < x_j\}.
\end{aligned}
$$

The interval tree consists of a root node v storing x_{mid}. Furthermore,

- the set I_{mid} is stored twice; once in a list $\mathcal{L}_{\text{left}}(v)$ that is sorted on the left endpoints of the intervals, and once in a list $\mathcal{L}_{\text{right}}(v)$ that is sorted on the right endpoints of the intervals,

- the left subtree of v is an interval tree for the set I_{left},

- the right subtree of v is an interval tree for the set I_{right}.

Lemma 10.2 *An interval tree on a set of n intervals uses $O(n)$ storage and has depth $O(\log n)$.*

Proof. The bound on the depth is trivial, so we prove the storage bound. Note that I_{left}, I_{mid}, and I_{right} are disjoint subsets. As a result, each interval is only stored in a set I_{mid} once and, hence, only appears once in the two sorted lists. This shows that the total amount of storage required for all associated lists is bounded by $O(n)$. The tree itself uses $O(n)$ storage as well. ⧠

The following recursive algorithm for building an interval tree follows directly from its definition. (Recall that $lc(v)$ and $rc(v)$ denote the left and right child, respectively, of a node v.)

Algorithm CONSTRUCTINTERVALTREE(I)
Input. A set I of intervals on the real line.
Output. The root of an interval tree for I.
1.   **if** $I = \emptyset$
2.   **then return** an empty leaf

3.   **else** Create a node v. Compute x_{mid}, the median of the set of interval endpoints, and store x_{mid} with v.

4.   Compute I_{mid} and construct two sorted lists for I_{mid}: a list $\mathcal{L}_{\mathrm{left}}(v)$ sorted on left endpoint and a list $\mathcal{L}_{\mathrm{right}}(v)$ sorted on right endpoint. Store these two lists at v.

5. $lc(v) \leftarrow \text{CONSTRUCTINTERVALTREE}(I_{\mathrm{left}})$

6. $rc(v) \leftarrow \text{CONSTRUCTINTERVALTREE}(I_{\mathrm{right}})$

7.   **return** v

Finding the median of a set of points takes linear time. Actually, it is better to compute the median by presorting the set of points, as in Chapter 5. It is easy to maintain these presorted sets through the recursive calls. Let $n_{\mathrm{mid}} :=$ card(I_{mid}). Creating the lists $\mathcal{L}_{\mathrm{left}}(v)$ and $\mathcal{L}_{\mathrm{right}}(v)$ takes $O(n_{\mathrm{mid}} \log n_{\mathrm{mid}})$ time. Hence, the time we spend (not counting the time needed for the recursive calls) is $O(n + n_{\mathrm{mid}} \log n_{\mathrm{mid}})$. Using similar arguments as in the proof of Lemma 10.2 we can conclude that the algorithm runs in $O(n \log n)$ time.

Lemma 10.3 *An interval tree on a set of n intervals can be built in $O(n \log n)$ time.*

It remains to show how to use the interval tree to find the intervals containing a query point q_x. We already sketched how to do this, but now we can give the exact algorithm.

Algorithm QUERYINTERVALTREE(v, q_x)
Input. The root v of an interval tree and a query point q_x.
Output. All intervals that contain q_x.
1.   **if** v is not a leaf
2.   **then if** $q_x < x_{\mathrm{mid}}(v)$
3.   **then** Walk along the list $\mathcal{L}_{\mathrm{left}}(v)$, starting at the interval with the leftmost endpoint, reporting all the intervals that contain q_x. Stop as soon as an interval does not contain q_x.
4.   QUERYINTERVALTREE$(lc(v), q_x)$
5.   **else** Walk along the list $\mathcal{L}_{\mathrm{right}}(v)$, starting at the interval with the rightmost endpoint, reporting all the intervals that contain q_x. Stop as soon as an interval does not contain q_x.
6.   QUERYINTERVALTREE$(rc(v), q_x)$

Analyzing the query time is not very difficult. At any node v that we visit we spend $O(1 + k_v)$ time, where k_v is the number of intervals that we report at v. The sum of the k_v's over the visited nodes is, of course, k. Furthermore, we visit at most one node at any depth of the tree. As noted above, the depth of the interval tree is $O(\log n)$. So the total query time is $O(\log n + k)$.

The following theorem summarizes the results about interval trees.

Theorem 10.4 *An interval tree for a set I of n intervals uses $O(n)$ storage and can be built in $O(n \log n)$ time. Using the interval tree we can report all intervals that contain a query point in $O(\log n + k)$ time, where k is the number of reported intervals.*

It's time to pause a moment and see where all this has brought us. The problem we originally wanted to solve is this: store a set S of axis-parallel segments in a data structure that allows us to find the segments intersecting a query window $W = [x : x'] \times [y : y']$. Finding the segments that have an endpoint inside W could be done using a data structure from Chapter 5, the range tree. The other segments intersecting W had to intersect the boundary of W twice. We planned to find these segments by querying with the left and top edges of W. So we needed a data structure that stores a set of horizontal segments such that the ones intersecting a vertical query segment can be reported efficiently, and a similar data structure storing the vertical segments that allows for intersection queries with horizontal segments. We started by developing a data structure that solves a slightly simpler problem, namely where the query object is a full line. This led to the interval tree. Now let's see how to extend the interval tree to the case where the query object is a vertical line segment.

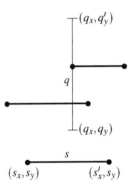

Let $S_H \subseteq S$ be the subset of horizontal segments in S, and let q be the vertical query segment $q_x \times [q_y : q'_y]$. For a segment $s := [s_x : s'_x] \times s_y$ in S_H, we call $[s_x : s'_x]$ the *x-interval* of the segment. Suppose we have stored the segments in S_H in an interval tree $\mathcal{T}$ according to their x-intervals. Let's go through the query algorithm QUERYINTERVALTREE to see what happens when we query $\mathcal{T}$ with the vertical query segment q. Suppose q_x lies to the left of the x_{mid}-value stored at the root of the interval tree $\mathcal{T}$. It is still correct that we only search recursively in the left subtree: segments completely to the right of x_{mid} cannot be intersected by q so we can skip the right subtree. The way the set I_{mid} is treated, however, is not correct anymore. For a segment $s \in I_{\text{mid}}$ to be intersected by q, it is not sufficient that its left endpoint lies to the left of q; it is also required that its y-coordinate lies in the range $[q_y : q'_y]$. Figure 10.4 illustrates this. So storing the endpoints in an ordered list is not

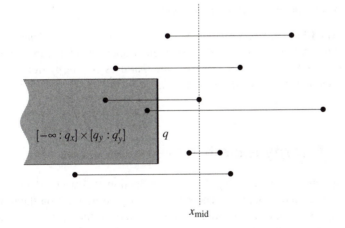

Figure 10.4
Segments intersected by q must have their left endpoint in the shaded region.

enough. We need a more elaborate associated structure: given a query range $(-\infty : q_x] \times [q_y : q'_y]$, we must be able to report all the segments whose left endpoint lies in that range. If q lies to the right of x_{mid} we want to report all segments whose right endpoint lies in the range $[q_x : +\infty) \times [q_y : q'_y]$, so for this case we need a second associated structure. How should we implement the associated structure? Well, the query that we wish to perform is nothing more than a rectangular range query on a set of points. A 2-dimensional range tree, described in Chapter 5, will therefore do the trick. This way the associated structure uses $O(n_{mid} \log n_{mid})$ storage, where $n_{mid} := \text{card}(I_{mid})$, and its query time is $O(\log n_{mid} + k)$.

The data structure that stores the set S_H of horizontal segments is now as follows. The main structure is an interval tree $\mathcal{T}$ on the x-intervals of the segments. Instead of the sorted lists $\mathcal{L}_{left}(v)$ and $\mathcal{L}_{right}(v)$ we have two range trees: a range tree $\mathcal{T}_{left}(v)$ on the left endpoints of the segments in $I_{mid}(v)$, and a range tree $\mathcal{T}_{right}(v)$ on the right endpoints of the segments in $I_{mid}(v)$. Because the storage required for a range tree is a factor $\log n$ larger than for sorted lists, the total amount of storage for the data structure becomes $O(n \log n)$. The pre-processing time remains $O(n \log n)$.

The query algorithm is the same as QUERYINTERVALTREE, except that, instead of walking along the sorted list $\mathcal{L}_{left}(v)$, say, we perform a query in the range tree $\mathcal{T}_{left}(v)$. So at each of the $O(\log n)$ nodes v on the search path we spend $O(\log n + k_v)$ time, where k_v is the number of reported segments. The total query time therefore becomes $O(\log^2 n + k)$.

We have proved the following theorem.

Theorem 10.5 *Let S be a set of n horizontal segments in the plane. The segments intersecting a vertical query segment can be reported in $O(\log^2 n + k)$ time with a data structure that uses $O(n \log n)$ storage, where k is the number of reported segments. The structure can be built in $O(n \log n)$ time.*

If we combine this with the result of Lemma 10.1 we get a solution to the windowing problem for axis-parallel segments.

Corollary 10.6 *Let S be a set of n axis-parallel segments in the plane. The segments intersecting an axis-parallel rectangular query window can be reported in $O(\log^2 n + k)$ time with a data structure that uses $O(n \log n)$ storage, where k is the number of reported segments. The structure can be built in $O(n \log n)$ time.*

10.2 Priority Search Trees

In the structure for windowing described in Section 10.1 we used a range tree for the associated structures. The range queries we perform on them have a special property: they are unbounded on one side. In this section we will describe a different data structure, the *priority search tree*, that uses this property to improve the bound on storage to $O(n)$. This data structure is also a lot simpler because it does not require fractional cascading. Using priority search trees instead of range trees in the data structure for windowing reduces the storage bound in Theorem 10.5 to $O(n)$. It does not improve the storage bound in

Corollary 10.6 because there we also need a range tree to report the endpoints that lie in the window.

Let $P := \{p_1, p_2, \ldots, p_n\}$ be a set of points in the plane. We want to design a structure for rectangular range queries of the form $(-\infty : q_x] \times [q_y : q'_y]$. To get some idea of how this special property can be used, let's look at the 1-dimensional case. A normal 1-dimensional range query would ask for the points lying in a range $[q'_x : q_x]$. To find these points efficiently we can store the set of points in a 1-dimensional range tree, as described in Chapter 5. If the range is unbounded to the left, we are asking for the points lying in $(-\infty : q_x]$. This can be solved by simply walking along an ordered list starting at the leftmost point, until we encounter a point that is not in the range. The query time is $O(1 + k)$, instead of $O(\log n + k)$ which we needed in the general case.

How can we extend this strategy to 2-dimensional range queries that are unbounded to the left? Somehow we must integrate information about the y-coordinate in the structure without using associated structures, so that, among the points whose x-coordinate is in $(-\infty : q_x]$, we can easily select the ones whose y-coordinate is in $[q_y : q'_y]$. A simple linear list doesn't lend itself well for this. Therefore we take a different structure to work with: the *heap*.

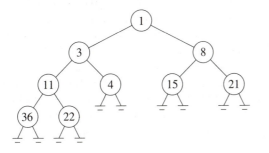

Figure 10.5
A heap for the set
$\{1, 3, 4, 8, 11, 15, 21, 22, 36\}$

A heap is normally used for priority queries that ask for the smallest (or largest) value in a set. But heaps can also be used to answer 1-dimensional range queries of the form $(-\infty : q_x]$. A heap has the same query time as a sorted list, namely $O(1 + k)$. Normally the advantage of a heap over a sorted list is that points can be inserted and the maximum deleted more efficiently. For us the tree structure of a heap has another advantage: it makes it easier to integrate information about the y-coordinate, as we shall see shortly. A heap is a binary tree defined as follows. The root of the tree stores the point with minimum x-value. The remainder of the set is partitioned into two subsets of almost equal size, and these subsets are stored recursively in the same way. Figure 10.5 gives an example of a heap. We can do a query with $(-\infty : q_x]$ by walking down the tree. When we visit a node we check if the x-coordinate of the point stored at the node lies in $(-\infty : q_x]$. If it does, we report the point and continue the search in both subtrees; otherwise, we abort the search in this part of the tree. For example, when we search with $(-\infty : 5]$ in the tree of Figure 10.5, we report the points 1, 3, and 4. We also visit the nodes with 8 and 11 but abort the search there.

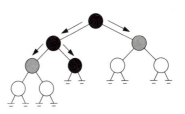

Heaps give us some freedom in how to partition the set into two subsets. If we also want to search on y-coordinate then the trick is to perform the partitioning not in an arbitrary way, as is the case for normal heaps, but according to y-coordinate. More precisely, we split the remainder of the set into two subsets of almost equal size such that the y-coordinate of any point in one subset is smaller than the y-coordinate of any point in the other subset. This is illustrated in Figure 10.6. The tree is drawn sideways to indicate that the partitioning is

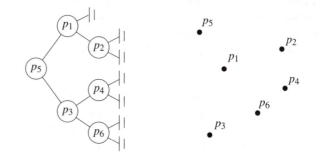

Figure 10.6
A set of points and the corresponding
priority search tree

on y-coordinate. In the example of Figure 10.6 the point p_5 has smallest x-coordinate and, hence, is stored in the root. The other points are partitioned in y-coordinate. The points p_3, p_4, and p_6 have smaller y-coordinate and are stored in the left subtree. Of these p_3 has smallest x-coordinate, so this point is places in the root of the subtree, and so on.

A formal definition of a priority search tree for a set P of points is as follows. We assume that all the points have distinct coordinates. In Chapter 5 (more precisely, in Section 5.5) we saw that this involves no loss of generality; by using composite numbers we can simulate that all coordinates are distinct.

- If $P = \emptyset$ then the priority search tree is an empty leaf.

- Otherwise, let p_{min} be the point in the set P with the smallest x-coordinate. Let y_{mid} be the median of the y-coordinates of the remaining points. Let

$$
\begin{aligned}
P_{\mathrm{below}} &:= \{p \in P \setminus \{p_{\mathrm{min}}\} : p_y < y_{\mathrm{mid}}\}, \\
P_{\mathrm{above}} &:= \{p \in P \setminus \{p_{\mathrm{min}}\} : p_y > y_{\mathrm{mid}}\}.
\end{aligned}
$$

The priority search tree consists of a root node v where the point $p(v) := p_{\mathrm{min}}$ and the value $y(v) := y_{\mathrm{mid}}$ are stored. Furthermore,

- the left subtree of v is a priority search tree for the set P_{below},

- the right subtree of v is a priority search tree for the set P_{above}.

It's straightforward to derive a recursive $O(n \log n)$ algorithm for building a priority search tree. Interestingly, priority search trees can even be built in linear time, if the points are already sorted on y-coordinate. The idea is to construct the tree bottom-up instead of top-down, in the same way heaps are normally constructed.

A query with a range $(-\infty : q_x] \times [q_y : q'_y]$ in a priority search tree is performed roughly as follows. We search with q_y and q'_y, as indicated in Figure 10.7. All the shaded subtrees in the figure store only points whose y-coordinate lies in the correct range. So we can search those subtrees based on x-coordinate only. This is done with the following subroutine, which is basically the query algorithm for a heap.

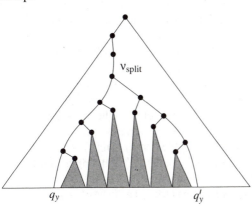

Figure 10.7
Querying a priority search tree

REPORTINSUBTREE(ν, q_x)
Input. The root ν of a subtree of a priority search tree and a value q_x.
Output. All points in the subtree with x-coordinate at most q_x.
1.   **if** ν is not a leaf and $(p(\nu))_x \leqslant q_x$
2.   **then** Report $p(\nu)$.
3.   REPORTINSUBTREE$(lc(\nu), q_x)$
4.   REPORTINSUBTREE$(rc(\nu), q_x)$

Lemma 10.7 REPORTINSUBTREE(ν, q_x) *reports in* $O(1 + k_\nu)$ *time all points in the subtree rooted at* ν *whose x-coordinate is at most* q_x, *where* k_ν *is the number of reported points.*

Proof. Let $p(\mu)$ be a point with $(p(\mu))_x \leqslant q_x$ that is stored at a node μ in the subtree rooted at ν. By definition of the data structure, the x-coordinates of the points stored at the path from μ to ν form a decreasing sequence, so all these points must have x-coordinates at most q_x. Hence, the search is not aborted at any of these nodes, which implies that μ is reached and $p(\mu)$ is reported. We conclude that all points with x-coordinate at most q_x are reported. Obviously, those are the only reported points.

At any node μ that we visit we spend $O(1)$ time. When we visit a node μ with $\mu \neq \nu$ we must have reported a point at the parent of μ. We charge the time we spend at μ to this point. This way any reported point gets charged twice, which means that the time we spend at nodes μ with $\mu \neq \nu$ is $O(k_\nu)$. Adding the time we spend at ν we get a total of $O(1 + k_\nu)$ time. $\quad\square$

If we call REPORTINSUBTREE at each of the subtrees that we select (the shaded subtrees of Figure 10.7), do we find all the points that lie in the query range? The answer is no. The root of the tree, for example, stores the point with smallest x-coordinate. This may very well be a point in the query range. In fact, any point stored at a node on the search path to q_y or q'_y may lie in the query range, so we should test them as well. This leads to the following query algorithm.

Algorithm QUERYPRIOSEARCHTREE$(\mathcal{T}, (-\infty : q_x] \times [q_y : q'_y])$
Input. A priority search tree and a range, unbounded to the left.
Output. All points lying in the range.
1. Search with q_y and q'_y in $\mathcal{T}$. Let v_{split} be the node where the two search paths split.
2. **for** each node v on the search path of q_y or q'_y
3.   **do if** $p(v) \in (-\infty : q_x] \times [q_y : q'_y]$ **then** report $p(v)$.
4. **for** each node v on the path of q_y in the left subtree of v_{split}
5.   **do if** the search path goes left at v
6.   **then** REPORTINSUBTREE$(rc(v), q_x)$
7. **for** each node v on the path of q'_y in the right subtree of v_{split}
8.   **do if** the search path goes right at v
9.   **then** REPORTINSUBTREE$(lc(v), q_x)$

Lemma 10.8 *Algorithm* QUERYPRIOSEARCHTREE *reports the points in a query range* $(-\infty : q_x] \times [q_y : q'_y]$ *in* $O(\log n + k)$ *time, where k is the number of reported points.*

Proof. First we prove that any point that is reported by the algorithm lies in the query range. For the points on the search paths to q_y and q'_y this is obvious: these points are tested explicitly for containment in the range. Consider a call REPORTINSUBTREE$(rc(v), q_x)$ in line 6. Let p be a point that is reported in this call. By Lemma 10.7 we have $p_x \leqslant q_x$. Furthermore, $p_y \leqslant q'_y$, because all the nodes visited in this call lie to the left of v_{split} and $q'_y > y(v_{\text{split}})$. Finally, $p_y \geqslant q_y$, because all the nodes visited in this call lie to the right of v and the search path to q_y went left at v. A similar argument applies for the points reported in line 9.

We have proved that all the reported points lie in the query range. Conversely, let $p(\mu)$ be a point in the range. Any point stored to the left of a node where the search path to q_y goes right must have a y-coordinate smaller than q_y. Similarly, any point stored to the right of a node where the search path to q'_y goes left must have a y-coordinate greater than q'_y. Hence, μ must either be on one of the search paths, or in one of the subtrees for which REPORTINSUBTREE is called. In both cases $p(\mu)$ will be reported.

It remains to analyze the time taken by the algorithm. This is linear in the number of nodes on the search paths to q_y and q'_y plus the time taken by all the executions of the procedure REPORTINSUBTREE. The depth of the tree and, hence, the number of nodes on the search paths, is $O(\log n)$. The time taken by all executions of REPORTINSUBTREE is $O(\log n + k)$ by Lemma 10.7. $\qquad\boxdot$

The performance of priority search trees is summarized in the following theorem.

Theorem 10.9 *A priority search tree for a set P of n points in the plane uses $O(n)$ storage and can be built in $O(n \log n)$ time. Using the priority search tree we can report all points in a query range of the form $(-\infty : q_x] \times [q_y : q'_y]$ in $O(\log n + k)$ time, where k is the number of reported points.*

10.3 Segment Trees

So far we have studied the windowing problem for a set of axis-parallel line segments. We developed a nice data structure for this problem using interval trees with priority search trees as associated structure. The restriction to axis-parallel segments was inspired by the application to printed circuit board design. When we are doing windowing queries in roadmaps, however, we must drop this restriction: roadmaps contain line segments at arbitrary orientation.

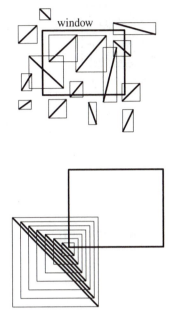

There is a trick that we can use to reduce the general problem to a problem on axis-parallel segments. We can replace each segment by its *bounding box*. Using the data structure for axis-parallel segments that we developed earlier, we can now find all the bounding boxes that intersect the query window W. We then check every segment whose bounding box intersects W to see if the segment itself also intersects W. In practice this technique usually works quite well: the majority of the segments whose bounding box intersects W will also intersect W themselves. In the worst case, however, the solution is quite bad: all bounding boxes may intersect W whereas none of the segments do. So if we want to guarantee a fast query time, we must look for another method.

As before we make a distinction between segments that have an endpoint in the window and segments that intersect the window boundary. The first type of segments can be reported using a range tree. To find the answers of the second type we perform an intersection query with each of the four boundary edges of the window. (Of course care has to be taken that answers are reported only once.) We will only show how to perform queries with vertical boundary edges. For the horizontal edges a similar approach can be used. So we are given a set S of line segments with arbitrary orientations in the plane, and we want to find those segments in S that intersect a vertical query segment $q := q_x \times [q_y : q'_y]$. We will assume that the segments in S don't intersect each other, but we allow them to touch. (For intersecting segments the problem is a lot harder to solve and the time bounds are worse. Techniques like the ones described in Chapter 16 are required in this case.)

Let's first see if we can adapt the solution of the previous sections to the case of arbitrarily oriented segments. By searching with q_x in the interval tree we select a number of subsets $I_{\text{mid}}(v)$. For a selected node v with $x_{\text{mid}}(v) > q_x$, the right endpoint of any segment in $I_{\text{mid}}(v)$ lies to the right of q. If the segment is horizontal, then it is intersected by the query segment if and only if its left endpoint lies in the range $(-\infty : q_x] \times [q_y : q'_y]$. If the segments have arbitrary orientation, however, things are not so simple: knowing that the right endpoint of a segment is to the right of q doesn't help us much. The interval tree is therefore not very useful in this case. Let's try to design a different data structure for

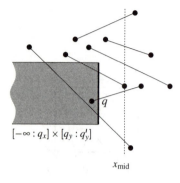

$[-\infty : q_x] \times [q_y : q_y']$

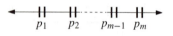

x_{mid}

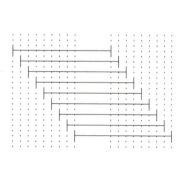

the 1-dimensional problem, one that is more suited for dealing with arbitrarily oriented segments.

One of the paradigms for developing data is structures is the *locus approach*. A query is described by a number of parameters; for the windowing problem, for example, there are four parameters, namely q_x, q_x', q_y, and q_y'. For each choice of the parameters we get a certain answer. Often nearby choices give the same answer; if we move the window slightly it will often still intersect the same collection of segments. Let the *parameter space* be the space of all possible choices for the parameters. For the windowing problem this space is 4-dimensional. The locus approach suggests partitioning the parameter space into regions such that queries in the same region have the same answer. Hence, if we locate the region that contains the query then we know the answer to it. Such an approach only works well when the number of regions is small. For the windowing problem this is not true. There can be $\Theta(n^4)$ different regions. But we can use the locus approach to create an alternative for the interval tree.

Let $I := \{[x_1 : x_1'], [x_2 : x_2'], \ldots, [x_n : x_n']\}$ be a set of n intervals on the real line. The data structure that we are looking for should be able to report the intervals containing a query point q_x. Our query has only one parameter, q_x, so the parameter space is the real line. Let $p_1, p_2, \ldots, p_m$ be the list of distinct interval endpoints, sorted from left to right. The partitioning of the parameter space is simply the partitioning of the real line induced by the points p_i. We call the regions in this partitioning *elementary intervals*. Thus the elementary intervals are, from left to right,

$$(-\infty : p_1), [p_1 : p_1], (p_1 : p_2), [p_2 : p_2], \ldots,$$

$$(p_{m-1} : p_m), [p_m : p_m], (p_m : +\infty).$$

The list of elementary intervals consists of open intervals between two consecutive endpoints p_i and p_{i+1}, alternated with closed intervals consisting of a single endpoint. The reason that we treat the points p_i themselves as intervals is, of course, that the answer to a query is not necessarily the same at the interior of an elementary interval and at its endpoints.

To find the intervals that contain a query point q_x, we must determine the elementary interval that contains q_x. To this end we build a binary search tree $\mathcal{T}$ whose leaves correspond to the elementary intervals. We denote elementary interval corresponding to a leaf μ by $\text{Int}(\mu)$.

If all the intervals in I containing $\text{Int}(\mu)$ would be stored at the leaf μ, then we could report the k intervals containing q_x in $O(\log n + k)$ time: first search in $O(\log n)$ time with q_x in $\mathcal{T}$, and then report all the intervals stored at μ in $O(1 + k)$ time. So queries can be answered efficiently. But what about the storage requirements of the data structure? Intervals that span a lot of elementary intervals are stored at many leaves in the data structure. Hence, the amount of storage will be high if there are many pairs of overlapping intervals. If we have bad luck the amount of storage can even become quadratic. Let's see if we can do something to reduce the amount of storage. In Figure 10.8 you

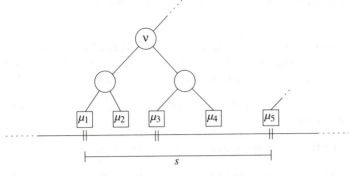

Figure 10.8
The segment s is stored at ν instead of at μ_1, μ_2, μ_3, and μ_4.

see an interval that spans five elementary intervals. Consider the elementary intervals corresponding to the leaves μ_1, μ_2, μ_3, and μ_4. When the search path to q_x ends in one of those leaves we must report the interval. The crucial observation is that a search path ends in μ_1, μ_2, μ_3, or μ_4 if and only if the path passes through the internal node ν. So why not store the interval at node ν (and at μ_5) instead of at the leaves μ_1, μ_2, μ_3, and μ_4 (and at μ_5)? In general, we store an interval at a number of nodes that together cover the interval, and we choose these nodes as high as possible. The data structure based on this principle is called a *segment tree*. We now describe the segment tree for a set I of intervals more precisely. Figure 10.9 shows a segment tree for a set of five intervals.

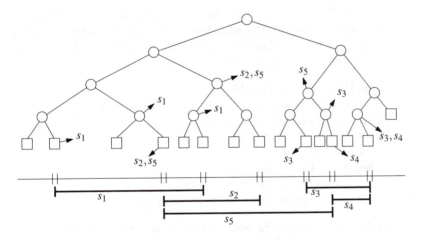

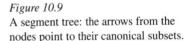

Figure 10.9
A segment tree: the arrows from the nodes point to their canonical subsets.

■ The skeleton of the segment tree is a balanced binary tree $\mathcal{T}$. The leaves of $\mathcal{T}$ correspond to the elementary intervals induced by the endpoints of the intervals in I in an ordered way: the leftmost leaf corresponds to the leftmost elementary interval, and so on. The elementary interval corresponding to leaf μ is denoted $\mathrm{Int}(\mu)$.

■ The internal nodes nodes of $\mathcal{T}$ correspond to intervals that are the union of elementary intervals: the interval $\mathrm{Int}(\nu)$ corresponding to node ν is the union of the elementary intervals $\mathrm{Int}(\mu)$ of the leaves in the subtree rooted

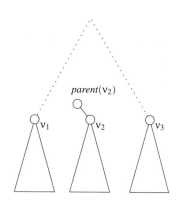

at v. (This implies that $Int(v)$ is the union of the intervals of its two children.)

■ Each node or leaf v in $\mathcal{T}$ stores the interval $Int(v)$ and a set $I(v) \subseteq I$ of intervals (for example, in a linked list). This *canonical subset* of node v contains the intervals $[x : x'] \in I$ such that $Int(v) \subseteq [x : x']$ and $Int(parent(v)) \not\subseteq [x : x']$.

Let's see if our strategy of storing intervals as high as possible has helped to reduce the amount of storage.

Lemma 10.10 *A segment tree on a set of n intervals uses $O(n \log n)$ storage.*

Proof. Because $\mathcal{T}$ is a balanced binary search tree with at most $4n + 1$ leaves, its height is $O(\log n)$. We claim that any interval $[x : x'] \in I$ is stored in the set $I(v)$ for at most two nodes at the same depth of the tree. To see why this is true, let v_1, v_2, v_3 be three nodes at the same depth, numbered from left to right. Suppose $[x : x']$ is stored at v_1 and v_3. This means that $[x : x']$ spans the whole interval from the left endpoint of $Int(v_1)$ to the right endpoint of $Int(v_3)$. Because v_2 lies between v_1 and v_3, $Int(parent(v_2))$ must be contained in $[x : x']$. Hence, $[x : x']$ will not be stored at v_2. It follows that any interval is stored at most twice at a given depth of the tree, so the total amount of storage is $O(n \log n)$. $\quad\square$

So the strategy has helped: we have reduced the worst-case amount of storage from quadratic to $O(n \log n)$. But what about queries: can they still be answered easily? The answer is yes. The following simple algorithm describes how this is done. It is first called with $v = root(\mathcal{T})$.

Algorithm QUERYSEGMENTTREE(v, q_x)
Input. The root of a (subtree of a) segment tree and a query point q_x.
Output. All intervals in the tree containing q_x.
1.   Report all the intervals in $I(v)$.
2.   **if** v is not a leaf
3.   **then if** $q_x \in Int(lc(v))$
4.   **then** QUERYSEGMENTTREE($lc(v), q_x$)
5.   **else** QUERYSEGMENTTREE($rc(v), q_x$)

The query algorithm visits one node per level of the tree, so $O(\log n)$ nodes in total. At a node v we spend $O(1 + k_v)$ time, where k_v is the number of reported intervals. This leads to the following lemma.

Lemma 10.11 *Using a segment tree, the intervals containing a query point q_x can be reported in $O(\log n + k)$ time, where k is the number of reported intervals.*

To construct a segment tree we proceed as follows. First we sort the endpoints of the intervals in I in $O(n \log n)$ time. This gives us the elementary intervals. We then construct a balanced binary tree on the elementary intervals, and we determine for each node v of the tree the interval $Int(v)$ it represents. This can be done bottom-up in linear time. It remains to compute the canonical subsets for the nodes. To this end we insert the intervals one by one into the segment tree. An interval is inserted into $\mathcal{T}$ by calling the following procedure with $v = root(\mathcal{T})$.

Algorithm INSERTSEGMENTTREE$(v, [x : x'])$
Input. The root of a (subtree of a) segment tree and an interval.
Output. The interval will be stored in the subtree.
1.   **if** Int$(v) \subseteq [x : x']$
2.   **then** store $[x : x']$ at v
3.   **else if** Int$(lc(v)) \cap [x : x'] \neq \emptyset$
4.   **then** INSERTSEGMENTTREE$(lc(v), q_x)$
5.   **if** Int$(rc(v)) \cap [x : x'] \neq \emptyset$
6.   **then** INSERTSEGMENTTREE$(rc(v), q_x)$

How much time does it take to insert an interval $[x : x']$ into the segment tree? At every node that we visit we spend constant time (assuming we store $I(v)$ in a simple structure like a linked list). When we visit a node v, we either store $[x : x']$ at v, or Int(v) contains an endpoint of $[x : x']$. We have already seen that an interval is stored at most twice at each level of $\mathcal{T}$. There is also at most one node at every level whose corresponding interval contains x and one node whose interval contains x'. So we visit at most 4 nodes per level. Hence, the time to insert a single interval is $O(\log n)$, and the total time to construct the segment tree is $O(n \log n)$.

The performance of segment trees is summarized in the following theorem.

Theorem 10.12 *A segment tree for a set I of n intervals uses $O(n \log n)$ storage and can be built in $O(n \log n)$ time. Using the segment tree we can report all intervals that contain a query point in $O(\log n + k)$ time, where k is the number of reported intervals.*

Recall that an interval tree uses only linear storage, and that it also allows us to report the intervals containing a query point in $O(\log n + k)$ time. So for this task an interval tree is to be preferred to a segment tree. When we want to answer more complicated queries, such as windowing in a set of line segments, then a segment tree is a more powerful structure to work with. The reason is that the set of intervals containing q_x is *exactly* the union of the canonical subsets that we select when we search in the segment tree. In an interval tree, on the other hand, we also select $O(\log n)$ nodes during a query, but not all intervals stored at those nodes contain the query point. We still have to walk along the sorted list to find the intersected intervals. So for segment trees, we have the possibility of storing the canonical subsets in an associated structure that allows for further querying.

Let's go back to the windowing problem. Let S be a set of arbitrarily oriented, disjoint segments in the plane. We want to report the segments intersecting a vertical query segment $q := q_x \times [q_y : q_y']$. Let's see what happens when we build a segment tree $\mathcal{T}$ on the x-intervals of the segments in S. A node v in $\mathcal{T}$ can now be considered to correspond to the vertical slab $\text{Int}(v) \times (-\infty : +\infty)$. A segment is in the canonical subset of v if it completely crosses the slab corresponding to v—we say that the segment *spans* the slab—but not the slab corresponding to the parent of v. We denote these subsets with $S(v)$. See Figure 10.10 for an illustration.

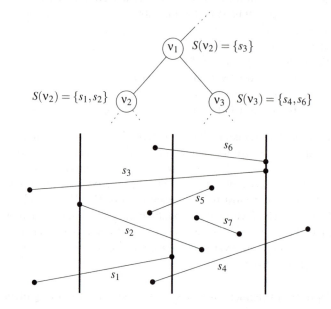

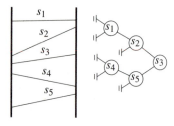

Figure 10.10
Canonical subsets contain segments
that span the slab of a node, but not the
slab of its parent.

When we search with q_x in $\mathcal{T}$ we find $O(\log n)$ canonical subsets—those of the nodes on the search path—that collectively contain all the segments whose x-interval contains q_x. A segment s in such a canonical subset is intersected by q if and only if the lower endpoint of q is below s and the upper endpoint of q is above s. How do we find the segments between the endpoints of q? Here we use the fact that the segments in the canonical subset $S(v)$ span the slab corresponding to v and that they do not intersect each other. This implies that the segments can be ordered vertically. Hence, we can store $S(v)$ in a search tree $\mathcal{T}(v)$ according to the vertical order. By searching in $\mathcal{T}(v)$ we can find the intersected segments in $O(\log n + k_v)$ time, where k_v is the number of intersected segments. The total data structure for the set S is thus as follows.

- The set S is stored in a segment tree $\mathcal{T}$ based on the x-intervals of the segments.

- The canonical subset of a node v in $\mathcal{T}$, which contains the segments spanning the slab corresponding to v but not the slab corresponding to the parent of v, is stored in a binary search tree $\mathcal{T}(v)$ based on the vertical order of the segments within the slab.

Because the associated structure of any node v uses storage linear in the size of $S(v)$, the total amount of storage remains $O(n \log n)$. The associated structures can be built in $O(n \log n)$ time, leading to a preprocessing time of $O(n \log^2 n)$. With a bit of extra work this can be improved to $O(n \log n)$. The idea is to maintain a (partial) vertical order on the segments while building the segment tree. With this order available the associated structures can be computed in linear time.

The query algorithm is quite simple: we search with q_x in the segment tree in the usual way, and at every node v on the search path we search with the upper and lower endpoint of q in $\mathcal{T}(v)$ to report the segments in $S(v)$ intersected by q. This basically is a 1-dimensional range query—see Section 5.1. The search in $\mathcal{T}(v)$ takes $O(\log n + k_v)$ time, where k_v is the number of reported segments at v. Hence, the total query time is $O(\log^2 n + k)$, and we obtain the following theorem.

Theorem 10.13 *Let S be a set of n disjoint segments in the plane. The segments intersecting a vertical query segment can be reported in $O(\log^2 n + k)$ time with a data structure that uses $O(n \log n)$ storage, where k is the number of reported segments. The structure can be built in $O(n \log n)$ time.*

Actually, it is only required that the segments have disjoint interiors. It is easily verified that the same approach can be used when the endpoints of segments are allowed to coincide with other endpoints or segments. This leads to the following result.

Corollary 10.14 *Let S be a set of n segments in the plane with disjoint interiors. The segments intersecting an axis-parallel rectangular query window can be reported in $O(\log^2 n + k)$ time with a data structure that uses $O(n \log n)$ storage, where k is the number of reported segments. The structure can be built in $O(n \log n)$ time.*

10.4 Notes and Comments

The query that asks for all intervals that contain a given point is often referred to as a *stabbing query*. The interval tree structure for stabbing queries is due to Edelsbrunner [128] and McCreight [235]. The priority search tree was designed by McCreight [236]. He observed that the priority search tree can be used for stabbing queries as well. The transformation is simple: map each interval $[a : b]$ to the point (a, b) in the plane. Performing a stabbing query with a value q_x can be done by doing a query with the range $(-\infty : q_x] \times [q_x : +\infty)$. Ranges of this type are a special case of the ones supported by priority search trees.

(a,b)

(q_x, q_x)

The segment tree was discovered by Bentley [33]. Used as a 1-dimensional data structure for stabbing queries it is less efficient than the interval tree since it requires $O(n \log n)$ storage. The importance of the segment tree is mainly that the sets of intervals stored with the nodes can be structured in any manner convenient for the problem at hand. Therefore, there are many extensions of segment trees that deal with 2- and higher-dimensional objects [82, 128, 134,

265, 325]. A second plus of the segment tree over the interval tree is that the segment tree can easily be adapted to stabbing *counting* queries: report the number of intervals containing the query point. Instead of storing the intervals in a list with the nodes, we store an integer representing their number. A query with a point is answered by adding the integers on one search path. Such a segment tree for stabbing counting queries uses only linear storage and queries require $O(\log n)$ time, so it is optimal.

There has been a lot of research in the past on the dynamization of interval trees and segment trees, that is, making it possible to insert and/or delete intervals. Priority search trees were initially described as a fully dynamic data structure, by replacing the binary tree with a red-black tree [167, 115] or other balanced binary search tree that requires only $O(1)$ rotations per update. Dynamization is important in situations where the input changes. Dynamic data structures are also important in many plane sweep algorithms where the status normally needs to be stored in a dynamic structure. In a number of problems a dynamic version of the segment tree is required here.

The notion of *decomposable searching problems* [34, 36, 137, 219, 234, 242, 268, 270, 271, 272, 299] gave a great push toward the dynamization of a large class of data structures. Let S be a set of objects involved in a searching problem, and let $A \cup B$ be any partition of S. The searching problem is *decomposable* if the solution to the searching problem on S can be obtained in constant time from the solutions to the searching problems on A and on B. The stabbing query problem is decomposable because if we have reported the stabbed intervals in the two subsets of a partition, we have automatically reported all intervals in the whole set. Similarly, the stabbing counting query problem is decomposable because we can add the integer answers yielded by the subsets to give the answer for the whole set. Many other searching problems, like the range searching problem, are decomposable as well. Static data structures for decomposable searching problems can be turned into dynamic structures using general techniques. For an overview see the book by Overmars [263].

The stabbing query problem can be generalized to higher dimensions. Here we are given a collection of axis-parallel (hyper-)rectangles and ask for those rectangles that contain a given query point. To solve these higher-dimensional stabbing queries we can use multi-level segment trees. The structure uses $O(n \log^{d-1} n)$ storage and stabbing queries can be answered in $O(\log^d n)$ time. The use of fractional cascading—see Chapter 5—lowers the query time bound by a logarithmic factor. The use of the interval tree on the deepest level of associated structures lowers the storage bound with a logarithmic factor. A higher-dimensional version of the interval tree and priority search tree doesn't exist, that is, there is no clear extension of the structure that solves the analogous query problem in higher dimensions. But the structures can be used as associated structure of segment trees and range trees. This is useful, for instance, to solve the stabbing query problem in sets of axis-parallel rectangles, and for range searching with ranges of the form $[x : x'] \times [y : y'] \times [z : +\infty)$.

10.5 Exercises

10.1 In Section 10.1 we solved the problem of finding all horizontal line segments in a set that intersect a vertical segment. For this we used an interval tree with priority search trees as associated structures. There is also an alternative approach. We can use a 1-dimensional range tree on the y-coordinate of the segments to determine those segments whose y-coordinate lies in the vertical query segment. The resulting segments cannot lie above or below the query segment, but they may lie completely to the left or to the right of it. We get those segments in $O(\log n)$ canonical subsets. For each of these subsets we use as an associated structure an interval tree on the x-coordinates to find the segments that actually intersect the query segment.

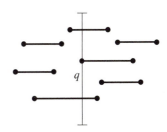

   a. Work out the details of this approach.
   b. Prove that the data structure correctly solves the queries.
   c. What are the bounds for preprocessing time, storage required, and query time of this structure? Prove your answers.

10.2 Let P be a set of n points in the plane, sorted on y-coordinate. Show that, because P is sorted, a priority search tree of the points in P can be constructed in $O(n)$ time.

10.3 In the description of the algorithms for priority search trees we assumed that all points have distinct coordinates. It was indicated that this restriction can be removed by using composite numbers as described in Section 5.5. Show that all basic operations required in building and querying priority search trees can indeed be performed with composite numbers.

10.4 Windowing queries in sets of non-intersecting segments are performed using a range query on the set of endpoints and intersection queries with the four boundaries of the window. Explain how to avoid that segments are reported more than once. To this end, make a list of all possible ways in which an arbitrarily oriented segment can intersect a query window.

Suppose that the segments are already sorted vertically. Describe a linear time algorithm for constructing the associated structure.

10.5 In this exercise you must show how the data structure of Theorem 10.13 can be built in $O(n \log n)$ time. The associated structures that were used in the solution are binary search trees on the vertical order of the segments. If the vertical order is given, then an associated structure can be built in linear time. So it remains to compute a sorted list of segments for each canonical subset in $O(n \log n)$ time in total.

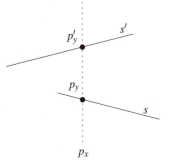

Let S be a set of n disjoint segments in the plane. For two segments $s, s' \in S$, we say that s lies below s', denoted $s \prec s'$, if and only if there are points $p \in s$, $p' \in s'$ with $p_x = p'_x$ and $p_y < p'_y$.

   a. Prove that the relation $\prec$ defines an acyclic relation on S. In other words, you must prove that an order $s_1, s_2, \ldots, s_n$ for the segments in S exists such that $i > j$ implies $s_i \nprec s_j$.

b. Describe an $O(n \log n)$ algorithm to compute such an order. *Hint:* Use plane sweep to find all segments that are vertically adjacent, construct a directed graph on these adjacencies, and apply topological sorting on the graph.

c. Explain how to obtain the sorted lists for the canonical subsets in the segment tree using this acyclic order.

10.6 Let I be a set of intervals on the real line. We want to be able to count the number of intervals containing a query point in $O(\log n)$ time. Thus, the query time must be independent of the number of segments containing the query point.

a. Describe a data structure for this problem based on segment trees, that uses only $O(n)$ storage. Analyze the amount of storage, preprocessing time, and the query time of the data structure.

b. Describe a data structure for this problem based on interval trees. You should replace the lists associated with the nodes of the interval tree with other structures. Analyze the amount of storage, preprocessing time, and the query time of the data structure.

10.7 We want to solve the following query problem: Given a set S of n disjoint line segments in the plane, determine those segments that intersect a vertical ray, running from a point (q_x, q_y) vertically upwards to infinity. Describe a data structure for this problem that uses $O(n \log n)$ storage and has a query time of $O(\log n + k)$, where k is the number of reported answers.

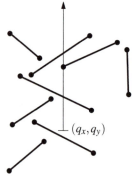

10.8 Segment trees can be used for multi-level data structures.

a. Let R be a set of n axis-parallel rectangles in the plane. Design a data structure for R such that the rectangles containing a query point q can be reported efficiently. Analyze the amount of storage and the query time of your data structure. *Hint:* Use a segment tree on the x-intervals of the rectangles, and store canonical subsets of the nodes in this segment tree in an appropriate associated structure.

b. Generalize this data structure to d-dimensional space. Here we are given a set of axis-parallel hyperrectangles—that is, polytopes of the form $[x_1 : x_1'] \times [x_2 : x_2'] \times \cdots \times [x_d : x_d']$—and we want to report the hyperrectangles containing a query point.

10.9 Let I be a set of intervals on the real line. We want to store these intervals such that we can efficiently determine those intervals that are completely contained in a given interval $[x : x']$. Describe a data structure that uses $O(n \log n)$ storage and solves such queries in $O(\log n + k)$ time, where k is the number of answers. *Hint:* Use a range tree.

10.10 Again we have a collection I of intervals on the real line, but this time we want to efficiently determine those intervals that contain a given interval $[x : x']$. Describe a data structure that uses $O(n)$ storage and solves such queries in $O(\log n + k)$ time, where k is the number of answers. *Hint:* Use a priority search tree.

10.11 Consider the following alternative approach to solve the 2-dimensional range searching problem: We construct a balanced binary search tree on the x-coordinate of the points. For a node v in the tree, let $P(v)$ be the set of points stored in the subtree rooted at v. For each node v we store two associated priority search trees of $P(v)$, a tree $\mathcal{T}_{\text{left}}$ allowing for range queries that are unbounded to the left, and a tree $\mathcal{T}_{\text{right}}$ for range queries that are unbounded to the right.

A query with a range $[x : x'] \times [y : y']$ is performed as follows. We search for the node v_{split} where the search paths toward x and x' split in the tree. Now we perform a query with range $[x : +\infty) \times [y : y']$ on $\mathcal{T}_{\text{right}}(lc(v))$ and a query with range $(-\infty : x'] \times [y : y']$ on $\mathcal{T}_{\text{left}}(rc(v))$. This gives all the answers (there is no need to search further down in the tree!).

a. Work out the details of this approach.
b. Prove that the data structure correctly solves range queries.
c. What are the bounds for preprocessing time, storage required, and query time of this structure? Prove your answers.

11 Convex Hulls

Mixing Things

The output of oil wells is a mixture of several different components, and the proportions of these components vary between different sources. This can sometimes be exploited: by mixing together the output of different wells, one can produce a mixture with proportions that are particularly favorable for the refining process.

Let's look at an example. For simplicity we assume that we are only interested in two of the components—call them A and B—of our product. Assume that we are given a mixture ξ_1 with 10% of component A and 35% of component B, and another mixture ξ_2 with 16% of A and 20% of B. Assume further that what we really need is a mixture that contains 12% of A and 30% of B. Can we produce this mixture from the given ones? Yes, mixing ξ_1 and ξ_2 in the ratio 2 : 1 gives the desired product. However, it is impossible to make a mixture of ξ_1 and ξ_2 that contains 13% of A and 22% of B. But if we have a third mixture ξ_3 containing 7% of A and 15% of B, then mixing ξ_1, ξ_2, and ξ_3 in the ratio of 1 : 3 : 1 will give the desired result.

What has all this to do with geometry? This becomes clear when we represent the mixtures ξ_1, ξ_2, and ξ_3 by points in the plane, namely by $p_1 := (0.1, 0.35)$, $p_2 := (0.16, 0.2)$, and $p_3 := (0.07, 0.15)$. Mixing ξ_1 and ξ_2 in the ratio 2 : 1 gives the mixture represented by the point $q := (2/3)p_1 + (1/3)p_2$. This is the point on the segment $\overline{p_1 p_2}$ such that $\operatorname{dist}(p_2, q) : \operatorname{dist}(q, p_1) = 2 : 1$, where $\operatorname{dist}(.,.)$ denotes the distance between two points. More generally, by mixing ξ_1 and ξ_2 in varying ratios, we can produce the mixtures represented by any point on the line segment $\overline{p_1 p_2}$. If we start with the three base mixtures ξ_1, ξ_2, and ξ_3, we can produce any point in the triangle $p_1 p_2 p_3$. For instance, mixing ξ_1, ξ_2, and ξ_3 in the ratio 1 : 3 : 1 gives the mixture represented by the point $(1/5)p_1 + (3/5)p_2 + (1/5)p_3 = (0.13, 0.22)$.

What happens if we don't have three but n base mixtures, for some $n > 3$, represented by points $p_1, p_2, \ldots, p_n$? Suppose that we mix them in the ratio $l_1 : l_2 : \cdots : l_n$. Let $L := \sum_{j=1}^{n} l_j$ and let $\lambda_i := l_i / L$. Note that

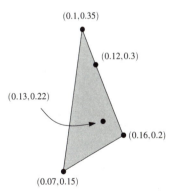

$$\lambda_i \geqslant 0 \text{ for all } i \quad \text{and} \quad \sum_{i=1}^{n} \lambda_i = 1.$$

The mixture we get by mixing the base mixtures in the given ratio is the one represented by

$$\sum_{i=1}^{n} \lambda_i p_i.$$

Such a linear combination of the points p_i where the λ_i satisfy the conditions stated above—each λ_i is non-negative, and the sum of the λ_i is one—is called a *convex combination*. In Chapter 1 we defined the *convex hull* of a set of points as the smallest convex set containing the points or, more precisely, as the intersection of all convex sets containing the points. One can show that the convex hull of a set of points is exactly the set of all possible convex combinations of the points. We can therefore test whether a mixture can be obtained from the base mixtures by computing the convex hull of their representative points, and checking whether the point representing the mixture lies inside it.

What if there are more than two interesting components in the mixtures? Well, what we have said above remains true; we just have to move to a space of higher dimension. More precisely, if we want to take d components into account we have to represent a mixture by a point in d-dimensional space. The convex hull of the points representing the base mixtures, which is a convex polytope, represents the set of all possible mixtures.

Convex hulls—in particular convex hulls in 3-dimensional space—are used in various applications. For instance, they are used to speed up collision detection in computer animation. Suppose that we want to check whether two objects $\mathcal{P}_1$ and $\mathcal{P}_2$ intersect. If the answer to this question is negative most of the time, then the following strategy pays off. Approximate the objects by simpler objects $\widehat{\mathcal{P}_1}$ and $\widehat{\mathcal{P}_2}$ that contain the originals. If we want to check whether $\mathcal{P}_1$ and $\mathcal{P}_2$ intersect, we first check whether $\widehat{\mathcal{P}_1}$ and $\widehat{\mathcal{P}_2}$ intersect; only if this is the case do we need to perform the—supposedly more costly—test on the original objects.

There is a trade-off in the choice of the approximating objects. On the one hand, we want them to be simple so that intersection tests are cheap. On the other hand, simple approximations most likely do not approximate the original objects very well, so there is a bigger chance we have to test the originals. Bounding spheres are on one side of the spectrum: intersection tests for spheres are quite simple, but for many objects spheres do not provide a good approximation. Convex hulls are more on the other side of the spectrum: intersection tests for convex hulls are more complicated than for spheres—but still simpler than for non-convex objects—but convex hulls can approximate most objects a lot better.

11.1 The Complexity of Convex Hulls in 3-Space

In Chapter 1 we have seen that the convex hull of a set P of n points in the plane is a convex polygon whose vertices are points in P. Hence, the convex hull has at most n vertices. In 3-dimensional space a similar statement is true:

the convex hull of a set P of n points is a convex polytope whose vertices are points in P and, hence, it has at most n vertices. In the planar case the bound on the number of vertices immediately implies that the complexity of the convex hull is linear, since the number of edges of a planar polygon is equal to the number of vertices. In 3-space this is no longer true; the number of edges of a polytope can be higher than the number of vertices. But fortunately the difference cannot be too large, as follows from the following theorem on the number of edges and facets of convex polytopes. (Formally, a *facet* of a convex polytope is defined to be a maximal subset of coplanar points on its boundary. A facet of a convex polytope is necessarily a convex polygon. An *edge* of a convex polytope is an edge of one of its facets.)

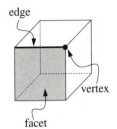

Theorem 11.1 *Let $\mathcal{P}$ be a convex polytope with n vertices. The number of edges of $\mathcal{P}$ is at most $3n - 6$, and the number of facets of $\mathcal{P}$ is at most $2n - 4$.*

Proof. Recall that Euler's formula states for a connected planar graph with n nodes, n_e arcs, and n_f faces the following relation holds:

$$n - n_e + n_f = 2.$$

Since we can interpret the boundary of a convex polytope as a planar graph—see Figure 11.1—the same relation holds for the numbers of vertices, edges, and facets in a convex polytope. (In fact, Euler's formula was originally stated

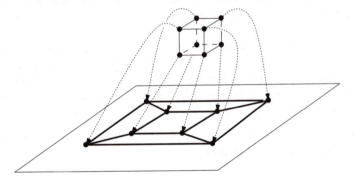

Figure 11.1
A cube interpreted as a planar graph: note that one facet maps to the unbounded face of the graph

in terms of polytopes, not in terms of planar graphs.) Every face of the graph corresponding to $\mathcal{P}$ has at least three arcs, and every arc is incident to two faces, so we have $2n_e \leqslant 3n_f$. Plugging this into Euler's formula we get

$$n + n_f - 2 \geqslant 3n_f/2,$$

so $n_f \leqslant 2n - 4$. This implies that $n_e \leqslant 3n - 6$. For the special case that every facet is a triangle—the case of a *simplicial polytope*—the bounds on the number of edges and facets of an n-vertex polytope are exact, because then $2n_e = 3n_f$. ▢

Theorem 11.1 also holds for non-convex polytopes whose so-called *genus* is zero, that is, polytopes without holes or tunnels; for polytopes of larger genus similar bounds hold. Since this chapter deals with convex hulls, however, we refrain from defining what a (non-convex) polytope exactly is, which we would need to do to prove the theorem in the non-convex case.

If we combine Theorem 11.1 with the earlier observation that the convex hull of a set of points in 3-space is a convex polytope whose vertices are points in P, we get the following result.

Corollary 11.2 *The complexity of the convex hull of a set of n points in 3-dimensional space is $O(n)$.*

11.2 Computing Convex Hulls in 3-Space

Let P be a set of n points in 3-space. We will compute $C\mathcal{H}(P)$, the convex hull of P, using a randomized incremental algorithm, following the paradigm we have met before in Chapters 4, 6, and 9.

The incremental construction starts by choosing four points in P that do not lie in a common plane, so that their convex hull is a tetrahedron. This can be done as follows. Let p_1 and p_2 be two points in P. We walk through the set P until we find a point p_3 that does not lie on the line through p_1 and p_2. We continue searching P until we find a point p_4 that does not lie in the plane through p_1, p_2, and p_3. (If we cannot find four such points, then all points in P lie in a plane. In this case we can use the planar convex hull algorithm of Chapter 1 to compute the convex hull.)

Next we compute a random permutation $p_5, \ldots, p_n$ of the remaining points. We will consider the points one by one in this random order, maintaining the convex hull as we go. For an integer $r \geqslant 1$, let $P_r := \{p_1, \ldots, p_r\}$. In a generic step of the algorithm, we have to add the point p_r to the convex hull of P_{r-1}, that is, we have to transform $C\mathcal{H}(P_{r-1})$ into $C\mathcal{H}(P_r)$. There are two cases.

- If p_r lies inside $C\mathcal{H}(P_{r-1})$, or on its boundary, then $C\mathcal{H}(P_r) = C\mathcal{H}(P_{r-1})$, and there is nothing to be done.

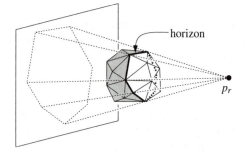

Figure 11.2
The horizon of a polytope

- Now suppose that p_r lies outside $C\mathcal{H}(P_{r-1})$. Imagine that you are standing at p_r, and that you are looking at $C\mathcal{H}(P_{r-1})$. You will be able to see some facets of $C\mathcal{H}(P_{r-1})$—the ones on the front side—but others will be invisible because they are on the back side. The visible facets form a connected region on the surface of $CH(P_{r-1})$, called the *visible region* of p_r on $CH(P_{r-1})$, which is enclosed by a closed curve consisting of edges of $CH(P_{r-1})$. We call this curve the *horizon* of p_r on $CH(P_{r-1})$. As you can see in Figure 11.2, the projection of the horizon is the boundary of the convex polygon obtained by projecting $CH(P_{r-1})$ onto a plane, with p_r as the center of projection. What exactly does "visible" mean geometrically? Consider the plane h_f containing a facet f of $CH(P_{r-1})$. By convexity, $CH(P_{r-1})$ is completely contained in one of the closed half-spaces defined by h_f. The face f is visible from a point if that point lies in the open half-space on the other side of h_f.

The horizon of p_r plays a crucial role when we want to transform $CH(P_{r-1})$ to $CH(P_r)$: it forms the border between the part of the boundary that can be kept—the invisible facets—and the part of the boundary that must be replaced—the visible facets. The visible facets must be replaced by facets connecting p_r to its horizon.

Before we go into more details, we should decide how we are going to represent the convex hull of points in space. As we observed before, the boundary of a 3-dimensional convex polytope can be interpreted as a planar graph. Therefore we store the convex hull in the form of a doubly-connected edge list, a data structure developed in Chapter 2 for storing planar subdivisions. The only difference is that vertices will now be 3-dimensional points. We will keep the convention that the half-edges are directed such that the ones bounding any face form a counterclockwise cycle *when seen from the outside* of the polytope.

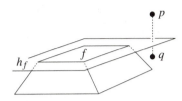
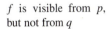

f is visible from p, but not from q

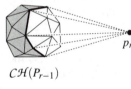

$CH(P_{r-1})$

p_r

$CH(P_r)$

p_r

Figure 11.3
Adding a point to the convex hull

Back to the addition of p_r to the convex hull. We have a doubly-connected edge list representing $CH(P_{r-1})$, which we have to transform to a doubly-connected edge list for $CH(P_r)$. Suppose that we knew all facets of $CH(P_{r-1})$ visible from p_r. Then it would be easy to remove all the information stored for these facets from the doubly-connected edge list, compute the new facets connecting p_r to the horizon, and store the information for the new facets in the doubly-connected edge list. All this will take linear time in the total complexity of the facets that disappear.

There is one subtlety we should take care of after the addition of the new facets: we have to check whether we have created any coplanar facets. This

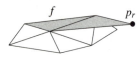

happens if p_r lies in the plane of a face of $C\mathcal{H}(P_{r-1})$. Such a face f is not visible from p_r by our definition of visibility above. Hence, f will remain unchanged, and we will add triangles connecting p_r to the edges of f that are part of the horizon. Those triangles are coplanar with f, and so they have to be merged with f into one facet.

In the discussion so far we have ignored the problem of finding the facets of $C\mathcal{H}(P_{r-1})$ that are visible to p_r. Of course this could be done by testing every facet. Since such a test takes constant time—we have to check to which side of a given plane the point p_r lies—we can find all visible facets in $O(r)$ time. This would lead to an $O(n^2)$ algorithm. Next we show how to do better.

The trick is that we are going to work ahead: besides the convex hull of the current point set we shall maintain some additional information, which will make it easy to find the visible facets. In particular, we maintain for each facet f of the current convex hull $C\mathcal{H}(P_r)$ a set $P_{\text{conflict}}(f) \subseteq \{p_{r+1}, p_{r+2}, \ldots, p_n\}$ containing the points that can see f. Conversely, we store for every point p_t, with $t > r$, the set $F_{\text{conflict}}(p_t)$ of facets of $C\mathcal{H}(P_r)$ visible from p_t. We will say that a point $p \in P_{\text{conflict}}(f)$ is *in conflict with* the facet f, because p and f cannot peacefully live together in the convex hull—once we add a point $p \in P_{\text{conflict}}(f)$ to the convex hull, the facet f must go. We call $P_{\text{conflict}}(f)$ and $F_{\text{conflict}}(p_t)$ *conflict lists*.

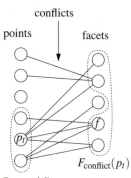

We maintain the conflicts in a so-called *conflict graph*, which we denote by $\mathcal{G}$. The conflict graph is a bipartite graph. It has one node set with a node for every point of P that has not been inserted yet, and one node set with a node for every facet of the current convex hull. There is an arc for every conflict between a point and a facet. In other words, there is an arc between a point $p_t \in P$ and facet f of $C\mathcal{H}(P_r)$ if $r < t$ and f is visible from p_t. Using the conflict graph $\mathcal{G}$, we can report the set $F_{\text{conflict}}(p_t)$ for a given point p_t (or $P_{\text{conflict}}(f)$ for a given facet f) in time linear in its size. This means that when we insert p_r into $C\mathcal{H}(P_{r-1})$, all we have to do is to look up $F_{\text{conflict}}(p_r)$ in $\mathcal{G}$ to get the visible facets, which we can then replace by the new convex hull facets connecting p_r to the horizon.

Initializing the conflict graph $\mathcal{G}$ for $C\mathcal{H}(P_4)$ can be done in linear time: we simply walk through the list of points P and determine which of the four faces of $C\mathcal{H}(P_4)$ they can see.

To update $\mathcal{G}$ after adding a point p_r, we first discard the nodes and incident arcs for all the facets of $C\mathcal{H}(P_{r-1})$ that disappear from the convex hull. These are the facets visible from p_r, which are exactly the neighbors of p_r in $\mathcal{G}$, so this is easy. We also discard the node for p_r. We then add nodes to $\mathcal{G}$ for the new facets we created, which connect p_r to the horizon. The essential step is to find the conflict lists of these new facets. No other conflicts have to be updated: the conflict set $P_{\text{conflict}}(f)$ of a facet f that is unaffected by the insertion of p_r remains unchanged.

The facets created by the insertion of p_r are all triangles, except for those that have been merged with existing coplanar facets. The conflict list of a facet of the latter type is trivial to find: it is the same as the conflict list of the existing facet, since the merging does not change the plane containing the facet. So let's look at one of the new triangles f incident to p_r in $CH(P_r)$. Suppose that a point p_t can see f. Then p_t can certainly see the edge e of f that is opposite p_r. This edge e is a horizon edge of p_r, and it was already present in $CH(P_{r-1})$. Since $CH(P_{r-1}) \subset CH(P_r)$, the edge e must have been visible from p_t in $CH(P_{r-1})$ as well. That can only be the case if one of the two facets incident to e in $CH(P_{r-1})$ is visible from p_t. This implies that the conflict list of f can be found by testing the points in the conflict lists of the two facets f_1 and f_2 that were incident to the horizon edge e in $CH(P_{r-1})$.

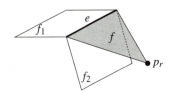

We stated earlier that we store the convex hull as a doubly-connected edge list, so changing the convex hull means changing the information in the doubly-connected edge list. To keep the code short, however, we have omitted all explicit references to the doubly-connected edge list in the pseudocode below, which summarizes the convex hull algorithm.

Algorithm CONVEXHULL(P)
Input. A set P of n points in three-space.
Output. The convex hull $CH(P)$ of P.
1.   Find four points p_1, p_2, p_3, p_4 in P that form a tetrahedron.
2. $C \leftarrow CH(\{p_1, p_2, p_3, p_4\})$
3.   Compute a random permutation $p_5, p_6, \ldots, p_n$ of the remaining points.
4.   Initialize the conflict graph G with all visible pairs (p_t, f), where f is a facet of C and $t > 4$.
5.   **for** $r \leftarrow 5$ **to** n
6.   **do** (* Insert p_r into C: *)
7.   **if** $F_{\text{conflict}}(p_r)$ is not empty (* that is, p_r lies outside C *)
8.   **then** Delete all facets in $F_{\text{conflict}}(p_r)$ from C.
9.   Walk along the boundary of the visible region of p_r (which consists exactly of the facets in $F_{\text{conflict}}(p_r)$) and create a list L of horizon edges in order.
10.   **for** all $e \in L$
11.   **do** Connect e to p_r by creating a triangular facet f.
12.   **if** f is coplanar with its neighbor facet f' along e
13.   **then** Merge f and f' into one facet, whose conflict list is the same as that of f'.
14.   **else** (* Determine conflicts for f: *)
15.   Create a node for f in G.
16.   Let f_1 and f_2 be the facets incident to e in the old convex hull.
17. $P(e) \leftarrow P_{\text{conflict}}(f_1) \cup P_{\text{conflict}}(f_2)$
18.   **for** all points $p \in P(e)$
19.   **do** If f is visible from p, add (p, f) to G.
20.   Delete the node corresponding to p_r and the nodes corresponding to the facets in $F_{\text{conflict}}(p_r)$ from G, together with their incident arcs.
21. **return** C

11.3* The Analysis

As usual when we analyse a randomized incremental algorithm, we first try to bound the expected structural change. For the convex hull algorithm this means we want to bound the total number of facets created by the algorithm.

Lemma 11.3 *The expected number of facets created by* CONVEXHULL *is at most* $6n - 20$.

Proof. The algorithm starts with a tetrahedron, which has four facets. In every stage r of the algorithm where p_r lies outside $\mathcal{CH}(P_{r-1})$, new triangular facets connecting p_r to its horizon on $\mathcal{CH}(P_{r-1})$ are created. What is the expected number of new facets? As in previous occasions where we analyzed randomized algorithms, we use backwards analysis. We look at $\mathcal{CH}(P_r)$ and imagine removing vertex p_r; the number of facets that disappear due to the removal of p_r from $\mathcal{CH}(P_r)$ is the same as the number of facets that were created due to the insertion of p_r into $\mathcal{CH}(P_{r-1})$. The disappearing facets are exactly the ones incident to p_r, and their number equals the number of edges incident to p_r in $\mathcal{CH}(P_r)$. We call this number the degree of p_r in $\mathcal{CH}(P_r)$, and we denote it by $\deg(p_r, \mathcal{CH}(P_r))$. We now want to bound the expected value of $\deg(p_r, \mathcal{CH}(P_r))$.

By Theorem 11.1 a convex polytope with r vertices has at most $3r - 6$ edges. This means that the sum of the degrees of the vertices of $\mathcal{CH}(P_r)$, which is a convex polytope with r or less vertices, is at most $6r - 12$. Hence, the average degree is bounded by $6 - 12/r$. Since we treat the vertices in random order, it seems that the expected degree of p_r is bounded by $6 - 12/r$. We have to be a little bit careful, though: the first four points are already fixed when we generate the random permutation, so p_r is a random element of $\{p_5, \ldots, p_r\}$, not of P_r. Because $p_1, \ldots, p_4$ have total degree at least 12, the expected value of $\deg(p_r, \mathcal{CH}(P_r))$ is bounded as follows:

$$
\begin{aligned}
\mathrm{E}[\deg(p_r, \mathcal{CH}(P_r))] &= \frac{1}{r-4} \sum_{i=5}^{r} \deg(p_i, \mathcal{CH}(P_r)) \\
&\leqslant \frac{1}{r-4} \left(\left\{ \sum_{i=1}^{r} \deg(p_i, \mathcal{CH}(P_r)) \right\} - 12 \right) \\
&\leqslant \frac{6r - 12 - 12}{r-4} = 6.
\end{aligned}
$$

The expected number of facets created by CONVEXHULL is the number of facets we start with (four) plus the expected total number of facets created during the additions of $p_5, \ldots, p_n$ to the hull. Hence, the expected number of created facets is

$$
4 + \sum_{r=5}^{n} \mathrm{E}[\deg(p_r, \mathcal{CH}(P_r))] \leqslant 4 + 6(n-4) = 6n - 20.
$$

$\square$

Now that we have bounded the total amount of structural change we can bound the expected running time of the algorithm.

Lemma 11.4 *Algorithm* CONVEXHULL *computes the convex hull of a set P of n points in $\mathbb{R}^3$ in $O(n \log n)$ expected time, where the expectation is with respect to the random permutation used by the algorithm.*

Proof. The steps before the main loop can certainly be done in $O(n \log n)$ time. Stage r of the algorithm takes constant time if $F_{\text{conflict}}(p_r)$ is empty, which is when p_r lies inside, or on the boundary of, the current convex hull.

If that is not the case, most of stage r takes $O(\text{card}(F_{\text{conflict}}(p_r)))$ time, where $\text{card}()$ denotes the cardinality of a set. The exceptions to this are the lines 17–19 and line 20. We shall bound the time spent in these lines later; first, we bound $\text{card}(F_{\text{conflict}}(p_r))$. Note that $\text{card}(F_{\text{conflict}}(p_r))$ is the number of facets deleted due to the addition of the point p_r. Clearly, a facet can only be deleted if it has been created before, and it is deleted at most once. Since the expected number of facets created by the algorithm is $O(n)$ by Lemma 11.3, this implies that the total number of deletions is $O(n)$ as well, so

$$\mathrm{E}[\sum_{r=5}^{n} \text{card}(F_{\text{conflict}}(p_r))] = O(n).$$

Now for lines 17–19 and line 20. Line 20 takes time linear in the number of nodes and arcs that are deleted from $\mathcal{G}$. Again, a node or arc is deleted at most once, and we can charge the cost of this deletion to the stage where we created it. It remains to look at lines 17–19. In stage r, these lines are executed for all horizon edges, that is, all edges in $\mathcal{L}$. For one edge $e \in \mathcal{L}$, they take $O(\text{card}(P(e)))$ time. Hence, the total time spent in these lines in stage r is $O(\sum_{e \in \mathcal{L}} \text{card}(P(e)))$. To bound the total expected running time, we therefore have to bound the expected value of

$$\sum_{e} \text{card}(P(e)),$$

where the summation is over all horizon edges that appear at any stage of the algorithm. We will prove below that this is $O(n \log n)$, which implies that the total running time is $O(n \log n)$. $\quad\square$

We use the framework of *configuration spaces* from Chapter 9 to supply the missing bound. The universe X is the set of P, and the configurations Δ correspond to convex hull edges. However, for technical reasons—in particular, to be able to deal correctly with degenerate cases—we attach a half-edge to both sides of the edge. To be more precise, a *flap* Δ is defined as an ordered four-tuple of points (p, q, s, t) that do not all lie in a plane. The defining set $D(\Delta)$ is simply the set $\{p, q, s, t\}$. The killing set $K(\Delta)$ is more difficult to visualize. Denote the line through p and q by ℓ. Given a point x, let $h(\ell, x)$ denote the half-plane bounded by ℓ that contains x. Given two points x, y, let $\rho(x, y)$ be the half-line starting in x and passing through y. A point $x \in X$ is in $K(\Delta)$ if and only if it lies in one of the following regions:

241

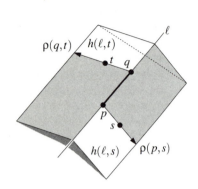

1. outside the closed convex 3-dimensional wedge defined by $h(\ell,s)$ and $h(\ell,t)$,

2. inside $h(\ell,s)$ but outside the closed 2-dimensional wedge defined by $\rho(p,q)$ and $\rho(p,s)$,

3. inside $h(\ell,t)$ but outside the closed 2-dimensional wedge defined by $\rho(q,t)$ and $\rho(q,p)$,

4. inside the line ℓ but outside the segment $\overline{pq}$,

5. inside the half-line $\rho(p,s)$ but outside the segment $\overline{ps}$,

6. inside the half-line $\rho(q,t)$ but outside the segment $\overline{qt}$.

For every subset $S \subseteq P$, we define the set $\mathcal{T}(S)$ of active configurations—this is what we want to compute—as prescribed in Chapter 9: $\Delta \in \mathcal{T}(S)$ if and only if $D(\Delta) \subseteq S$ and $K(\Delta) \cap S = \emptyset$.

Lemma 11.5 *A flap $\Delta = (p,q,s,t)$ is in $\mathcal{T}(S)$ if and only if $\overline{pq}$, $\overline{ps}$, and $\overline{qt}$ are edges of the convex hull $\mathcal{CH}(S)$, there is a facet f_1 incident to $\overline{pq}$ and $\overline{ps}$, and a different facet f_2 incident to $\overline{pq}$ and $\overline{qt}$. Furthermore, if one of the facets f_1 or f_2 is visible from a point $x \in P$ then $x \in K(\Delta)$.*

We leave the proof—which involves looking precisely at the cases when points are collinear or coplanar, but which is otherwise not difficult—to the reader.

As you may have guessed, the flaps take over the role of the horizon edges.

Lemma 11.6 *The expected value of $\sum_e \mathrm{card}(P(e))$, where the summation is over all horizon edges that appear at some stage of the algorithm, is $O(n \log n)$.*

Proof. Consider an edge e of the horizon of p_r on $\mathcal{CH}(P_{r-1})$. Let $\Delta = (p,q,s,t)$ be one of the two flaps with $\overline{pq} = e$. By Lemma 11.5, $\Delta \in \mathcal{T}(P_{r-1})$, and the points in $P \setminus P_r$ that can see one of the facets incident to e are all in $K(\Delta)$, so $P(e) \subseteq K(\Delta)$. By Theorem 9.15, it follows that the expected value of

$$\sum_{\Delta} \mathrm{card}(K(\Delta)),$$

where the summation is over all flaps Δ appearing in at least one $\mathcal{T}(P_r)$, is bounded by

$$\sum_{r=1}^{n} 16 \left(\frac{n-r}{r} \right) \left(\frac{\mathrm{E}\left[\mathrm{card}(\mathcal{T}(P_r))\right]}{r} \right).$$

The cardinality of $\mathcal{T}(P_r)$ is twice the number of edges of $\mathcal{CH}(P_r)$. Therefore it is at most $6r - 12$, so we get the bound

$$\sum_{e} \mathrm{card}(P(e)) \leqslant \sum_{\Delta} \mathrm{card}(K(\Delta)) \leqslant \sum_{r=1}^{n} 16 \left(\frac{n-r}{r} \right) \left(\frac{6r-12}{r} \right) \leqslant 96 n \ln n.$$

$\boxdot$

This finishes the last piece of the analysis of the convex hull algorithm. We get the following result:

Theorem 11.7 *The convex hull of a set of n points in $\mathbb{R}^3$ can be computed in $O(n \log n)$ randomized expected time.*

11.4* Convex Hulls and Half-Space Intersection

In Chapter 8 we have met the concept of duality. The strenth of duality lies in that it allows us to look at a problem from a new perspective, which can lead to more insight in what is really going on. Recall that we denote the line that is the dual of a point p by p^*, and the point that is the dual of a line ℓ by ℓ^*. The duality transform is incidence and order preserving: $p \in \ell$ if and only if $\ell^* \in p^*$, and p lies above ℓ if and only if ℓ^* lies above p^*.

Let's have a closer look at what convex hulls correspond to in dual space. We will do this for the planar case. Let P be a set of points in the plane. For technical reasons we focus on its *upper convex hull*, denoted $\mathcal{UH}(P)$, which consists of the convex hull edges that have P *below* their supporting line—see the left side of Figure 11.4. The upper convex hull is a polygonal chain that connects the leftmost point in P to the rightmost one. (We assume for simplicity that no two points have the same x-coordinate.)

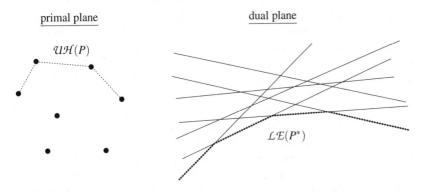

primal plane dual plane

$\mathcal{UH}(P)$

$\mathcal{LE}(P^*)$

Figure 11.4
Upper hulls correspond to lower
envelopes.

When does a point $p \in P$ appear as a vertex of the upper convex hull? That is the case if and only if there is a non-vertical line ℓ through p such that all other points of P lie below ℓ. In the dual plane this statement translates to the following condition: there is a point ℓ^* on the line $p^* \in P^*$ such that ℓ^* lies below all other lines of P^*. If we look at the arrangement $\mathcal{A}(P^*)$, this means that p^* contributes an edge to the unique bottom cell of the arrangement. This cell is the intersection of the half-planes bounded by a line in P^* and lying below that line. The boundary of the bottom cell is an x-monotone chain. We can define this chain as the minimum of the linear functions whose graphs are the lines in P^*. For this reason, the boundary of the bottom cell in an arrangement is often called the *lower envelope* of the set of lines. We denote the lower envelope of P^* by $\mathcal{LE}(P^*)$—see the right hand side of Figure 11.4.

The points in P that appear on $\mathcal{UH}(P)$ do so in order of increasing x-coordinate. The lines of P^* appear on the boundary of the bottom cell in order of decreasing slope. Since the slope of the line p^* is equal to the x-coordinate of p, it follows that the left-to-right list of points on $\mathcal{UH}(P)$ corresponds exactly to the right-to-left list of edges of $\mathcal{LE}(P^*)$. So the upper convex hull of a set of points is essentially the same as the lower envelope of a set of lines.

Let's do one final check. Two points p and q in P form an upper convex hull edge if and only if all other points in P lie below the line ℓ through p and q. In the dual plane, this means that all lines r^*, with $r \in P \setminus \{p, q\}$, lie above the intersection point ℓ^* of p^* and q^*. This is exactly the condition under which $p^* \cap q^*$ is a vertex of $\mathcal{LE}(P^*)$.

What about the *lower convex hull* of P and the *upper envelope* of P^*? (We leave the precise definitions to the reader.) By symmetry, these concepts are dual to each other as well.

We now know that the intersection of *lower half-planes*—half-planes bounded from above by a non-vertical line—can be computed by computing an upper convex hull, and that the intersection of *upper half-planes* can be computed by computing a lower convex hull. But what if we want to compute the intersection of an arbitrary set H of half-planes? Of course, we can split the set H into a set H_+ of upper half-planes and a set H_- of lower half-planes, compute $\bigcup H_+$ by computing the lower convex hull of H_+^* and $\bigcup H_-$ by computing the upper convex hull of H_-^*, and then compute $\bigcap H$ by intersecting $\bigcup H_+$ and $\bigcup H_-$.

But is this really necessary? If lower envelopes correspond to upper convex hulls, and upper envelopes correspond to lower convex hulls, shouldn't then the intersection of arbitrary half-planes correspond to full convex hulls? In a sense, this is true. The problem is that our duality transformation cannot handle vertical lines, and lines that are close to vertical but have opposite slope are mapped to very different points. This explains why the dual of the convex hull consists of two parts that lie rather far apart.

It is possible to define a different duality transformation that allows vertical lines. However, to apply this duality to a given set of half-planes, we need a point in the intersection of the half-planes. But that was to be expected. As long as we do not want to leave the Euclidean plane, there cannot be any general duality that turns the intersection of a set of half-planes into a convex hull, because the intersection of half-planes can have one special property: it can be empty. What could that possibly correspond to in the dual? The convex hull of a set of points in Euclidean space is always well defined: there is no such thing as "emptiness." (This problem is nicely solved if one works in oriented projective space, but this concept is beyond the scope of this book.) Only once you know that the intersection is not empty, and a point in the interior is known, can you define a duality that relates the intersection with a convex hull.

We shall leave it at this for the moment. The important thing is that—although there are some technical complications—convex hulls and intersections of half-planes (or half-spaces in three dimensions) are essentially the dual of each other. As a consequence, an algorithm to compute the intersection of half-planes in the plane (or the intersection of half-spaces in three dimensions) can be given by dualizing the algorithm described in this chapter.

11.5* Voronoi Diagrams Revisited

In Chapter 7 we introduced the Voronoi diagram of a set of points in the plane. It may come as a surprise that there is a close relationship between planar Voronoi diagrams and the intersection of upper half-spaces in 3-dimensional space. By the result on duality of the previous section, this implies a close relation between planar Voronoi diagrams and lower convex hulls in 3-space.

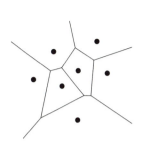

This has to do with an amazing property of the unit paraboloid in 3-space. Let $\mathcal{U} := (z = x^2 + y^2)$ denote the unit paraboloid, and let $p := (p_x, p_y, 0)$ be a point in the plane $z = 0$. Consider the vertical line through p. It intersects $\mathcal{U}$ in the point $p' := (p_x, p_y, p_x^2 + p_y^2)$. Let $h(p)$ be the non-vertical plane $z = 2p_x x + 2p_y y - (p_x^2 + p_y^2)$. Notice that $h(p)$ contains the point p'. Now consider any other point $q := (q_x, q_y, 0)$ in the plane $z = 0$. The vertical line through q intersects $\mathcal{U}$ in the point $q' := (q_x, q_y, q_x^2 + q_y^2)$, and it intersects $h(p)$ in

$$q(p) := (q_x, q_y, 2p_x q_x + 2p_y q_y - (p_x^2 + p_y^2)).$$

The vertical distance between q' and $q(p)$ is

$$q_x^2 + q_y^2 - 2p_x q_x - 2p_y q_y + p_x^2 + p_y^2 = (q_x - p_x)^2 + (q_y - p_y)^2 = \text{dist}(p,q)^2.$$

Hence, the plane $h(p)$ encodes—together with the unit paraboloid—the distance between p and any other point in the plane $z = 0$. (Since $\text{dist}(p,q)^2 \geqslant 0$ for any point q, and $p' \in h(p)$, this also implies that $h(p)$ is the tangent plane to $\mathcal{U}$ at p'.)

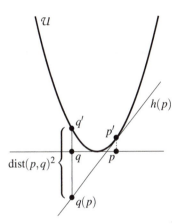

The fact that the plane $h(p)$ encodes the distance of other points to p leads to a correspondence between Voronoi diagrams and upper envelopes, as explained next. Let P be a planar point set, which we imagine to lie in the plane $z = 0$ of 3-dimensional space. Consider the set $H := \{h(p) \mid p \in P\}$ of planes, and let $\mathcal{UE}(H)$ be the upper envelope of the planes in H. We claim that the projection of $\mathcal{UE}(H)$ on the plane $z = 0$ is the Voronoi diagram of P. Figure 11.5 illustrates this one dimension lower: the Voronoi diagram of the points p_i on the line $y = 0$ is the projection of the upper envelope of the lines $h(p_i)$.

Theorem 11.8 *Let P be a set of points in 3-dimensional space, all lying in the plane $z = 0$. Let H be the set of planes $h(p)$, for $p \in P$, defined as above. Then the projection of $\mathcal{UE}(H)$ on the plane $z = 0$ is the Voronoi diagram of P.*

Proof. To prove the theorem, we will show that the Voronoi cell of a point $p \in P$ is exactly the projection of the facet of $\mathcal{UE}(H)$ that lies on the plane $h(p)$. Let q be a point in the plane $z = 0$ lying in the Voronoi cell of p. Hence, we have $\text{dist}(q, p) < \text{dist}(q, r)$ for all $r \in P$ with $r \neq p$. We have to prove that the vertical line through q intersects $\mathcal{UE}(H)$ at a point lying on $h(p)$. Recall that for a point $r \in P$, the plane $h(r)$ is intersected by the vertical line through q at the point $q(p) := (q_x, q_y, q_x^2 + q_y^2 - \text{dist}(q, r)^2)$. Of all points in P, the point p has the smallest distance to q, so $h(p)$ is the highest intersection point. Hence, the vertical line through q intersects $\mathcal{UE}(H)$ at a point lying on $h(p)$, as claimed. $\square$

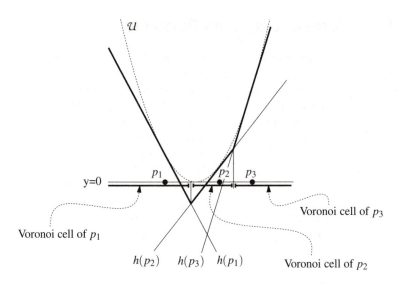

Figure 11.5
The correspondence between Voronoi
diagrams and upper envelopes

This theorem implies that we can compute a Voronoi diagram in the plane by computing the upper envelope of a set of planes in 3-space. By Exercise 11.9 (see also the previous section), the upper envelope of a set of planes in 3-space is in one-to-one correspondence to the lower convex hull of the points H^*, so we can immediately use our algorithm CONVEXHULL.

Not surprisingly, the lower convex hull of H^* has a geometric meaning as well: its projection on the plane $z = 0$ is the Delaunay triangulation of P.

11.6 Notes and Comments

The early convex hull algorithms worked only for points in the plane—see the notes and comments of Chapter 1 for a discussion of these algorithms. Computing convex hulls in 3-dimensional space turns out to be considerably more difficult. One of the first algorithms was the "gift wrapping" algorithm due to Chand and Kapur [63]. It finds facet after facet by "rotating" a plane over known edges of the hull until the first point is found. The running time is $O(nf)$ for a convex hull with f facets, which is $O(n^2)$ in the worst case. The first algorithm to achieve $O(n \log n)$ running time was a divide-and-conquer algorithm by Preparata and Hong [285, 286]. Early incremental algorithms run in time $O(n^2)$ [307, 189]. The randomized version presented here is due to Clarkson and Shor [111]. The version we presented needs $O(n \log n)$ space; the original paper gives a simple improvement to linear space. The idea of a conflict graph, used here for the first time in this book, also comes from the paper of Clarkson and Shor. Our analysis, however, is due to Mulmuley [256].

In this chapter we have concentrated on 3-dimensional space, where convex hulls have linear complexity. The so-called *Upper Bound Theorem* states that the worst-case combinatorial complexity of the convex hull of n points in d-dimensional space—phrased in dual space: the intersection of n half-spaces—is $\Theta(n^{\lfloor d/2 \rfloor})$. (We proved this result for the case $d = 3$, using Euler's relation.) The algorithm described in this chapter generalizes to higher dimensions, and is optimal in the worst case: its expected running time is $\Theta(n^{\lfloor d/2 \rfloor})$. Interestingly, the best known deterministic convex hull algorithm for odd-dimensional spaces is based on a (quite complicated) derandomization of this algorithm [76]. Since the convex hull in dimensions greater than three can have non-linear complexity, output-sensitive algorithms may be useful. The best known output-sensitive algorithm for computing convex hulls in $\mathbb{R}^d$ is due to Chan [62]. Its running time is $O(n \log k + (nk)^{1-1/(\lfloor d/2 \rfloor + 1)} \log^{O(1)})$, where k denotes the complexity of the convex hull. Readers who want to know more about the mathematical aspects of polytopes in higher dimensions can consult Grünbaum's book [162], which is a classical reference for polytope theory, or Ziegler's book [345], which treats the combinatorial aspects.

In Section 11.5 we have seen that the Voronoi diagram of a planar point set is the projection of the upper envelope of a certain set of planes in 3-dimensional space. A similar statement is true in higher dimensions: the Voronoi diagram of a set of points in $\mathbb{R}^d$ is the projection of the upper envelope of a certain set of hyperplanes in $\mathbb{R}^{d+1}$. Not all sets of (hyper)planes define an upper envelope whose projection is the Voronoi diagram of some point set. Interestingly, any upper envelope *does* project onto a so-called power diagram, a generalization of the Voronoi diagram where the sites are spheres rather than points [13].

11.7 Exercises

11.1 In Chapter 1 we defined the convex hull of a set P of points as the intersection of all convex sets containing the points. In the current chapter we saw another definition: the convex hull of P is the set of all convex combinations of the points in P. Prove that these two definitions are equivalent, that is, prove that a point q is a convex combination of points in P if and only if q lies in every convex set containing P.

11.2 Prove that the worst case running time of algorithm CONVEXHULL is $O(n^2)$, and that there are sets of points where a bad choice of the random permutation makes the algorithm actually need $\Theta(n^2)$ time.

11.3 Describe a randomized incremental algorithm to compute the convex hull of n points in the plane. Describe how to deal with degeneracies. Analyze the expected running time of your algorithm.

11.4 In many applications, only a small percentage of the points in a given set P of n points are extreme. In such a case, the convex hull of P has less than n vertices. This can actually make our algorithm CONVEXHULL run faster than $\Theta(n \log n)$.

Assume, for instance, that the number of extreme points in a random sample of P of size r is only $O(r^\alpha)$, for some constant $\alpha < 1$. (This is true when the set P has been created by uniformly picking points at random in a ball.) Prove that under this condition, the running time of the algorithm is $O(n)$.

11.5 Describe a data structure that allows you to test whether a query point q lies inside a convex polytope with n vertices in $\mathbb{R}^3$. (*Hint:* Use the results from Chapter 6.)

11.6 Define a simple polytope to be a region in 3-space that is topologically equivalent to a ball (but not necessarily convex) and whose boundary consists of planar polygons. Describe how to test in $O(n)$ time whether a point lies inside a simple polytope with n vertices in 3-dimensional space.

11.7 Describe a randomized incremental algorithm to compute the intersection of half-planes, and analyze its expected running time. Your algorithm should maintain the intersection of the current set of half-planes. To figure out where to insert a new half-plane, maintain a conflict graph between the vertices of the current intersection and the half-planes that are still to be inserted.

11.8 Describe a randomized incremental algorithm to compute the intersection of half-spaces in 3-dimensional space, and analyze its expected running time. Maintain a conflict graph analogous to the previous exercise.

11.9 In this exercise you have to work out the details of a 3-dimensional duality transformation. Given a point $p := (p_x, p_y, p_z)$ in $\mathbb{R}^3$, let p^* be the plane $z = p_x x + p_y y - (p_x^2 + p_y^2)/2$. For a non-vertical plane h, define h^* such that $(h^*)^* = h$. Define the upper convex hull $\mathcal{UH}(P)$ of a set of points P and the lower envelope $\mathcal{LE}(H)$ of a set H of planes in 3-space.

Show the following properties.

- A point p lies on a plane h if and only if h^* lies on p^*.

- A point p lies above h if and only if h^* lies above p^*.

- A point $p \in P$ is a vertex of $\mathcal{UH}(P)$ if and only if p^* appears on $\mathcal{LE}(P^*)$.

- A segment $\overline{pq}$ is an edge of $\mathcal{UH}(P)$ if and only if p^* and q^* share an edge on $\mathcal{LE}(P^*)$.

- Points $p_1, p_2, \ldots, p_k$ are the vertices of a facet f of $\mathcal{UH}(P)$ if and only if $p_1^*, p_2^*, \ldots, p_k^*$ support facets of $\mathcal{LE}(P^*)$ that share a common vertex.

12 Binary Space Partitions
The Painter's Algorithm

These days pilots no longer have their first flying experience in the air, but on the ground in a flight simulator. This is cheaper for the air company, safer for the pilot, and better for the environment. Only after spending many hours in the simulator are pilots allowed to operate the control stick of a real airplane. Flight simulators must perform many different tasks to make the pilot forget that she is sitting in a simulator. An important task is visualization: pilots must be able

Figure 12.1
The Microsoft flight simulator in Windows 95

to see the landscape above which they are flying, or the runway on which they are landing. This involves both modeling landscapes and rendering the models. To render a scene we must determine for each pixel on the screen the object that is visible at that pixel; this is called *hidden surface removal*. We must also perform shading calculations, that is, we must compute the intensity of the light that the visible object emits in the direction of the view point. The latter task is very time-consuming if highly realistic images are desired: we must compute how much light reaches the object—either directly from light sources or indirectly via reflections on other objects—and consider the interaction of the light with the surface of the object to see how much of it is reflected in the direction of the view point. In flight simulators rendering must be performed in real-time, so there is no time for accurate shading calculations. Therefore a fast and simple shading technique is employed and hidden surface removal becomes an important factor in the rendering time.

The *z-buffer algorithm* is a very simple method for hidden surface removal. This method works as follows. First, the scene is transformed such that the viewing direction is the positive z-direction. Then the objects in the scene are scan-converted in arbitrary order. Scan-converting an object amounts to determining which pixels it covers in the projection; these are the pixels where the object is potentially visible. The algorithm maintains information about the already processed objects in two buffers: a frame buffer and a z-buffer. The frame buffer stores for each pixel the intensity of the currently visible object, that is, the object that is visible among those already processed. The z-buffer stores for each pixel the z-coordinate of the currently visible object. (More precisely, it stores the z-coordinate of the point on the object that is visible at the pixel.) Now suppose that we select a pixel when scan-converting an object. If the z-coordinate of the object at that pixel is smaller than the z-coordinate stored in the z-buffer, then the new object lies in front of the currently visible object. So we write the intensity of the new object to the frame buffer, and its z-coordinate to the z-buffer. If the z-coordinate of the object at that pixel is larger than the z-coordinate stored in the z-buffer, then the new object is not visible, and the frame buffer and z-buffer remain unchanged. The z-buffer algorithm is easily implemented in hardware and quite fast in practice. Hence, this is the most popular hidden surface removal method. Nevertheless, the algorithm has

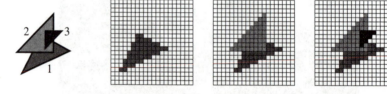

Figure 12.2
The painter's algorithm in action

some disadvantages: a large amount of extra storage is needed for the z-buffer, and an extra test on the z-coordinate is required for every pixel covered by an object. The *painter's algorithm* avoids these extra costs by first sorting the objects according to their distance to the view point. Then the objects are scan-converted in this so-called *depth order*, starting with the object farthest from the view point. When an object is scan-converted we do not need to perform any test on its z-coordinate, we always write its intensity to the frame buffer. Entries in the frame buffer that have been filled before are simply overwritten. Figure 12.2 illustrates the algorithm on a scene consisting of three triangles. On the left, the triangles are shown with numbers corresponding to the order in which they are scan-converted. The images after the first, second, and third triangle have been scan-converted are shown as well. This approach is correct because we scan-convert the objects in back-to-front order: for each pixel the last object written to the corresponding entry in the frame buffer will be the one closest to the viewpoint, resulting in a correct view of the scene. The process resembles the way painters work when they put layers of paint on top of each other, hence the name of the algorithm.

To apply this method successfully we must be able to sort the objects quickly. Unfortunately this is not so easy. Even worse, a depth order may not always exist: the in-front-of relation among the objects can contain cycles. When such a *cyclic overlap* occurs, no ordering will produce a correct view of this scene. In this case we must break the cycles by splitting one or more of the objects, and hope that a depth order exists for the pieces that result from the splitting. When there is a cycle of three triangles, for instance, we can always split one of them into a triangular piece and a quadrilateral piece, such that a correct displaying order exists for the resulting set of four objects. Computing which objects to split, where to split them, and then sorting the object fragments is an expensive process. Because the order depends on the position of the view point, we must recompute the order every time the view point moves. If we want to use the painter's algorithm in a real-time environment such as flight simulation, we should preprocess the scene such that a correct displaying order can be found quickly for any view point. An elegant data structure that makes this possible is the binary space partition tree, or BSP tree for short.

12.1 The Definition of BSP Trees

To get a feeling for what a BSP tree is, take a look at Figure 12.3. This figure shows a binary space partition (BSP) for a set of objects in the plane, together with the tree that corresponds to the BSP. As you can see, the binary space partition is obtained by recursively splitting the plane with a line: first we split the entire plane with ℓ_1, then we split the half-plane above ℓ_1 with ℓ_2 and the half-plane below ℓ_1 with ℓ_3, and so on. The splitting lines not only partition the plane, they may also cut objects into fragments. The splitting continues until there is only one fragment left in the interior of each region. This process is

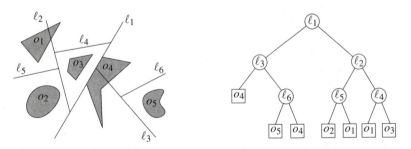

Figure 12.3
A binary space partition and the corresponding tree

naturally modeled as a binary tree. Each leaf of this tree corresponds to a face of the final subdivision; the object fragment that lies in the face is stored at the leaf. Each internal node corresponds to a splitting line; this line is stored at the node. When there are 1-dimensional objects (line segments) in the scene then objects could be contained in a splitting line; in that case the corresponding internal node stores these objects in a list.

For a hyperplane $h : a_1 x_1 + a_2 x_2 + \cdots + a_d x_d + a_{d+1} = 0$, we let h^+ be the open positive half-space bounded by h and we let h^- be the open negative half-space:

$$h^+ := \{(x_1, x_2, \ldots, x_d) : a_1 x_1 + a_2 x_2 + \cdots + a_d x_d + a_{d+1} > 0\}$$

and

$$h^- := \{(x_1, x_2, \ldots, x_d) : a_1 x_1 + a_2 x_2 + \cdots + a_d x_d + a_{d+1} < 0\}.$$

A binary space partition tree, or BSP tree, for a set S of objects in d-dimensional space is now defined as a binary tree $\mathcal{T}$ with the following properties:

- If $\mathrm{card}(S) \leqslant 1$ then $\mathcal{T}$ is a leaf; the object fragment in S (if it exists) is stored explicitly at this leaf. If the leaf is denoted by v, then the (possibly empty) set stored at the leaf is denoted by $S(v)$.

- If $\mathrm{card}(S) > 1$ then the root v of $\mathcal{T}$ stores a hyperplane h_v, together with the set $S(v)$ of objects that are fully contained in h_v. The left child of v is the root of a BSP tree $\mathcal{T}^-$ for the set $S^- := \{h_v^- \cap s : s \in S\}$, and the right child of v is the root of a BSP tree $\mathcal{T}^+$ for the set $S^+ := \{h_v^+ \cap s : s \in S\}$.

The *size* of a BSP tree is the total size of the sets $S(v)$ over all nodes v of the BSP tree. In other words, the size of a BSP tree is the total number of object fragments that are generated. If the BSP does not contain useless splitting lines—lines that split off an empty subspace—then the number of nodes of the tree is at most linear in the size of the BSP tree. Strictly speaking, the size of the BSP tree does not say anything about the amount of storage needed to store it, because it says nothing about the amount of storage needed for a single fragment. Nevertheless, the size of a BSP tree as we defined it is a good measure to compare the quality of different BSP trees for a given set of objects.

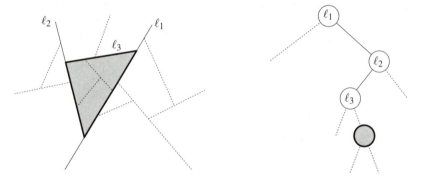

Figure 12.4
The correspondence between nodes
and regions

The leaves in a BSP tree represent the faces in the subdivision that the BSP induces. More generally, we can identify a convex region with each node v in a BSP tree $\mathcal{T}$: this region is the intersection of the half-spaces $h_\mu^\diamond$, where μ is an ancestor of v and $\diamond = +$ when v is in the left subtree of μ, and $\diamond = -$ when it is in the right subtree. The region corresponding to the root of $\mathcal{T}$ is the whole space. Figure 12.4 illustrates this: the grey node corresponds to the grey region $\ell_1^+ \cap \ell_2^- \cap \ell_3^+$.

The splitting hyperplanes used in a BSP can be arbitrary. For computational purposes, however, it can be convenient to restrict the set of allowable splitting hyperplanes. A usual restriction is the following. Suppose we want to construct a BSP for a set of line segments in the plane. An obvious set of candidates for the splitting lines is the set of extensions of the input segments. A BSP that only uses such splitting lines is called an *auto-partition*. For a set of planar polygons in 3-space, an auto-partition is a BSP that only uses planes through the input polygons as splitting planes. It seems that the restriction to auto-partitions is a severe one. But, although auto-partitions cannot always produce minimum-size BSP trees, we shall see that they can produce reasonably small ones.

12.2 BSP Trees and the Painter's Algorithm

Suppose we have built a BSP tree $\mathcal{T}$ on a set S of objects in 3-dimensional space. How can we use $\mathcal{T}$ to get the depth order we need to display the set S with the painter's algorithm? Let p_{view} be the view point and suppose that p_{view} lies above the splitting plane stored at the root of $\mathcal{T}$. Then clearly none of the objects below the splitting plane can obscure any of the objects above it. Hence, we can safely display all the objects (more precisely, object fragments) in the subtree $\mathcal{T}^-$ before displaying those in $\mathcal{T}^+$. The order for the object fragments in the two subtrees $\mathcal{T}^+$ and $\mathcal{T}^-$ is obtained recursively in the same way. This is summarized in the following algorithm.

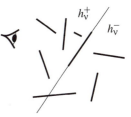

Algorithm PAINTERSALGORITHM($\mathcal{T}, p_{\text{view}}$)
1.   Let v be the root of $\mathcal{T}$.
2.   **if** v is a leaf
3.   **then** Scan-convert the object fragments in $S(v)$.
4.   **else if** $p_{\text{view}} \in h_v^+$
5.   **then** PAINTERSALGORITHM($\mathcal{T}^-, p_{\text{view}}$)
6.   Scan-convert the object fragments in $S(v)$.
7.   PAINTERSALGORITHM($\mathcal{T}^+, p_{\text{view}}$)
8.   **else if** $p_{\text{view}} \in h_v^-$
9.   **then** PAINTERSALGORITHM($\mathcal{T}^+, p_{\text{view}}$)
10.   Scan-convert the object fragments in $S(v)$.
11.   PAINTERSALGORITHM($\mathcal{T}^-, p_{\text{view}}$)
12.   **else** (* $p_{\text{view}} \in h_v$ *)
13.   PAINTERSALGORITHM($\mathcal{T}^+, p_{\text{view}}$)
14.   PAINTERSALGORITHM($\mathcal{T}^-, p_{\text{view}}$)

Note that we do not draw the polygons in $S(v)$ when p_{view} lies on the splitting plane h_v, because polygons are flat 2-dimensional objects and therefore not visible from points that lie in the plane containing them.

    The efficiency of this algorithm—indeed, of any algorithm that uses BSP trees—depends largely on the size of the BSP tree. So we must choose the splitting planes in such a way that fragmentation of the objects is kept to a

minimum. Before we can develop splitting strategies that produce small BSP trees, we must decide on which types of objects we allow. We became interested in BSP trees because we needed a fast way of doing hidden surface removal for flight simulators. Because speed is our main concern, we should keep the type of objects in the scene simple: we should not use curved surfaces, but represent everything in a polyhedral model. We assume that the facets of the polyhedra have been triangulated. So we want to construct a BSP tree of small size for a given set of triangles in 3-dimensional space.

12.3 Constructing a BSP Tree

When you want to solve a 3-dimensional problem, it is usually not a bad idea to gain some insight by first studying the planar version of the problem. This is also what we do in this section.

Let S be a set of n non-intersecting line segments in the plane. We will restrict our attention to auto-partitions, that is, we only consider lines containing one of the segments in S as candidate splitting lines. The following recursive algorithm for constructing a BSP immediately suggests itself. Let $\ell(s)$ denote the line that contains a segment s.

Algorithm 2DBSP(S)
Input. A set $S = \{s_1, s_2, \ldots, s_n\}$ of segments.
Output. A BSP tree for S.
1. **if** card(S) $\leqslant 1$
2.   **then** Create a tree $\mathcal{T}$ consisting of a single leaf node, where the set S is stored explicitly.
3.   **return** $\mathcal{T}$
4.   **else** (∗ Use $\ell(s_1)$ as the splitting line. ∗)
5. $S^+ \leftarrow \{s \cap \ell(s_1)^+ : s \in S\};$ $\mathcal{T}^+ \leftarrow$ 2DBSP(S^+)
6. $S^- \leftarrow \{s \cap \ell(s_1)^- : s \in S\};$ $\mathcal{T}^- \leftarrow$ 2DBSP(S^-)
7.   Create a BSP tree $\mathcal{T}$ with root node v, left subtree $\mathcal{T}^-$, right subtree $\mathcal{T}^+$, and with $S(v) = \{s \in S : s \subset \ell(s_1)\}$.
8.   **return** $\mathcal{T}$

The algorithm clearly constructs a BSP tree for the set S. But is it a small one? Perhaps we should spend a little more effort in choosing the right segment to do the splitting, instead of blindly taking the first segment, s_1. One approach that comes to mind is to take the segment $s \in S$ such that $\ell(s)$ cuts as few segments as possible. But this is too greedy: there are configurations of segments where this approach doesn't work well. Furthermore, finding this segment would be time consuming. What else can we do? Perhaps you already guessed: as in previous chapters where we had to make a difficult choice, we simply make a random choice. That is to say, we use a random segment to do the splitting. As we shall see later, the resulting BSP is expected to be fairly small.

To implement this, we put the segments in random order before we start the construction:

Algorithm 2DRANDOMBSP(S)
1. Generate a random permutation $S' = s_1, \ldots, s_n$ of the set S.
2. $\mathcal{T} \leftarrow$ 2DBSP(S')
3. **return** $\mathcal{T}$

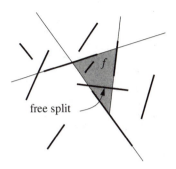

free split

Before we analyze this randomized algorithm, we note that one simple opti- mization is possible. Suppose that we have chosen the first few partition lines. These lines induce a subdivision of the plane whose faces correspond to nodes in the BSP tree that we are constructing. Consider one such face f. There can be segments that cross f completely. Selecting one of these crossing seg- ments to split f will not cause any fragmentation of other segments inside f, while the segment itself can be excluded from further consideration. It would be foolish not to take advantage of such *free splits*. So our improved strategy is to make free splits whenever possible, and to use random splits otherwise. To implement this optimization, we must be able to tell whether a segment is a free split. To this end we maintain two boolean variables with each segment, which indicate whether the left and right endpoint lie on one of the already added splitting lines. When both variables become true, then the segment is a free split.

We now analyze the performance of algorithm 2DRANDOMBSP. To keep it simple, we will analyze the version without free splits. (In fact, free splits do not make a difference asympotically.)

We start by analyzing the size of the BSP tree or, in other words, the number of fragments that are generated. Of course, this number depends heavily on the particular permutation generated in line 1: some permutations may give small BSP trees, while others give very large ones. As an example, consider the

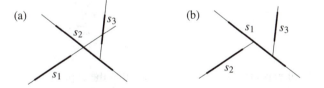

Figure 12.5
Different orders give different BSPs

collection of three segments depicted in Figure 12.5. If the segments are treated as illustrated in part (a) of the figure, then five fragments result. A different order, however, gives only three fragments, as shown in part (b). Because the size of the BSP varies with the permutation that is used, we will analyze the *expected* size of the BSP tree, that is, the average size over all $n!$ permutations.

Lemma 12.1 *The expected number of fragments generated by the algorithm* 2DRANDOMBSP *is* $O(n\log n)$.

Proof. Let s_i be a fixed segment in S. We shall analyze the expected number of other segments that are cut when $\ell(s_i)$ is added by the algorithm as the next splitting line.

In Figure 12.5 we can see that whether or not a segment s_j is cut when $\ell(s_i)$ is added—assuming it can be cut at all by $\ell(s_i)$—depends on segments that are

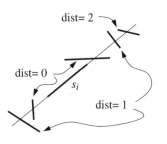

also cut by $\ell(s_i)$ and are 'in between' s_i and s_j. In particular, when the line through such a segment is used before $\ell(s_i)$, then it shields s_j from s_i. This is what happened in Figure 12.5(b): the segment s_1 shielded s_3 from s_2. These considerations lead us to define the distance of a segment with respect to the fixed segment s_i:

$$\mathrm{dist}(s_j) = \begin{cases} \text{the number of segments intersect-} & \text{if } \ell(s_i) \text{ intersects } s_j \\ \text{ing } \ell(s_i) \text{ in between } s_i \text{ and } s_j \\ +\infty & \text{otherwise} \end{cases}$$

For any finite distance, there are at most two segments at that distance, one on either side of s_i.

Let $k := \mathrm{dist}(s_j)$, and let $s_{j_1}, s_{j_2}, \dots, s_{j_k}$ be the segments in between s_i and s_j. What is the probability that $\ell(s_i)$ cuts s_j when added as a splitting line? For this to happen, s_i must come before s_j in the random ordering and, moreover, it must come before any of the segments in between s_i and s_j, which shield s_j from s_i. In other words, of the set $\{i, j, j_1, \dots, j_k\}$ of indices, i must be the smallest one. Because the order of the segments is random, this implies

$$\Pr[\ell(s_i) \text{ cuts } s_j] \leqslant \frac{1}{\mathrm{dist}(s_j) + 2}.$$

Notice that there can be segments that are not cut by $\ell(s_i)$ but whose *extension* shields s_j. This explains why the expression above is not an equality.

We can now bound the expected total number of cuts generated by s:

$$\begin{aligned} \mathrm{E}[\text{number of cuts generated by } s] &\leqslant \sum_{s' \neq s} \frac{1}{\mathrm{dist}_s(s') + 2} \\ &\leqslant 2 \sum_{i=0}^{n-2} \frac{1}{i + 2} \\ &\leqslant 2 \ln n. \end{aligned}$$

By linearity of expectation, we can conclude that the expected total number of cuts generated by all segments is at most $2n \ln n$. Since we start with n segments, the expected total number of fragments is bounded by $n + 2n \ln n$. $\quad\square$

We have shown that the expected size of the BSP that is generated by 2DRANDOMBSP is $n + 2n \ln n$. As a consequence, we have proven that a BSP of size $n + 2n \ln n$ *exists* for any set of n segments. Furthermore, at least half of all permutations lead to a BSP of size $n + 4n \ln n$. We can use this to find a BSP of that size: After running 2DRANDOMBSP we test the size of the tree, and if it exceeds that bound, we simply start the algorithm again with a fresh random permutation. The expected number of trials is two.

We have analyzed the size of the BSP that is produced by 2DRANDOMBSP. What about the running time? Again, this depends on the random permutation that is used, so we look at the expected running time. Computing the random permutation takes linear time. If we ignore the time for the recursive calls, then the time taken by algorithm 2DBSP is linear in the number of fragments in S. This number is never larger than n—in fact, it gets smaller with each recursive call. Finally, the number of recursive calls is obviously bounded by the total number of generated fragments, which is $O(n \log n)$. Hence, the total construction time is $O(n^2 \log n)$, and we get the following result.

Theorem 12.2 *A BSP of size $O(n \log n)$ can be computed in expected time $O(n^2 \log n)$.*

Although the expected size of the BSP that is constructed by 2DRANDOMBSP is fairly good, the running time of the algorithm is somewhat disappointing. In many applications this is not so important, because the construction is done off-line. Moreover, the construction time is only quadratic when the BSP is very unbalanced, which is rather unlikely to occur in practice. Nevertheless, from a theoretical point of view the construction time is disappointing. Using an approach based on segment trees—see Chapter 10—this can be improved: one can construct a BSP of size $O(n \log n)$ in $O(n \log n)$ time with a deterministic algorithm. This approach does not give an auto-partition, however, and in practice it produces BSPs that are slightly larger.

A natural question is whether it is also possible to improve the size of the BSP generated by 2DRANDOMBSP: is there an $O(n)$ size BSP for any set of segments in the plane, or are there perhaps sets for which any BSP must have size $\Omega(n \log n)$? The answer to this question is currently unknown.

The algorithm we described for the planar case immediately generalizes to 3-dimensional space. Let S be a set of n non-intersecting triangles in $\mathbb{R}^3$. Again we restrict ourselves to auto-partitions, that is, we only use partition planes that contain a triangle of S. For a triangle t we denote the plane containing it by $h(t)$.

Algorithm 3DBSP(S)
Input. A set $S = \{t_1, t_2, \ldots, t_n\}$ of triangles in $\mathbb{R}^3$.
Output. A BSP tree for S.
1. **if** card(S) $\leqslant 1$
2. **then** Create a tree $\mathcal{T}$ consisting of a single leaf node, where the set S is stored explicitly.
3. **return** $\mathcal{T}$
4. **else** (* Use $h(t_1)$ as the splitting plane. *)
5. $S^+ \leftarrow \{t \cap h(t_1)^+ : t \in S\}; \qquad \mathcal{T}^+ \leftarrow$ 3DBSP(S^+)
6. $S^- \leftarrow \{t \cap h(t_1)^- : t \in S\}; \qquad \mathcal{T}^- \leftarrow$ 3DBSP(S^-)
7. Create a BSP tree $\mathcal{T}$ with root node v, left subtree $\mathcal{T}^-$, right subtree $\mathcal{T}^+$, and with $S(v) = \{t \in T : t \subset h(t_1)\}$.
8. **return** $\mathcal{T}$

The size of the resulting BSP again depends on the order in of the triangles; some orders give more fragments than others. As in the planar case, we can try to get a good expected size by first putting the triangles in a random order.

This usually gives a good result in practice. However, it is not known how to analyze the expected behavior of this algorithm theoretically. Therefore we will analyze a variant of the algorithm in the next section, although the algorithm described above is probably superior in practice.

12.4* The Size of BSP Trees in 3-Space

The randomized algorithm for constructing a BSP tree in 3-space that we analyze in this section is almost the same as the improved algorithm described above: it treats the triangles in random order, and it makes free splits whenever possible. A free split now occurs when a triangle of S splits a cell into two disconnected subcells. The only difference is that when we use some plane $h(t)$ as a splitting plane, we use it in all cells intersected by that plane, not just in the cells that are intersected by t. (And therefore a simple recursive implementation is no longer possible.) There is one exception to the rule that we split all cells with $h(t)$: when the split is completely useless for a cell, because all the triangles in that cell lie completely to one side of it, then we do not split it.

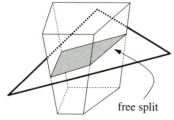

free split

Figure 12.6 illustrates this on a 2-dimensional example. In part (a) of the figure, the subdivision is shown that is generated by the algorithm of the previous section after treating segments s_1, s_2, and s_3 (in that order). In part (b) the subdivision is shown as generated by the modified algorithm. Note that the modified algorithm uses $\ell(s_2)$ as a splitting line in the subspace below $\ell(s_1)$, and that $\ell(s_3)$ is used as a splitting line in the subspace to the right of $\ell(s_2)$. The line $\ell(s_3)$ is not used in the subspace between $\ell(s_1)$ and $\ell(s_2)$, however, because it is useless there.

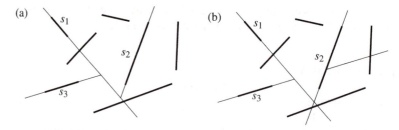

Figure 12.6
The original and the modified algorithm

The modified algorithm can be summarized as follows. Working out the details is left as an exercise.

Algorithm 3DRANDOMBSP2(S)
Input. A set $S = \{t_1, t_2, \dots, t_n\}$ of triangles in $\mathbb{R}^3$.
Output. A BSP tree for S.
1.   Generate a random permutation $t_1, \dots, t_n$ of the set S.
2.   **for** $i \leftarrow 1$ **to** n
3.   **do** Use $h(t_i)$ to split every cell where the split is useful.
4.   Make all possible free splits.

The next lemma analyzes the expected number of fragments generated by the algorithm.

Lemma 12.3 *The expected number of object fragments generated by algorithm* 3DRANDOMBSP2 *over all* $n!$ *possible permutations is* $O(n^2)$.

Proof. We shall prove a bound on the expected number of fragments into which a fixed triangle $t_k \in S$ is cut. For a triangle t_i with $i < k$ we define $\ell_i := h(t_i) \cap h(t_k)$. The set $L := \{\ell_1, \ldots, \ell_{k-1}\}$ is a set of at most $k-1$ lines lying in the plane $h(t_k)$. Some of these lines intersect t_k, others miss t_k. For a line ℓ_i that intersects t_k we define $s_i := \ell_i \cap t_k$. Let I be the set of all such intersections s_i. Due to free splits the number of fragments into which t_k is cut is in general *not* simply the number of faces in the arrangement that I induces on t_k. To understand this, consider the moment that t_{k-1} is treated. Assume that ℓ_{k-1} intersects t_k; otherwise t_{k-1} does not cause any fragmentation on t_k. The segment s_{k-1} can intersect several of the faces of the arrangement on t_k induced by $I \setminus \{s_k\}$. If, however, such a face f is not incident to the one of the edges of t_k—we call f an interior face—then a free split already has been made through this part of t_k. In other words, $h(t_{k-1})$ only causes cuts in exterior faces, that is, faces that are incident to one of the three edges of t_k. Hence, the number of splits on t_k caused by $h(t_{k-1})$ equals the number of edges that s_{k-1} contributes to exterior faces of the arrangement on t_k induced by I. (In the analysis that follows, it is important that the collection of exterior faces is independent of the order in which $t_1, \ldots, t_{k-1}$ have been treated. This is not the case for the algorithm in the previous section, which is the reason for the modification.) What is the expected number of such edges? To answer this question we first bound the total number of edges of the exterior faces.

In Chapter 8 we defined the *zone* of a line ℓ in an arrangement of lines in the plane as the set of faces of the arrangement intersected by ℓ. You may recall that for an arrangement of m lines the complexity of the zone is $O(m)$. Now let e_1, e_2, and e_3 be the edges of t_k and let $\ell(e_i)$ be the line through e_i, for $i = 1, 2, 3$. The edges that we are interested in must be in the zone of either $\ell(e_1)$, $\ell(e_2)$, or $\ell(e_3)$ in the arrangement induced by the set L on the plane $h(t_k)$. Hence, the total number of edges of exterior faces is $O(k)$.

If the total number of edges of exterior faces is $O(k)$, then the average number of edges lying on a segment s_i is $O(1)$. Because $t_1, \ldots, t_n$ is a random permutation, so is $t_1, \ldots, t_{k-1}$. Hence, the expected number of edges on segment s_{k-1} is constant, and therefore the expected number of extra fragments on t_k caused by $h(t_{k-1})$ is $O(1)$. The same argument shows that the expected number of fragmentations on t_k generated by each of the splitting planes $h(t_1)$ through $h(t_{k-2})$ is constant. This implies that the expected number of fragments into which t_k is cut is $O(k)$. The total number of fragments is therefore

$$O(\sum_{k=1}^{n} k) = O(n^2).$$

$\square$

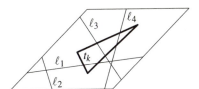

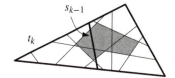

The quadratic bound on the expected size of the partitioning generated by 3DRANDOMBSP immediately proves that a BSP tree of quadratic size exists.

You may be a bit disappointed by the bound that we have achieved. A quadratic size BSP tree is not what you are hoping for when you have a set of 10,000 triangles. The following theorem tells us that we cannot hope to prove anything better if we restrict ourselves to auto-partitions.

Lemma 12.4 *There are sets of n non-intersecting triangles in 3-space for which any auto-partition has size $\Omega(n^2)$.*

Proof. Consider a collection of rectangles consisting of a set R_1 of rectangles parallel to the xy-plane and a set R_2 of rectangles parallel to the yz-plane, as illustrated in the margin. (The example also works with a set of triangles, but with rectangles it is easier to visualize.) Let $n_1 := \operatorname{card}(R_1)$, let $n_2 := \operatorname{card}(R_2)$, and let $G(n_1, n_2)$ be the minimum size of an auto-partition for such a configuration. We claim that $G(n_1, n_2) = (n_1 + 1)(n_2 + 1) - 1$. The proof is by induction on $n_1 + n_2$. The claim is obviously true for $G(1,0)$ and $G(0,1)$, so now consider the case where $n_1 + n_2 > 1$. Without loss of generality, assume that the auto-partition chooses a rectangle r from the set R_1. The plane through r will split all the rectangles in R_2. Moreover, the configurations in the two subscenes that must be treated recursively have exactly the same form as the initial configuration. If m denotes the number of rectangles of R_1 lying above r, then we have

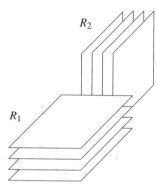

$$
\begin{aligned}
G(n_1, n_2) &= 1 + G(m, n_2) + G(n_1 - m - 1, n_2) \\
&= 1 + ((m+1)(n_2+1) - 1) + ((n_1 - m)(n_2+1) - 1) \\
&= (n_1 + 1)(n_2 + 1) - 1.
\end{aligned}
$$

So perhaps we should not restrict ourselves to auto-partitions. In the lower bound in the proof of Lemma 12.4 the restriction to auto-partitions is definitely a bad idea: we have shown that such a partition necessarily has quadratic size,

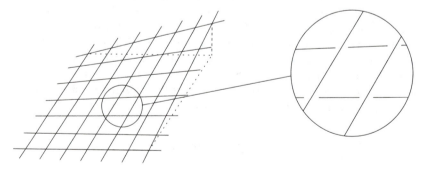

Figure 12.7
The general lower bound construction

whereas we can easily get a linear BSP if we first separate the set R_1 from the set R_2 with a plane parallel to the xz-plane. But even unrestricted partitions fail to give a small BSP for the configuration of Figure 12.7. This configuration is obtained as follows. We start by taking a grid in the plane made up of $n/2$

lines parallel to the x-axis and $n/2$ lines parallel to the y-axis. (Instead of the lines we could also take very long and skinny triangles.) We skew these lines a little to get the configuration of Figure 12.7; the lines now lie on a so-called hyperbolic paraboloid. Finally we move the lines parallel to the y-axis slightly upward so that the lines no longer intersect. What we get is the set of lines

$$\{y = i, z = ix : 1 \leqslant i \leqslant n/2\} \cup \{x = i, z = iy + \varepsilon : 1 \leqslant i \leqslant n/2\},$$

where ε is a small positive constant. If ε is sufficiently small then any BSP must cut at least one of the four lines that bound a grid cell in the immediate neighborhood of that cell. The formal proof of this fact is elementary, but tedious and not very instructive. The idea is to show that the lines are skewed in such a way that no plane fits simultaneously through the four "openings" at its corners. Since there is a quadratic number of grid cells, this will result in $\Theta(n^2)$ fragments.

Theorem 12.5 *For any set of n non-intersecting triangles in $\mathbb{R}^3$ a BSP tree of size $O(n^2)$ exists. Moreover, there are configurations for which the size of any BSP is $\Omega(n^2)$.*

The quadratic lower bound on the size of BSP trees might give you the idea that they are useless in practice. Fortunately this is not the case. The configurations that yield the lower bound are quite artificial. In many practical situations BSP trees perform just fine.

12.5 Notes and Comments

BSP trees are popular in many application areas, in particular in computer graphics. The application mentioned in this chapter is to perform hidden surface removal with the painter's algorithm [154]. Other application include shadow generation [102], set operations on polyhedra [257, 323], and visibility preprocessing for interactive walkthroughs [322]. They have also been used for cell decomposition methods in motion planning [24].

The study of BSP trees from a theoretical point of view was initiated by Paterson and Yao [280]; the results of this chapter come from their paper. They also proved bounds on BSPs in higher dimensions: any set of $(d-1)$-dimensional simplices in $\mathbb{R}^d$, with $d \geqslant 3$, admits a BSP of size $O(n^{d-1})$.

Paterson and Yao also studied the special case of line segments in the plane that are all either horizontal or vertical. They showed that in such a case a linear BSP is possible. The same result was achieved by D'Amore and Franciosa [116]. Paterson and Yao generalized the results to orthogonal objects in higher dimensions [281]. For instance, they proved that any set of orthogonal rectangles in $\mathbb{R}^3$ admits a BSP of size $O(n\sqrt{n})$, and that this bound is tight in the worst case.

The discrepancy between the quadratic lower bound on the worst-case size of a BSP in 3-dimensional space and their size in practice led de Berg et al. [39]

to study BSPs for scenes with special properties. In particular they showed that any set of *fat objects* in the plane—which means that long and skinny objects are forbidden—admits a BSP of linear size. A set of line segments where the ratio between the length of the longest and the shortest one is bounded admits a linear size BSP as well. These results strengthen the belief that any set of segments in the plane should admit a linear size BSP, but the best known worst-case bound is still the $O(n \log n)$ bound of Paterson and Yao. The result on fat objects was extended to higher dimensions and to a larger class of objects by de Berg [37].

Finally, we remark that two other well-known structures, kd-trees and quad trees, are in fact special cases of BSP trees, where only orthogonal splitting planes are used. Kd-trees were discussed extensively in Chapter 5 and quad trees will be discussed in Chapter 14.

12.6 Exercises

12.1 Prove that PAINTERSALGORITHM is correct. That is, prove that if (some part of) an object A is scan-converted before (some part of) object B is scan-converted, then A cannot lie in front of B.

12.2 Let S be a set of m polygons in the plane with n vertices in total. Let $\mathcal{T}$ be a BSP tree for S of size k. Prove that the total complexity of the fragments generated by the BSP is $O(n + k)$.

12.3 Give an example of a set of line segments in the plane where the greedy method of constructing an auto-partition (where the splitting line $\ell(s)$ is taken that induces the least number of cuts) results in a BSP of quadratic size.

12.4 Give an example of a set S of n non-intersecting line segments in the plane for which a BSP tree of size n exists, whereas any auto-partition of S has size at least $\lfloor 4n/3 \rfloor$.

12.5 We have shown that the expected size of the partitioning produced by 2DRANDOMBSP is $O(n \log n)$. What is the worst-case size?

12.6 Suppose we apply 2DRANDOMBSP to a set of *intersecting* line segments in the plane. Can you say anything about the expected size of the resulting BSP tree?

12.7 In 3DRANDOMBSP2, it is not described how to find the cells that must be split when a splitting plane is added, nor is it described how to perform the split efficiently. Work out the details for this step, and analyze the running time of your algorithm.

12.8 Give a deterministic divide-and-conquer algorithm that constructs a BSP tree of size $O(n \log n)$ for a set of n line segments in the plane. *Hint:* Use as many free splits as possible and use vertical splitting lines otherwise.

12.9 In Chapter 5 kd-trees were introduced. These trees are in fact a special type of BSP tree, where the splitting lines for nodes at even depth in the tree are horizontal and the splitting lines at odd levels are vertical.

    a.   Discuss the advantages and/or disadvantages of BSP trees over kd-trees.

    b.   For any set of two non-intersecting line segments in the plane there exists a BSP tree of size 2. Prove that there is no constant c such that for any set of two non-intersecting line segments there exists a kd-tree of size at most c.

12.10* Prove or disprove: For any set of n non-intersecting line segments in the plane, there exists a BSP tree of size $O(n)$. (This problem is still open and probably difficult.)

13 Robot Motion Planning
Getting Where You Want to Be

One of the ultimate goals in robotics is to design autonomous robots: robots that you can tell *what* to do without having to say *how* to do it. Among other things, this means a robot has to be able to plan its own motion.

To be able to plan a motion, a robot must have some knowledge about the environment in which it is moving. For example, a mobile robot moving around in a factory must know where obstacles are located. Some of this information—where walls and machines are located—can be provided by a floor plan. For other information the robot will have to rely on its sensors. It should be able to detect obstacles that are not on the floor plan—people, for instance. Using the information about the environment, the robot has to move to its goal position without colliding with any of the obstacles.

This *motion planning problem* has to be solved whenever any kind of robot wants to move in physical space. The description above assumed that we have an autonomous robot moving around in some factory environment. That kind of robot is still quite rare compared to the robot arms that are now widely employed in industry.

A robot arm, or *articulated robot*, consists of a number of links, connected by joints. Normally, one end of the arm—its *base*—is firmly connected to the ground, while the other end carries a *hand* or some kind of tool. The number of links varies between three to six or even more. The joints are usually of two types, the *revolute joint* type that allows the links to rotate around the joint, much like an elbow, or the *prismatic joint* type that allows one of the links to slide in and out. Robot arms are mostly used to assemble or manipulate parts of an object, or to perform tasks like welding or spraying. To do this, they must be able to move from one position to another, without colliding with the environment, the object they are operating on, or—an interesting complication—with themselves.

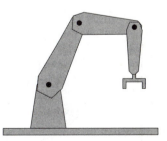

In this chapter we introduce some of the basic notions and techniques used in motion planning. The general motion planning problem is quite difficult, and we shall make some simplifying assumptions.

The most drastic simplification is that we will look at a 2-dimensional mo-

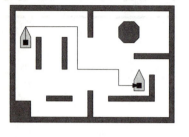

tion planning problem. The environment will be a planar region with polygonal obstacles, and the robot itself will also be polygonal. We also assume that the environment is static—there are no people walking in the way of our robot—and known to the robot. The restriction to planar robots is not as severe as it appears at first sight: for a robot moving around on a work floor, a floor plan showing the location of walls, tables, machines, and so on, is often sufficient to plan a motion.

The types of motions a robot can execute depend on its mechanics. Some robots can move in any direction, while others are constrained in their motions. *Car-like robots*, for instance, cannot move sideways—otherwise parallel parking would be less challenging. In addition, they often have a certain minimum turning radius. The geometry of the motions of car-like robots is quite complicated, so we will restrict ourselves to robots that can move in arbitrary directions. In fact, we will mainly look at robots that can translate only; at the end of the chapter we'll briefly consider the case of robots that can also change their orientation by rotation.

13.1 Work Space and Configuration Space

Let $\mathcal{R}$ be a robot moving around in a 2-dimensional environment, or *work space*, consisting of a set $S = \{\mathcal{P}_1, \ldots, \mathcal{P}_t\}$ of obstacles. We assume that $\mathcal{R}$ is a simple polygon. A *placement*, or *configuration*, of the robot can now be specified by a translation vector. We denote the robot translated over a vector (x, y) by $\mathcal{R}(x, y)$. For instance, if the robot is the polygon with vertices $(1, -1)$, $(1, 1)$, $(0, 3)$, $(-1, 1)$, and $(-1, -1)$, then the vertices of $\mathcal{R}(6, 4)$ are $(7, 3)$, $(7, 5)$, $(6, 7)$, $(5, 5)$, and $(5, 3)$. With this notation, a robot can be specified by listing the vertices of $\mathcal{R}(0, 0)$.

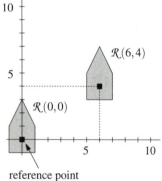

reference point

An alternative way to view this is in terms of a *reference point*. This is most intuitive if the origin $(0, 0)$ lies in the interior of $\mathcal{R}(0, 0)$. By definition, this point is then called the reference point of the robot. We can specify a placement of $\mathcal{R}$ by simply stating the coordinates of the reference point if the robot is in the given placement. Thus $\mathcal{R}(x, y)$ specifies that the robot is placed with its reference point at (x, y). In general, the reference point does not have to be inside the robot; it can also be a point outside the robot, which we might imagine to be attached to the robot by an invisible stick. By definition, this point is at origin for $\mathcal{R}(0, 0)$.

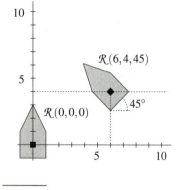

Now suppose the robot can change its orientation by rotation, say around its reference point. We then need an extra parameter, ϕ, to specify the orientation of the robot. We let $\mathcal{R}(x, y, \phi)$ denote the robot with its reference point at (x, y) and rotated through an angle ϕ. So what is specified initially is $\mathcal{R}(0, 0, 0)$.

In general, a placement of a robot is specified by a number of parameters that corresponds to the number of *degrees of freedom (DOF)* of the robot. This number is two for planar robots that can only translate, and it is three for planar robots that can rotate as well as translate. The number of parameters we need for a robot in 3-dimensional space is higher, of course: a translating robot in $\mathbb{R}^3$ has three degrees of freedom, and a robot that is free to translate and rotate in $\mathbb{R}^3$ has six degrees of freedom.

The parameter space of a robot $\mathcal{R}$ is usually called its *configuration space*. It is denoted by $C(\mathcal{R})$. A point p in this configuration space corresponds to a certain placement $\mathcal{R}(p)$ of the robot in the work space.

In the example of a translating and rotating robot in the plane the configuration space is 3-dimensional. A point (x, y, ϕ) in this space corresponds to the placement $\mathcal{R}(x, y, \phi)$ in the work space. The configuration space is not the Euclidean 3-dimensional space; it is the space $\mathbb{R}^2 \times [0 : 360)$. Because rotations over zero and 360 degrees are equivalent, the configuration space of a rotating robot has a special topology, which is like a cylinder.

The configuration space of a translating robot in the plane is the 2-dimensional Euclidean plane, and therefore identical to the work space. Still, it is useful to distinguish the two notions: the work space is the space where the robot actually moves around—the real world, so to speak—and the configuration space is the parameter space of the robot. A polygonal robot in the work space is represented by a point in configuration space, and any point in configuration space corresponds to some placement of an actual robot in work space.

We now have a way to specify a placement of the robot, namely by specifying values for the parameters determining the placement or, in other words, by specifying a point in configuration space. But clearly not all points in configuration space are possible; points corresponding to placements where the robot intersects one of the obstacles in S are forbidden. We call the part of the configuration space consisting of these points the *forbidden configuration space*, or *forbidden space* for short. It is denoted by $C_{\mathrm{forb}}(\mathcal{R}, S)$. The rest of the configuration space, which consists of the points corresponding to *free placements*—placements where the robot does not intersect any obstacle—is called the *free configuration space*, or *free space*, and it is denoted by $C_{\mathrm{free}}(\mathcal{R}, S)$.

A path for the robot maps to a curve in the configuration space, and vice versa: every placement along the path simply maps to the corresponding point in configuration space. A collision-free path maps to a curve in the free space. Figure 13.1 illustrates this for a translating planar robot. On the left the work space is shown, with a collision-free path from the initial position to the goal position of the robot. On the right the configuration space is shown, with the grey area indicating the forbidden part of it. The unshaded area in between the grey area is the free space. For clarity, the obstacles are still shown in the configuration space, although they have no meaning there. The curve corresponding to the collision-free path is also shown.

We have seen how to map placements of the robot to points in the configuration space, and paths of the robot to curves in that space. Can we also map obstacles to configuration space? The answer is yes: an obstacle $\mathcal{P}$ is mapped to the set of points p in configuration space such that $\mathcal{R}(p)$ intersects $\mathcal{P}$. The resulting set is called the *configuration-space obstacle*, or *C-obstacle* for short, of $\mathcal{P}$.

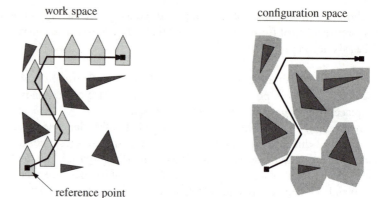

work space   configuration space

Figure 13.1
A path in the work space and the
corresponding curve in the
configuration space

reference point

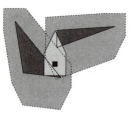

C-obstacles may overlap even when the obstacles in the work space are disjoint. This happens when there are placements of the robot where it intersects more than one obstacle at the same time.

There is one subtle issue that we have ignored so far: does the robot collide with an obstacle when it touches that obstacle? In other words, do we define the obstacles to be topologically open or closed sets? In the remainder we will choose the first option: obstacles are open sets, so that the robot is allowed to touch them. This is of little importance in this chapter, but it will become useful in Chapter 15. In practice a movement where the robot passes very close to an obstacle cannot be considered safe because of possible errors in robot control. Such movements can be avoided by slightly enlarging all the obstacles before the computation of a path.

13.2 A Point Robot

Before we try to plan the motion of a polygonal robot in the plane, let's have a look at point robots. Given the mapping from work space to configuration space that we saw in the previous section this is not such a strange idea. Furthermore, it's always good to start with a simple case. As before, we denote the robot by $\mathcal{R}$ and we denote the obstacles by $\mathcal{P}_1, \ldots, \mathcal{P}_t$. The obstacles are polygons with disjoint interiors, whose total number of vertices is denoted by n. For a point robot, the work space and the configuration space are identical. (That is to say, if we make the natural assumption that its reference point is the point robot itself. Otherwise the configuration space is a translated copy of the work space.)

Rather than finding a path from a particular start position to a particular goal position we will construct a data structure storing a representation of the free space. This data structure can then be used to compute a path between any two given start and goal positions. Such an approach is useful if the work space of the robot does not change and many paths have to be computed.

To simplify the description we restrict the motion of the robot to a large bounding box B that contains the set of polygons. In other words, we add one extra infinitely large obstacle, which is the area outside B. The free configuration space C_{free} now consists of the part of B not covered by any obstacle:

$$C_{\text{free}} = B \setminus \bigcup_{i=1}^{t} \mathcal{P}_i.$$

The free space is a possibly disconnected region, which may have holes. Our goal is to compute a representation of the free space that allows us to find a path for any start and goal position. We will use the *trapezoidal map* for this. Recall from Chapter 6 that the trapezoidal map of a set of non-intersecting line segments inside a bounding box is obtained by drawing two vertical extensions from every segment endpoint, one going upward until a segment (or the bounding box) is hit, and one going downward until a segment (or the bounding box) is hit. In Chapter 6 we developed a randomized algorithm, TRAPEZOIDALMAP, that computes the trapezoidal map of a set of n segments in $O(n \log n)$ expected time. The following algorithm, which computes a representation of the free space, uses this algorithm as a subroutine.

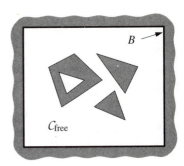

Algorithm COMPUTEFREESPACE(S)

Input. A set S of disjoint polygons.
Output. A trapezoidal map of $C_{\text{free}}(\mathcal{R}, S)$ for a point robot $\mathcal{R}$.
1. Let E be the set of edges of the polygons in S.
2. Compute the trapezoidal map $\mathcal{T}(E)$ with algorithm TRAPEZOIDALMAP described in Chapter 6.
3. Remove the trapezoids that lie inside one of the polygons from $\mathcal{T}(E)$ and return the resulting subdivision.

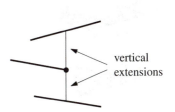

vertical
extensions

The algorithm is illustrated in Figure 13.2. Part (a) of the figure shows the trapezoidal map of the obstacle edges inside the bounding box; this is what is computed in line 2 of the algorithm. Part (b) shows the map after the trapezoids inside the obstacles have been removed in line 3.

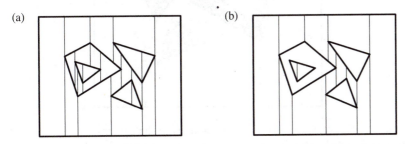

(a)   (b)

Figure 13.2
Computing a trapezoidal map of the free space

There is one detail left: how do we find the trapezoids inside the obstacles, which have to be removed? This is not so difficult, because after running TRAPEZOIDALMAP we know for each trapezoid the edge that bounds it from the top, and we known to which obstacle that edge belongs. To decide whether or not to remove the trapezoid, it suffices to check whether the edge bounds the obstacle from above or from below. The latter test takes only constant time, because the edges of the obstacles are listed in order along the boundary so that the obstacle lies to a specific, known side of the edges.

The expected time taken by TRAPEZOIDALMAP is $O(n \log n)$, so we get the following result.

Lemma 13.1 *A trapezoidal map of the free configuration space for a point robot moving among a set of disjoint polygonal obstacles with n edges in total can be computed by a randomized algorithm in $O(n \log n)$ expected time.*

In what follows, we will denote the trapezoidal map of the free space by $\mathcal{T}(C_{\text{free}})$.

How do we use $\mathcal{T}(C_{\text{free}})$ to find a path from a start position p_{start} to a goal position p_{goal}?

If p_{start} and p_{goal} are in the same trapezoid of the map, this is easy: the robot can simply move to its goal in a straight line.

If the start and goal position are in different trapezoids, however, then things are not so easy. In this case the path will cross a number of trapezoids and it may have to make turns in some of them. To guide the motion across trapezoids we construct a *road map* through the free space. The road map is a graph $\mathcal{G}_{\text{road}}$, which is embedded in the plane. More precisely, it is embedded in the free space. Except for an initial and final portion, paths will always follow the road map. Notice that any two neighboring trapezoids share a vertical edge that is a vertical extension of a segment endpoint. This leads us to define the road map as follows. We place one node in the center of each trapezoid, and we place one node in the middle of each vertical extension. There is an arc

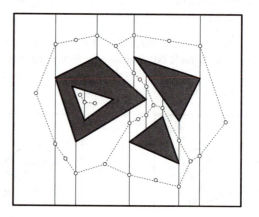

Figure 13.3
A road map

between two nodes if and only if one node is in the center of a trapezoid and the other node is on the boundary of that same trapezoid. The arcs are embedded in the plane as straight line segments, so following an arc in the road map corresponds to a straight-line motion of the robot. Figure 13.3 illustrates this. The road map $\mathcal{G}_{\text{road}}$ can be constructed in $O(n)$ time by traversing the doubly-connected edge list of $\mathcal{T}(C_{\text{free}})$. Using the arcs in the road map we can go from the node in the center of one trapezoid to the node in the center of a neighboring trapezoid via the node on their common boundary.

We can use the road map, together with the trapezoidal map, to plan a motion from a start to a goal position. To this end we first determine the trapezoids Δ_{start} and Δ_{goal} containing these points. If they are the same trapezoid, then we move from p_{start} to p_{goal} in a straight line. Otherwise, let v_{start} and v_{goal} be the nodes of G_{road} that have been placed in the center of these trapezoids. The path from p_{start} to p_{goal} that we will construct now consists of three parts: the first part is a straight-line motion from p_{start} to v_{start}, the second part is a path from v_{start} to v_{goal} along the arcs of the road map, and the final part is a straight-line motion from v_{goal} to p_{goal}. Figure 13.4 illustrates this.

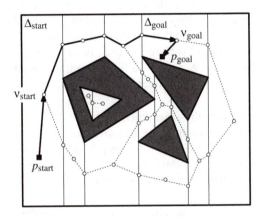

Figure 13.4
A path computed from the road map of Figure 13.3

The following algorithm summarizes how a path is found.

Algorithm COMPUTEPATH($\mathcal{T}(C_{free}), G_{road}, p_{start}, p_{goal}$)
Input. The trapezoidal map $\mathcal{T}(C_{free})$ of the free space, the road map G_{road}, a start position p_{start}, and goal position p_{goal}.
Output. A path from p_{start} to p_{goal} if it exists. If a path does not exist, this fact is reported.
1. Find the trapezoid Δ_{start} containing p_{start} and the trapezoid Δ_{goal} containing p_{goal}.
2. **if** Δ_{start} or Δ_{goal} does not exist
3.   **then** Report that the start or goal position is in the forbidden space.
4.   **else** Let v_{start} be the node of G_{road} in the center of Δ_{start}.
5.   Let v_{goal} be the node of G_{road} in the center of Δ_{goal}.
6.   Compute a path in G_{road} from v_{start} to v_{goal} using breadth-first search.
7.   **if** there is no such path
8.   **then** Report that there is no path from p_{start} to p_{goal}.
9.   **else** Report the path consisting of a straight-line motion from p_{start} to v_{start}, the path found in G_{road}, and a straight-line motion from v_{goal} to p_{goal}.

Before we analyze the time complexity of algorithm, let's think about its correctness. Are the paths we report always collision-free, and do we always find a collision-free path if one exists?

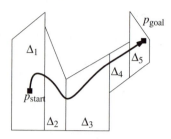

The first question is easy to answer: any path we report must be collision-free, since it consists of segments inside trapezoids and all trapezoids are in the free space.

To answer the second question, suppose that there is a collision-free path from p_{start} to p_{goal}. Obviously p_{start} and p_{goal} must lie in one of the trapezoids covering the free space, so it remains to show that there is a path in $\mathcal{G}_{road}$ from v_{start} to v_{goal}. The path from p_{start} to p_{goal} must cross a sequence of trapezoids. Denote the sequence of trapezoids by $\Delta_1, \Delta_2, \ldots, \Delta_k$. By definition, $\Delta_1 = \Delta_{start}$ and $\Delta_k = \Delta_{goal}$. Let v_i be the node of $\mathcal{G}_{road}$ that is in the center of Δ_i. If the path goes from Δ_i to Δ_{i+1}, then Δ_i and Δ_{i+1} must be neighbors, so they share a vertical extension. But $\mathcal{G}_{road}$ is constructed such that the nodes of such trapezoids are connected through the node on their common boundary. Hence, there is a path (consisting of two arcs) in $\mathcal{G}_{road}$ from v_i to v_{i+1}. This means that there is a path from v_1 to v_k as well. It follows that the breadth-first search in $\mathcal{G}_{road}$ will find some (possibly different) path from v_{start} to v_{goal}.

We now analyze the time the algorithm takes.

Finding the trapezoids containing the start and goal can be done in $O(\log n)$ using the point location structure of Chapter 6. Alternatively, we can simply check all trapezoids in linear time; we shall see that the rest of the algorithm takes linear time anyway, so this does not increase the time bound asymptotically.

The breadth-first search takes linear time in the size of the graph $\mathcal{G}_{road}$. This graph has one node per trapezoid plus one node per vertical extension. Both the number of vertical extensions and the number of trapezoids are linear in the total number of vertices of the obstacles. The number of arcs in the graph is linear as well, because it is planar. Hence, the breadth-first search takes $O(n)$ time.

The time to report the path is bounded by the maximum number of arcs on a path in $\mathcal{G}_{road}$, which is $O(n)$.

We get the following theorem.

Theorem 13.2 *Let $\mathcal{R}$ be a point robot moving among a set S of polygonal obstacles with n edges in total. We can preprocess S in $O(n \log n)$ expected time, such that between any start and goal position a collision-free path for $\mathcal{R}$ can be computed in $O(n)$ time, if it exists.*

The path computed by the algorithm of this section is collision-free, but we can give no guarantee that the path does not make large detours. In Chapter 15 we will develop an algorithm that actually computes the shortest possible path. That algorithm, however, will be slower by an order of magnitude.

13.3 Minkowski Sums

In the previous section we solved the motion planning problem for a point robot; we computed a trapezoidal map of its free space and used that map to plan its motions. The same approach can be used if the robot is a polygon. There is one difference that makes dealing with a polygonal robot more difficult: the configuration-space obstacles are no longer the same as the obstacles in work space. Therefore we start by studying the free configuration space of a translating polygonal robot. In the next section we will then describe how to compute it, so that we can use it to plan the motion of the robot.

We assume that the robot $\mathcal{R}$ is convex, and for the moment we also assume that the obstacles are convex. Recall that we use $\mathcal{R}(x,y)$ to denote the placement of $\mathcal{R}$ with its reference point at (x,y). The configuration-space obstacle, or C-obstacle, of an obstacle $\mathcal{P}$ and the robot $\mathcal{R}$ is defined as the set of points in configuration space such that the corresponding placement of $\mathcal{R}$ intersects $\mathcal{P}$. So if we denote the C-obstacle of $\mathcal{P}$ by $C\mathcal{P}$, then we have

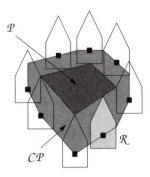

$$CP := \{(x,y) \ : \ \mathcal{R}(x,y) \cap \mathcal{P} \neq \emptyset\}.$$

You can visualize the shape of $C\mathcal{P}$ by sliding $\mathcal{R}$ along the boundary of $\mathcal{P}$; the curve traced by the reference point of $\mathcal{R}$ is the boundary of $C\mathcal{P}$.

We can describe this in a different way using the notion of *Minkowski sums*. The Minkowski sum of two sets $S_1 \subset \mathbb{R}^2$ and $S_2 \subset \mathbb{R}^2$, denoted by $S_1 \oplus S_2$, is defined as

$$S_1 \oplus S_2 := \{p+q \ : \ p \in S_1, q \in S_2\},$$

where $p+q$ denotes the vector sums of the vectors p and q, that is, if $p = (p_x, p_y)$ and $q = (q_x, q_y)$ then we have

$$p+q := (p_x + q_x, p_y + q_y).$$

Because a polygon is a planar set the definition of Minkowski sums also applies to them.

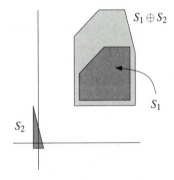

To be able to express the C-obstacles as Minkowski sums, we need one more piece of notation. For a point $p = (p_x, p_y)$ we define $-p := (-p_x, -p_y)$, and for a set S we define $-S := \{-p \ : \ p \in S\}$. In other words, we get $-S$ by reflecting S about the origin. We now have the following theorem.

Theorem 13.3 *Let $\mathcal{R}$ be a planar, translating robot and let $\mathcal{P}$ be an obstacle. Then the C-obstacle of $\mathcal{P}$ is $\mathcal{P} \oplus (-\mathcal{R}(0,0))$.*

Proof. We have to prove that $\mathcal{R}(x,y)$ intersects $\mathcal{P}$ if and only if we have that $(x,y) \in \mathcal{P} \oplus (-\mathcal{R}(0,0))$.

First, suppose that $\mathcal{R}(x,y)$ intersects $\mathcal{P}$, and let $q = (q_x, q_y)$ be a point in the intersection. It follows from $q \in \mathcal{R}(x,y)$ that we have $(q_x - x, q_y - y) \in \mathcal{R}(0,0)$ or, equivalently, that $(-q_x + x, -q_y + y) \in -\mathcal{R}(0,0)$. Because we also have $q \in \mathcal{P}$, this implies that $(x,y) \in \mathcal{P} \oplus (-\mathcal{R}(0,0))$.

Conversely, let $(x,y) \in \mathcal{P} \oplus (-\mathcal{R}(0,0))$. Then there are points $(r_x, r_y) \in \mathcal{R}(0,0)$ and $(p_x, p_y) \in \mathcal{P}$ such that $(x,y) = (p_x - r_x, p_y - r_y)$ or, in other words, such that $p_x = r_x + x$ and $p_y = r_y + y$, which implies that $\mathcal{R}(x,y)$ intersects $\mathcal{P}$. $\square$

So for a planar translating robot $\mathcal{R}$ the C-obstacles are the Minkowski sums of the obstacles and $-\mathcal{R}(0,0)$. (Sometimes $\mathcal{P} \oplus (-\mathcal{R}(0,0))$ is referred to as the *Minkowski difference* of $\mathcal{P}$ and $\mathcal{R}(0,0)$. Since Minkowski differences are defined differently in the mathematics literature we shall avoid this term.)

In the remainder of this section we will derive some useful properties of Minkowski sums and develop an algorithm to compute them.

We start with a simple observation about extreme points on Minkowski sums.

Observation 13.4 *Let* $\mathcal{P}$ *and* $\mathcal{R}$ *be two objects on the plane, and let* $C\mathcal{P} := \mathcal{P} \oplus \mathcal{R}$. *An extreme point in direction* $\vec{d}$ *on* $C\mathcal{P}$ *is the sum of extreme points in direction* $\vec{d}$ *on* $\mathcal{P}$ *and* $\mathcal{R}$.

Figure 13.5 illustrates the observation. Using this observation we now prove that the Minkowski sum of two convex polygons has linear complexity.

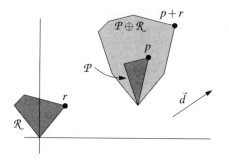

Figure 13.5
An extreme point on a Minkowski sum
is the sum of extreme points.

Theorem 13.5 *Let* $\mathcal{P}$ *and* $\mathcal{R}$ *be two convex polygons with* n *and* m *edges, respectively. Then the Minkowski sum* $\mathcal{P} \oplus \mathcal{R}$ *is a convex polygon with at most* $n + m$ *edges.*

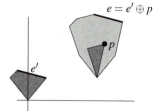

Proof. The convexity of the Minkowski sum of two convex sets follows directly from the definition.

To see that the complexity of the Minkowski sum is linear, consider an edge e of $\mathcal{P} \oplus \mathcal{R}$. This edge is extreme in the direction of its outer normal. Hence, it must be generated by points on $\mathcal{P}$ and $\mathcal{R}$ that are extreme in the same direction. Moreover, at least one of $\mathcal{P}$ and $\mathcal{R}$ must have an edge that is extreme in that direction. We charge e to this edge. This way each edge is charged at most once, so the total number of edges is at most $n + m$. (If $\mathcal{P}$ and $\mathcal{R}$ don't have parallel edges, then the number of edges of the Minkowski sum is exactly $n + m$.) ☐

So the Minkowski sum of two convex polygons is convex and has linear complexity. But there is more: the boundaries of two Minkowski sums can only intersect in a very special manner. To make this precise, we need one more piece of terminology.

We call a pair o_1, o_2 of planar objects a pair of *pseudodiscs* if it satisfies the *pseudodisc property*: the sets $o_1 \setminus o_2$ and $o_2 \setminus o_1$ are connected.

Figure 13.6 illustrates this definition. A collection of objects is called a collection of pseudodiscs if every pair of objects is a pair of pseudodiscs. Notice that any collection of discs forms a collection of pseudodiscs. Also a collection of axis-parallel squares is a collection of pseudodiscs.

pseudodiscs   not pseudodiscs

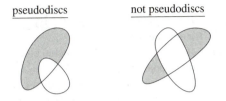

Figure 13.6
The pseudodisc property

Define a proper intersection of two curves to be an intersection consisting of a single point that is not a point where the curves touch. The boundaries of pseudodiscs have the following important property.

Observation 13.6 *For any pair of pseudodiscs there are at most two proper intersections between the boundaries.*

Below we will prove that a collection of Minkowski sums forms a collection of pseudodiscs. But first we need one more observation about directions and extreme points on pairs of convex polygons with disjoint interiors. We will say that one polygon is more extreme in a direction $\vec{d}$ than another polygon if its extreme points lie further in that direction than the extreme points of the other polygon. For instance, a polygon is more extreme in the positive x-direction if its rightmost points lie to the right of the rightmost points of the other polygon.

We will look at extreme points for various directions. To this end we model the set of all directions by the unit circle centered at the origin: a point p on the unit circle represents the direction given by the vector from the origin to p. The range from a direction $\vec{d}_1$ to a direction $\vec{d}_2$ is defined as the the directions corresponding to points in the counter-clockwise circle segment from the point representing $\vec{d}_1$ to the point representing $\vec{d}_2$. Note that the range from $\vec{d}_1$ to $\vec{d}_2$ is not the same as the range from $\vec{d}_2$ to $\vec{d}_1$. The following observation is illustrated in Figure 13.7.

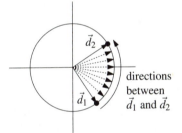

directions between $\vec{d}_1$ and $\vec{d}_2$

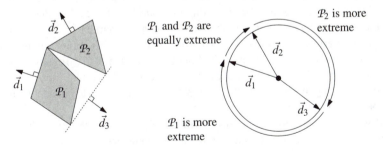

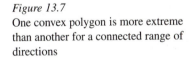

Observation 13.7 *Let $\mathcal{P}_1$ and $\mathcal{P}_2$ be convex polygons with disjoint interiors. Let $\vec{d}_1$ and $\vec{d}_2$ be directions in which $\mathcal{P}_1$ is more extreme than $\mathcal{P}_2$. Then either $\mathcal{P}_1$ is more extreme than $\mathcal{P}_2$ in all directions in the range from $\vec{d}_1$ to $\vec{d}_2$, or it is more extreme in all directions in the range from $\vec{d}_2$ to $\vec{d}_1$.*

We are now ready to prove that Minkowski sums are pseudodiscs.

Theorem 13.8 *Let P_1 and P_2 be two convex polygons with disjoint interiors, and let R be another convex polygon. Then the two Minkowski sums $P_1 \oplus R$ and $P_2 \oplus R$ are pseudodiscs.*

Proof. Define $CP_1 := P_1 \oplus R$ and $CP_2 := P_2 \oplus R$. By symmetry, it suffices to show that $CP_1 \setminus CP_2$ is connected.

Since CP_1 and CP_2 are both convex, a connected component of $CP_1 \setminus CP_2$ must contain a point on the convex hull of $CP_1 \cup CP_2$. If there is more than one connected component, that means that there must be two directions $\vec{d_1}$ and $\vec{d_2}$ such that CP_1 is more extreme in those directions than CP_2. From Observation 13.4 it follows that P_1 is more extreme than P_2 in those two directions. By Observation 13.7 this means that either P_1 is more extreme than P_2 for all directions in the range from $\vec{d_1}$ to $\vec{d_2}$, or for all directions in the range from $\vec{d_2}$ to $\vec{d_1}$. Applying Observation 13.4 once more we conclude that the same holds for CP_1 with respect to CP_2. In other words, the set of directions for which the extreme points on CP_1 lie further out than those on CP_2 is connected, a contradiction to the assumption that there are two connected components. $\square$

This result is useful in combination with the following theorem.

Theorem 13.9 *Let S be a collection of polygonal pseudodiscs with n edges in total. Then the complexity of their union is $O(n)$.*

Proof. We prove this by charging every vertex of the union to a pseudodisc vertex in such a way that any pseudodisc vertex is charged at most two times. This leads to a bound of $2n$ on the maximal complexity of the union.

The charging is done as follows. There are two types of vertices in the union boundary: pseudodisc vertices and intersections of pseudodisc edges.

Vertices of the former type are simply charged to themselves.

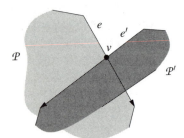

The charging for intersection points is a little more difficult. Consider a vertex v of the union that is the intersection of an edge e of a pseudodisc $P \in S$ and an edge e' of a pseudodisc $P' \in S$. Imagine that these edges are directed such that they enter the interior of the other pseudodisc at v. Follow the edge e into the interior of P'. If it ends in the interior, then we charge v to the endpoint of e in the interior of P'. Otherwise e leaves the interior. The point where this happens is an intersection point of the boundaries of P and P'. Together with v this makes two intersection points, so that e' cannot leave the interior of P without violating the pseudodisc property. Hence, e' has a vertex inside P, and we charge v to this vertex.

If we do the charging in this way, then every pseudodisc vertex is charged at most twice. For vertices that are on the union boundary this is obvious: they only get charged by themselves. For the other vertices this can be seen as follows. From such a vertex, follow its incident edges until the union boundary is reached at an intersection point with some other edge; these two intersection points, if they exist, are the only ones that get charged to the vertex. $\square$

The proof of this theorem depends heavily on the pseudodiscs being polygonal, but the theorem itself generalizes to arbitrary pseudodiscs: the complexity of the union of any set of pseudodiscs is linear in the total complexity of the pseudodiscs. This implies for instance that the union of n discs in the plane has $O(n)$ complexity. This more general theorem is a lot more difficult to prove.

Before we return to our motion planning application, we will give an algorithm to compute the Minkowski sum of two convex polygons P and R. A very simple algorithm is the following. For each pair v, w of vertices, with $v \in P$ and $w \in R$, compute $v + w$. Next, compute the convex hull of all these sums. Although the algorithm is very simple, it is inefficient when the polygons have a lot of vertices, because it looks at every pair of vertices. Below we give an alternative algorithm, which is as easy to implement. It only looks at pairs of vertices that are extreme in the same direction—this is allowed because of Observation 13.4—which makes it run in linear time. In the algorithm we use the notation $angle(pq)$ to denote the angle that the vector $\vec{pq}$ makes with the positive x-axis.

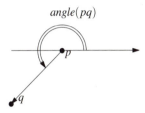

$angle(pq)$

Algorithm MINKOWSKISUM(P, R)
Input. A convex polygon P with vertices $v_1, \ldots, v_n$, and a convex polygon R with vertices $w_1, \ldots, w_m$. The lists of vertices are assumed to be in counter-clockwise order, with v_1 and w_1 being the vertices with smallest y-coordinate (and smallest x-coordinate in case of ties).
Output. The Minkowski sum $P \oplus R$.
1. $i \leftarrow 1; j \leftarrow 1$
2. $v_{n+1} \leftarrow v_1; w_{m+1} \leftarrow w_1$
3.   **repeat**
4.   Add $v_i + w_j$ as a vertex to $P \oplus R$.
5.   **if** $angle(v_i v_{i+1}) < angle(w_j w_{j+1})$
6.   **then** $i \leftarrow (i+1)$
7.   **else if** $angle(v_i v_{i+1}) > angle(w_j w_{j+1})$
8.   **then** $j \leftarrow (j+1)$
9.   **else** $i \leftarrow (i+1)$
10. $j \leftarrow (j+1)$
11. **until** $i = n+1$ **and** $j = m+1$

MINKOWSKISUM runs in linear time, because at each execution of the **repeat** loop either i or j is incremented and—as is not difficult to prove—they will not be incremented after reaching the values $n + 1$ and $m + 1$. The fact that the correct pairs of vertices are taken is similar to the proof of Theorem 13.5; one just has to observe that any vertex of the Minkowski sum is the sum of two original vertices that are extreme in a common direction, and argue that the angle test ensures that all extreme pairs are found.

We conclude with the following theorem:

Theorem 13.10 *The Minkowski sum of two convex polygons with n and m vertices, respectively, can be computed in $O(n + m)$ time.*

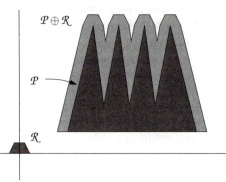

What happens if one or both of the polygons are not convex? This question is not so hard to answer if we realize that the following equality holds for any sets S_1, S_2, and S_3:

$$S_1 \oplus (S_2 \cup S_3) = (S_1 \oplus S_2) \cup (S_1 \oplus S_3).$$

Now consider the Minkowski sum of a non-convex polygon $\mathcal{P}$ and a convex polygon $\mathcal{R}$ with n and m vertices respectively. What is the complexity of $\mathcal{P} \oplus \mathcal{R}$? We know from Chapter 3 that the polygon $\mathcal{P}$ can be triangulated into $n-2$ triangles $t_1, \ldots, t_{n-2}$, where n is its number of vertices. From the equality above we can conclude that

$$\mathcal{P} \oplus \mathcal{R} = \bigcup_{i=1}^{n-2} t_i \oplus \mathcal{R}.$$

Since t_i is a triangle and $\mathcal{R}$ a convex polygon with m vertices, we know that $t_i \oplus \mathcal{R}$ is a convex polygon with at most $m+3$ vertices. Moreover, the triangles have disjoint interiors, so the collection of Minkowski sums is a collection of pseudodiscs. Hence, the complexity of their union is linear in the sum of their complexities. This implies that the complexity of $\mathcal{P} \oplus \mathcal{R}$ is $O(nm)$.

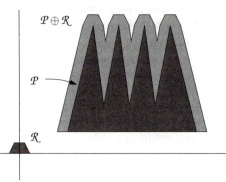

Figure 13.8
The Minkowski sum of a non-convex
and a convex polygon

This upper bound on the complexity of a non-convex and a convex polygon is tight in the worst case. To see this, consider a polygon $\mathcal{P}$ with $\lfloor n/2 \rfloor$ spikes pointing upward, and a much smaller polygon $\mathcal{R}$ that is the top half of a regular $(2m-2)$-gon. The Minkowski sum of these polygons will also have $\lfloor n/2 \rfloor$ spikes, each of which has m vertices at its top. Figure 13.8 illustrates the construction.

To bound the complexity of the Minkowski sum of two non-convex polygons $\mathcal{P}$ and $\mathcal{R}$, we triangulate both polygons. We get a collection of $n-2$ triangles t_i, and a collection of $m-2$ triangles u_j. The Minkowski sum of $\mathcal{P}$ and $\mathcal{R}$ is now the union of the Minkowski sums of the pairs t_i, u_j. Each sum $t_i \oplus u_j$ has constant complexity. Hence, $\mathcal{P} \oplus \mathcal{R}$ is the union of $(n-2)(m-2)$ constant-complexity polygons. This implies that the total complexity of $\mathcal{P} \oplus \mathcal{R}$ is $O(n^2 m^2)$. Again, this bound is tight in the worst case: there are non-convex

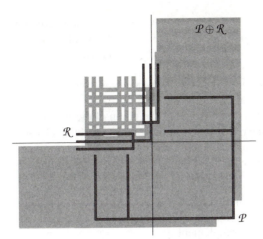

Figure 13.9
The Minkowski sum of two
non-convex polygons

polygons whose Minkowski sum really has $\Theta(n^2m^2)$ complexity. Figure 13.9 illustrates this.

The following theorem summarizes the results on the complexity of Minkowski sums. For completeness the complexity in the case of two convex polygons is given as well.

Theorem 13.11 *Let $\mathcal{P}$ and $\mathcal{R}$ be polygons with n and m vertices, respectively. The complexity of the Minkowski sum $\mathcal{P} \oplus \mathcal{R}$ is bounded as follows:*
(i)   it is $O(n+m)$ if both polygons are convex;
(ii) it is $O(nm)$ if one of the polygons is convex and one is non-convex;
(iii) it is $O(n^2m^2)$ if both polygons are non-convex.
These bounds are tight in the worst case.

Computing Minkowski sums of non-convex polygons is not very difficult: triangulate both polygons, compute the Minkowski sum of each pair of triangles, and take their union. This approach is basically the same as the approach described in the next section for computing the forbidden space of a translating robot, so we omit the details here.

13.4   Translational Motion Planning

It is time to return to the planar motion planning problem. Recall that our robot $\mathcal{R}$ can only translate and that the obstacles are disjoint polygons. Early in the previous section we have shown that the C-obstacle corresponding to an obstacle $\mathcal{P}_i$ is the Minkowski sum $\mathcal{P}_i \oplus (-\mathcal{R})$. Moreover, we have seen that Minkowski sums of convex polygons are pseudodiscs. We use this to prove our first major result on the motion planning problem, which states that the complexity of the free space of a translating planar robot is linear.

Theorem 13.12 *Let $\mathcal{R}$ be a convex robot of constant complexity, translating among a set S of non-intersecting polygonal obstacles with a total of n edges. Then the complexity of the free configuration space $C_{\text{free}}(\mathcal{R}, S)$ is $O(n)$.*

Proof. First we triangulate each obstacle polygon. We get a set of $O(n)$ triangular, and hence convex, obstacles with disjoint interiors. The free configuration space is the complement of the union of the C-obstacles of these triangles. Because the robot has constant complexity, the C-obstacles have constant complexity, and according to Theorem 13.8 they form a set of pseudodiscs. Theorem 13.9 now implies that the union has linear complexity. ☐

It remains to find an algorithm to construct the free space. Rather than computing the free space C_{free}, we shall compute the forbidden space C_{forb}; the free space is simply its complement.

Let $\mathcal{P}_1, \dots, \mathcal{P}_n$ denote the triangles that we get when we triangulate the obstacles. We want to compute

$$C_{\text{forb}} = \bigcup_{i=1}^{n} C\mathcal{P}_i = \bigcup_{i=1}^{n} \mathcal{P}_i \oplus (-\mathcal{R}(0,0)).$$

In Section 13.3 we saw how to compute the individual Minkowski sums $C\mathcal{P}_i$. To compute their union we use a simple divide-and-conquer approach.

Algorithm FORBIDDENSPACE($C\mathcal{P}_1, \dots, C\mathcal{P}_n$)
Input. A collection of C-obstacles.
Output. The forbidden space $C_{\text{forb}} = \bigcup_{i=1}^{n} C\mathcal{P}_i$.
1.   **if** $n = 1$
2.   **then return** $C\mathcal{P}_1$
3.   **else** $C_{\text{forb}}^1 \leftarrow$ FORBIDDENSPACE($\mathcal{P}_1, \dots, \mathcal{P}_{\lceil n/2 \rceil}$)
4. $C_{\text{forb}}^2 \leftarrow$ FORBIDDENSPACE($\mathcal{P}_{\lceil n/2 \rceil + 1}, \dots, \mathcal{P}_n$)
5.   Compute $C_{\text{forb}} = C_{\text{forb}}^1 \cup C_{\text{forb}}^2$.
6.   **return** C_{forb}

The heart of this algorithm is the subroutine to compute the union of two planar regions, which we need to perform the merge step (line 5). If we represent these regions by doubly-connected edge lists, this can be done by the overlay algorithm described in Chapter 2.

The following lemma summarizes the result.

Lemma 13.13 *The free configuration space C_{free} of a convex robot of constant complexity translating among a set of polygons with n edges in total can be computed in $O(n \log^2 n)$ time.*

Proof. In Chapter 3 we saw that a polygon with m vertices can be triangulated in $O(m \log m)$ time. (In fact, it can even be done in $O(m)$ time with a very complicated algorithm, as stated in the notes and comments of Chapter 3.) Hence, if m_i denotes the complexity of obstacle $\mathcal{P}_i$, then triangulating all the obstacles takes time proportional to

$$\sum_{i=1}^{t} m_i \log m_i \;\leqslant\; \sum_{i=1}^{t} m_i \log n \;=\; n \log n.$$

Computing the C-obstacles of each of the resulting triangles takes linear time in total. It remains to bound the time that FORBIDDENSPACE needs to compute the union of the C-obstacles.

Using the results from Chapter 2, the merge step (line 5) can be done in $O((n_1 + n_2 + k) \log(n_1 + n_2))$ where n_1, n_2, and k denote the complexity of C_{forb}^1, C_{forb}^2, and $C_{\text{forb}}^1 \cup C_{\text{forb}}^2$. Theorem 13.12 states that the complexity of the free space—and, hence, of the forbidden space—is linear in the sum of the complexities of the obstacles. In our case this means that n_1, n_2, and k are all $O(n)$, so the time for the merge step is $O(n \log n)$. We get the following recurrence for $T(n)$, the time the algorithm needs when applied to a set of n constant-complexity C-obstacles:

$$T(n) = T(\lceil n/2 \rceil) + T(\lfloor n/2 \rfloor) + O(n \log n).$$

The solution of this recurrence is $O(n \log^2 n)$. ⊟

The result of this theorem is not the best possible—see the notes and comments of this chapter.

Now that we have computed the free space, we can continue in exactly the same way as in Section 13.2: we compute a trapezoidal map of the free space, together with a roadmap. Given a start and a goal placement of the robot $\mathcal{R}$, we find a path as follows. First, we map the start and goal placement to points in the configuration space. Then we compute a path between these two points through the free space using the trapezoidal map and the road map, as described in Section 13.2. Finally, we map the path back to a path for $\mathcal{R}$ in the work space.

The next theorem summarizes the result of our efforts.

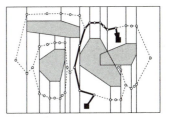

Theorem 13.14 *Let $\mathcal{R}$ be a convex robot of constant complexity translating among a set S of disjoint polygonal obstacles with n edges in total. We can preprocess S in $O(n \log^2 n)$ expected time, such that between any start and goal position a collision-free path for $\mathcal{R}$, if it exists, can be computed in $O(n)$ time.*

13.5* Motion Planning with Rotations

In the previous sections the robot was only allowed to translate. When the robot is circular this does not limit its possible motion. On the other hand, when it is long and skinny, translational motion is often not enough: it may have to change its orientation to be able to pass through a narrow passage or to go around a corner. In this section we sketch a method to plan motion for robots that can rotate as well as translate.

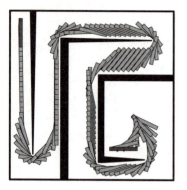

Let $\mathcal{R}$ be a convex polygonal robot that can translate and rotate in a planar work space that contains a set $\mathcal{P}_1, \ldots, \mathcal{P}_t$ of disjoint polygonal obstacles. The robot $\mathcal{R}$ has three degrees of freedom: two translational and one rotational degree of freedom. Hence, we can specify a placement for $\mathcal{R}$ by three parameters: the x- and y-coordinate of a reference point of $\mathcal{R}$, and an angle ϕ that specifies its orientation. As in Section 13.1, we use $\mathcal{R}(x, y, \phi)$ to denote the robot placed with its reference point at (x, y) and rotated over an angle ϕ.

The configuration space that we get is the 3-dimensional space $\mathbb{R}^2 \times [0 : 360)$, with a topology where points $(x, y, 0)$ and $(x, y, 360)$ are identified. Recall that $C\mathcal{P}_i$, the C-obstacle of an obstacle $\mathcal{P}_i$, is defined as follows:

$$C\mathcal{P}_i := \{(x, y, \phi) \in \mathbb{R}^2 \times [0 : 360) \ : \ \mathcal{R}(x, y, \phi) \cap \mathcal{P}_i \neq \emptyset\}.$$

What do these C-obstacles look like? This question is difficult to answer directly, but we can get an idea by looking at cross-sections with planes of constant ϕ. In such a plane the rotation angle is fixed, so we are dealing with a purely translational problem. Hence, we know the shape of the cross-section: it is a Minkowski sum. More precisely, the cross-section of $C\mathcal{P}_i$ with the plane $h : \phi = \phi_0$ is equal to $\mathcal{P}_i \oplus \mathcal{R}(0, 0, \phi_0)$. (More precisely, it is a copy of the

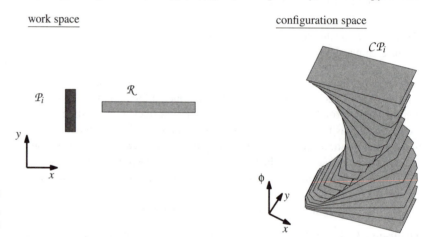

work space   configuration space

Figure 13.10
The C-obstacle of a rotating and translating robot

Minkowski sum placed at height ϕ_0.) Now imagine sweeping a horizontal plane upwards through configuration space, starting at $\phi = 0$ and ending at $\phi = 360$. At any time during the sweep the cross-section of the plane with $C\mathcal{P}_i$ is a Minkowski sum. The shape of the Minkowski sum changes continuously: at $\phi = \phi_0$ the cross-section is $\mathcal{P}_i \oplus \mathcal{R}(0, 0, \phi_0)$, and at $\phi = \phi_0 + \varepsilon$ the cross-section is $\mathcal{P}_i \oplus \mathcal{R}(0, 0, \phi_0 + \varepsilon)$. This means that $C\mathcal{P}_i$ looks like a twisted pillar, as in Figure 13.10. The edges and facets of this twisted pillar, except for the top facet and bottom facet, are curved.

So we know more or less what C-obstacles look like. The free space is the complement of the union of these C-obstacles. Due to the nasty shape of the C-obstacles, the free space is rather complicated: its boundary is no longer polygonal, but curved. Moreover, the combinatorial complexity of the free space can be quadratic for a convex robot, and even cubic for a non-convex robot. Nevertheless, we can solve the motion planning problem using the same approach we took before: compute a decomposition of the free space into simple cells, and construct a road map to guide the motions between neighboring cells. Given a start and goal placement of the robot we then find a path as follows. We map these placements to points in configuration space, find the cells that contain the points, and construct a path consisting of three parts: a path from the start position to the node of the road map in the center of the start cell, a path along the road map to the node in the center of the goal cell, and a path inside the goal cell to the final destination. It remains to map the path in configuration space back to a motion in the work space.

Because of the complex shape of the C-obstacles, it is difficult to compute a suitable cell decomposition, especially when it comes to an actual implementation. Therefore we shall describe a different, simpler approach. As we shall see, however, this approach has its drawbacks as well. Our approach is based on the same observation we used to study the shape of the C-obstacles, namely that the motion planning problem reduces to a purely translational problem if we restrict the attention to a horizontal cross-section of the configuration space. We will call such a cross-section a *slice*. The idea is to compute a finite number of slices. A path for the robot now consists of two types of motion: motions within a slice—these are purely translational—and motions from one slice to the next or previous one—these will be purely rotational.

Let's formalize this. Let z denote the number of slices we take. For every integer i with $0 \leqslant i \leqslant z-1$, let $\phi_i = i \times (360/z)$. We compute a slice of the free space for each ϕ_i. Since within the slice we are dealing with a purely translational problem for the robot $\mathcal{R}(0,0,\phi_i)$, we can compute the slice using the methods of the previous section. This will give us the trapezoidal map $\mathcal{T}_i$ of the free space within the slice. For each $\mathcal{T}_i$ we compute a road map $\mathcal{G}_i$. These road maps are used to plan the motions within a slice, as in Section 13.2.

It remains to connect consecutive slices. More precisely, we connect every pair of roadmaps $\mathcal{G}_i$, $\mathcal{G}_{i+1}$ to obtain a roadmap $\mathcal{G}_{\text{road}}$ of the entire configuration space. This is done as follows. We take the trapezoidal maps of each pair of consecutive slices, and compute their overlay with the algorithm of Chapter 2. (Strictly speaking, we should say that we compute the overlay of the projections of $\mathcal{T}_i$ and $\mathcal{T}_{i+1}$ onto the plane $h : \phi = 0$.) This tells us all the pairs Δ_1, Δ_2 with $\Delta_1 \in \mathcal{T}_i$ and $\Delta_2 \in \mathcal{T}_{i+1}$ such that Δ_1 intersects Δ_2. Let $(x,y,0)$ be a point in $\Delta_1 \cap \Delta_2$. We then add an extra node to $\mathcal{G}_{\text{road}}$ at (x,y,ϕ_i) and at (x,y,ϕ_{i+1}), which we connect by an arc. Moving from one slice to the other along this arc corresponds to a rotation from ϕ_i to ϕ_{i+1}, or back. Furthermore, the node at (x,y,ϕ_i) is connected to the node at the center of Δ_1, and the node at (x,y,ϕ_{i+1}) is connected to the node at the center of Δ_2. These connections stay within a slice, so they correspond to purely translational motions. We connect $\mathcal{G}_{z-1}$ and $\mathcal{G}_0$ in the same way. Note that paths in the graph $\mathcal{G}_{\text{road}}$ correspond to paths of the robot that are composed of purely translational motion (when we move along an arc connecting nodes in the same slice) and purely rotational motion (when we move along an arc connecting nodes in different slices).

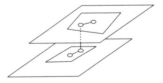

Once we have constructed this road map, we can use it to plan a motion for $\mathcal{R}$ from any start placement $\mathcal{R}(x_{\text{start}}, y_{\text{start}}, \phi_{\text{start}})$ to any goal placement $\mathcal{R}(x_{\text{goal}}, y_{\text{goal}}, \phi_{\text{goal}})$. To do this, we first determine the slices closest to the start and goal placement by rounding the orientations ϕ_{start} and ϕ_{goal} to the nearest orientation ϕ_i for which we constructed a slice. Within those slices we determine the trapezoids Δ_{start} and Δ_{goal} containing the start and goal position. If one of these trapezoids does not exist because the start or goal position lies in the forbidden space within the slice, then we report that we cannot compute a path. Otherwise, let v_{start} and v_{goal} be the nodes of the road map that have been placed in their center. We try to find a path in G_{road} from v_{start} to v_{goal} using breadth-first search. If there is no path in the graph, we report that we cannot compute a motion. Otherwise we report a motion consisting of five parts: a purely rotational motion from the start position to the nearest slice, a purely translational motion within that slice to the node v_{start}, a motion that corresponds to the path from v_{start} to v_{goal} in G_{road}, a purely translational motion from v_{goal} to the final position within that slice (which is the slice nearest to the goal position), and finally a purely rotational motion to the real goal position.

This method is a generalization of the method we used for translating motions, but is has a major problem: it is not always correct. Sometimes it may erroneously report that a path does not exist. For instance, the start position can be in the free space, whereas the start position within the nearest slice is not. In this case we report that there is no path, which need not be true. Even worse is that the paths we report need not be collision-free. The translational motions within a slice are okay, because we solved the problem within a slice exactly, but the rotational motions from one slice to the next may cause problems: the placements within the two slices are collision-free, but halfway the robot could collide with an obstacle. Both problems are less likely to occur when we increase the number of slices, but we can never be certain of the correctness of the result. This is especially bothersome for the second problem: we definitely don't want our possibly very expensive robot to have a collision.

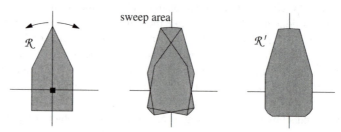

Figure 13.11
Enlarging the robot

Therefore we use the following trick. We make the robot slightly larger, and use the method described above on the enlarged robot $\mathcal{R}'$. This is done in such a way that although $\mathcal{R}'$ can collide during rotations, the original robot $\mathcal{R}$ cannot. To achieve this, the robot is enlarged as follows. Rotate $\mathcal{R}$ clockwise and counter-clockwise over an angle of $(180/z)°$. During this rotation $\mathcal{R}$ sweeps a part of the plane. We use for the enlarged robot $\mathcal{R}'$ a convex polygon that contains the sweep area—see Figure 13.11. We now compute the trapezoidal maps and the road map for $\mathcal{R}'$ instead of $\mathcal{R}$. It is not difficult to prove that $\mathcal{R}$ cannot collide with an obstacle during a purely rotational motion between two adjacent slices, even though $\mathcal{R}'$ can. By enlarging the robot we have introduced another way to incorrectly decide that there is no path. Again, this becomes less likely when the number of slices increases. So with a large number of slices, the method probably performs reasonably well in practical situations.

13.6 Notes and Comments

The motion planning problem has received a lot of attention over the years, both from people working in computational geometry and from people working in robotics, and this chapter only scratches the surface of all the research. A much more extensive treatment of the problem is given by Latombe [207]. Nevertheless, the concepts we have introduced—configuration space, decomposition of the free space, road maps that transform the geometric problem into a graph searching problem—underly the majority of approaches that have been proposed.

These concepts date back to the work of Lozano-Pérez [221, 222, 223]. An important difference between his method and the method of this chapter is that he used an approximate decomposition of the free space. The approach of Section 13.2, which uses an exact decomposition of the free space of a planar translating robot into trapezoids, is based on more recent work by Kedem et al. [195, 196]. An improved algorithm, which runs in $O(n \log n)$ time, was given by Bhattacharya and Zorbas [48].

A very general method that is based on finding an exact cell decomposition of the free space was given by Schwartz and Sharir [304]. It is based on a decomposition method of Collins [114]. Unfortunately, this method takes time doubly exponential in the dimension of the configuration space. This can be improved using a decomposition method of Chazelle et al. [81].

In this chapter we have seen that the cell decomposition approach leads to an $O(n \log^2 n)$ algorithm when applied to a convex robot translating in the plane. The bottleneck in the algorithm was the computation of the union of a collection of Minkowski sums. Using a randomized incremental algorithm, instead of a divide-and-conquer algorithm, this step can be done in $O(n \log n)$ time [42, 246].

The translational motion planning problem in 3-dimensional space can be solved in $O(n^2 \log^3 n)$ time [10].

The approach we sketched for robots that can translate and rotate is approximate: it isn't guaranteed to find a path if it exists. It is possible to find an exact solution by computing an exact cell decomposition of the free space in $O(n^3)$ time [21]. For a convex robot, the running time can be reduced to $O(n^2 \log^2 n)$ [197].

The free space of a robot may consist of a number of disconnected components. Of course, the motions of the robot are confined to the component where it starts; to go to another component it would have to pass through the forbidden space. Hence, it is sufficient to compute only a *single cell* in the free space, instead of the entire free space. Usually the worst-case complexity of a single cell is one order of magnitude lower than the complexity of the entire free space. This can be used to speed up the asymptotic running time of the motion planning algorithms. The book by Agarwal and Sharir [313] and the thesis by Halperin [173] discuss single cells and their connection to motion planning at length.

The theoretical complexity of the motion planning is exponential in the number of degrees of freedom of the robot, which makes the problem appear intractable for high DOF robots. Under some mild restrictions on the shape of the robot and the obstacles—which are likely to be satisfied in practical situations—one can show that the complexity of the free space is only linear [318, 319].

Cell decomposition methods are not the only exact methods for motion planning. Another approach is the so-called *retraction method*. Here a road map is constructed directly, without decomposing the free space. Furthermore, a retraction function is defined, which maps any point in the free space to a point on the road map. Once this has been done, paths can be found be retracting both start and goal to the road map and next following a path along the road map. Different types of road maps and retraction functions have been proposed. A nice road map is the Voronoi diagram, because it stays as far away from the obstacles as possible. If the robot is a disc, then we can use the normal Voronoi diagram; otherwise one has to use a different distance function in the definition of the Voronoi diagram, which depends on the shape of the robot. Still, such a diagram can often be computed in $O(n \log n)$ time [220, 260], leading to another $O(n \log n)$ time algorithm for translational motion planning. A very general road map method has been proposed by Canny [60]. It can solve almost any motion planning problem in time $O(n^d \log n)$, where d is the dimension of the configuration space, that is, the number of degrees of freedom of the robot. Unfortunately, the method is very complicated, and it has the disadvantage that most of the time the robot moves in contact with an obstacle. This is often not the preferred type of motion.

In this chapter we have concentrated mainly on exact motion planning. There are also a number of heuristic approaches.

For instance, one can use *approximate cell decompositions* [53, 221, 222, 344] instead of exact ones. These are often based on quad trees.

Another heuristic is the *potential field method* [27, 199, 327]. Here one defines a potential field on the configuration space by making the goal position attract the robot and making the obstacles repel it. The robot then moves in the direction dictated by the potential field. The problem with this approach is that the robot can get stuck in local minima of the potential field. Various techniques have been proposed to escape the minima.

Another heuristic which has recently become popular is the probabilistic road map method [274, 194]. It computes a number of random placements for the robot, which are connected in some way to form a road map of the free space. The road map can then be used to plan paths between any given start and goal placement.

Minkowski sums not only play an important role in motion planning, but also in other problems. An example is the problem of placing one polygon inside another [97]; this can be useful if one wants to cut out some shape from a piece of fabric. For some basic results on properties of Minkowski sums and their computation we refer to [31, 165].

In this chapter we concentrated on finding some path for the robot, but we didn't try to find a short path. This is the topic of Chapter 15.

Finally, we note that we allowed paths where the robot touches an obstacle. Such paths are sometimes called "semi-free" [207, 303]. Paths that do not touch any obstacle are then called "free". It is useful to be aware of these terms when studying the motion planning literature.

13.7 Exercises

13.1 Let $\mathcal{R}$ be a robotic arm with a fixed base and seven links. The last joint of $\mathcal{R}$ is a prismatic joint, the other ones are revolute joints. Give a set of parameters that determines a placement of $\mathcal{R}$. What is the dimension of the configuration space resulting from your choice of parameters?

13.2 In the road map $\mathcal{G}_{\text{road}}$ that was constructed on the trapezoidal decomposition of the free space we added a node in the center of each trapezoid and on each vertical wall. It is possible to avoid the nodes in the center of each trapezoid. Show how the graph can be changed such that only nodes on the vertical walls are required. (Avoid an increase in the number of edges in the graph.) Explain how to adapt the query algorithm.

13.3 Prove that the shape of $\mathcal{CP}_i$ is independent of the choice of the reference point in the robot $\mathcal{R}$.

13.4 What does the Minkowski sum of two circles with radius r_1 and r_2 look like?

13.5 Let $\mathcal{P}_1$ and $\mathcal{P}_2$ be two convex polygons. Let S_1 be the collection of vertices of $\mathcal{P}_1$ and S_2 be the collection of vertices of $\mathcal{P}_2$. Prove that

$$P_1 \oplus P_2 = ConvexHull(S_1 \oplus S_2).$$

13.6 Prove Observation 13.4.

13.7 In Theorem 13.9 we gave an $O(n)$ bound on the complexity of the union of a set of polygonal pseudodiscs with n vertices in total. We are interested in the precise bound.

a. Assume that the union boundary contains m original vertices of the polygons. Show that the complexity of the union boundary is bounded by $2n - m$. Use this to prove an upper bound of $2n - 3$.

b. Prove a lower bound of $2n - 6$ by constructing an example that has this complexity.

14 Quadtrees

Non-Uniform Mesh Generation

Almost all electrical devices, from shavers and telephones to televisions and computers, contain some electronic circuitry to control their functioning. This circuitry—VLSI circuits, resistors, capacitors, and other electric components—is placed on a printed circuit board. To design printed circuit boards one has to decide where to place the components, and how to connect them. This raises a number of interesting geometric problems, of which this chapter tackles one: mesh generation.

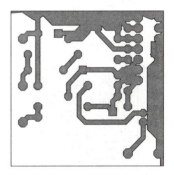

Many components on a printed circuit board emit heat during operation. In order for the board to function properly, the emission of heat should be below a certain threshold. It is difficult to predict in advance whether the heat emission will cause problems, since this depends on the relative positions of the components and the connections. In former days one therefore made a prototype of the board to determine the heat emission experimentally. If it turned out to be too high, an alternative layout had to be designed. Today the experiments can often be simulated. Because the design is largely automated, a computer model of the board is readily available and simulation is a lot faster than building a prototype. Simulation also allows testing during the initial stages of the design phase, so that faulty designs can be rejected as early as possible.

The heat transfer between different materials on the printed circuit board is a quite complicated process. To simulate the heat processes on the board one therefore has to resort to approximation using *finite element methods*. Such methods first divide the board into many small regions, or *elements*. The elements are usually triangles or quadrilaterals. The heat that each element emits by itself is assumed to be known. It is also assumed to be known how neighboring elements influence each other. This leads to a big system of equations, which is then solved numerically.

The accuracy of finite element methods depends heavily on the mesh: the finer the mesh, the better the solution. The other side of the coin is that the computation time for the numerical process increases drastically when the number of elements increases. So we would like to use a fine mesh only where necessary. Often this is at the border between regions of different material. It is also important that the mesh elements respect the borders, that is, that any mesh el-

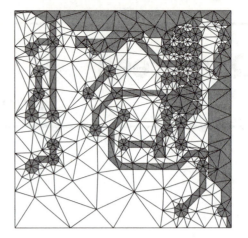

Figure 14.1
Triangular mesh of a part of a printed
circuit board

ement is contained in only one region. Finally, the shape of the mesh elements plays an important role: irregularly shaped elements such as very thin triangles often lead to a slower convergence of the numerical process.

14.1   Uniform and Non-Uniform Meshes

We'll study the following variant of the mesh generation problem. The input is a square—the printed circuit board—with a number of disjoint polygonal components inside it. The square together with the components is sometimes called the *domain* of the mesh. The vertices of the square are at $(0,0)$, $(0,U)$, $(U,0)$, (U,U), where $U = 2^j$ for a positive integer j. The coordinates of the vertices of the components are assumed to be integers between 0 and U. We make one more assumption, which is satisfied in a number of applications: the edges of the components have only four different orientations. In particular, the angle that an edge makes with the x-axis is either $0°$, $45°$, $90°$, or $135°$.

Our goal is to compute a *triangular mesh* of the square, that is, a subdivision of the square into triangles. We require the mesh to have the following properties:

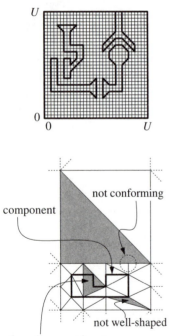

- The mesh must be *conforming*: a triangle is not allowed to have a vertex of another triangle in the interior of one of its edges.

- The mesh must *respect the input*: the edges of the components must be contained in the union of the edges of the mesh triangles.

- The mesh triangles must be *well-shaped*: the angles of any mesh triangle should not be too large nor too small. In particular, we require them to be in the range from $45°$ to $90°$.

Finally, we would like the mesh to be fine only where necessary. Where this is depends on the application. The exact property we shall require is as follows.

- The mesh must be *non-uniform*: it should be fine near the edges of the components and coarse far away from the edges.

We have already seen triangulations earlier in this book. Chapter 3 presented an algorithm for triangulating a simple polygon, and Chapter 9 presented an algorithm for triangulating a point set. The latter algorithm computes the Delaunay triangulation, a triangulation that maximizes the minimum angle over all possible triangulations. Given our restriction on the angles of the mesh triangles this seems quite useful, but there are two problems.

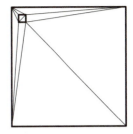

First of all, a triangulation of the vertices of the components need not respect the edges of the components. Even if it did, there is a second problem: there can still be angles that are too small. Suppose that the input is a square with side length 16 and one component, which is a small square with side length 1 placed in the top left corner at distance 1 from the left and the top side of the square. Then the Delaunay triangulation contains triangles that have an angle of less than 5°. Since the Delaunay triangulation maximizes the minimum angle, it seems impossible to generate a mesh with only well-shaped triangles. But there is a catch: unlike in a triangulation, the triangles in a mesh are not required to have their vertices at the input points. We are allowed to add extra points, called *Steiner points*, to help us obtain well-shaped triangles. A triangulation that uses Steiner points is sometimes called a *Steiner triangulation*. In our example, if we add a Steiner point at every grid point inside the square, then we can easily obtain a mesh consisting only of triangles with two 45° angles and one 90° angle—see the mesh shown on the left in Figure 14.2. Unfortunately, this mesh suffers from another problem: it uses small triangles

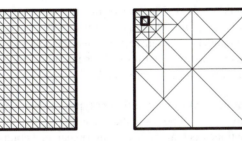

Figure 14.2
A uniform and a non-uniform mesh

everywhere, not only near the edges of the input, so it is a *uniform* mesh. As a result, it has many triangles. We cannot simply replace all the triangles in, say, the bottom right quarter of the square by two big triangles, because then the mesh would no longer be conforming. Nevertheless, if we gradually increase the size of the triangles when we get farther away from top left corner, then it is possible to get a conforming mesh with only well-shaped triangles, as shown on the right in Figure 14.2. This leads to a significantly smaller number of triangles: the uniform mesh has 512 triangles, whereas the non-uniform one only has 52.

14.2 Quadtrees for Point Sets

The non-uniform mesh generation method we shall describe in the next section is based on *quadtrees*. A quadtree is a rooted tree in which every internal node has four children. Every node in the quadtree corresponds to a square. If a node ν has children, then their corresponding squares are the four quadrants of the square of ν—hence the name of the tree. This implies that the squares of the leaves together form a subdivision of the square of the root. We call this subdivision the *quadtree subdivision*. Figure 14.3 gives an example of a quadtree and the corresponding subdivision. The children of the root are

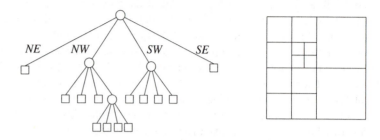

Figure 14.3
A quadtree and the corresponding
subdivision

labelled NE, NW, SW, and SE to indicate to which quadrant they correspond; NE stands for the north-east quadrant, NW for the north-west quadrant, and so on.

Before we continue, we introduce some terminology related to quadtree subdivisions. The faces in a quadtree subdivision have the shape of a square. Although they can have more than four vertices, we shall call them squares anyway. The four vertices at the corners of the square are called *corner vertices*, or *corners* for short. The line segments connecting consecutive corners are the *sides of the square*. The edges of the quadtree subdivision that are contained in the boundary of a square are called the *edges of the square*. Hence, a side contains at least one, but possibly many more, edges. We say that two squares are *neighbors* if they share an edge.

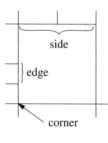

Quadtrees can be used to store different types of data. We will describe the variant that stores a set of points in the plane. In this case the recursive splitting of squares continues as long as there is more than one point in a square. So the definition of a quadtree for a set P of points inside a square σ is as follows. Let $\sigma := [x_\sigma : x'_\sigma] \times [y_\sigma : y'_\sigma]$.

- If $card(P) \leqslant 1$ then the quadtree consists of a single leaf where the set P and the square σ are stored.

- Otherwise, let $\sigma_{NE}, \sigma_{NW}, \sigma_{SW}$, and σ_{SE} denote the four quadrants of σ. Let $x_{mid} := (x_\sigma + x'_\sigma)/2$ and $y_{mid} := (y_\sigma + y'_\sigma)/2$, and define

$$
\begin{aligned}
P_{NE} &:= \{p \in P : p_x > x_{mid} \text{ and } p_y > y_{mid}\}, \\
P_{NW} &:= \{p \in P : p_x \leqslant x_{mid} \text{ and } p_y > y_{mid}\}, \\
P_{SW} &:= \{p \in P : p_x \leqslant x_{mid} \text{ and } p_y \leqslant y_{mid}\}, \\
P_{SE} &:= \{p \in P : p_x > x_{mid} \text{ and } p_y \leqslant y_{mid}\}.
\end{aligned}
$$

The quadtree now consists of a root node ν where the square σ is stored. Below we shall denote the square stored at ν by $\sigma(ν)$. Furthermore, ν has four children:

- the NE-child is the root of a quadtree for the set P_{NE} inside the square σ_{NE},

- the NW-child is the root of a quadtree for the set P_{NW} inside the square σ_{NW},

- the SW-child is the root of a quadtree for the set P_{SW} inside the square σ_{SW},

- the SE-child is the root of a quadtree for the set P_{SE} inside the square σ_{SE}.

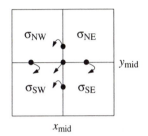

The choice of using less-than-or-equal-to and greater-than in the definition of the sets P_{NE}, P_{NW}, P_{SW}, and P_{SE} means that we define the vertical splitting line to belong to the left two quadrants, and the horizontal splitting line to the lower two quadrants.

Every node v of the quadtree stores its corresponding square $\sigma(v)$. This is not necessary; we could store only the square of the root of the tree. When we walk down the tree we would then have to maintain the square of the current node. This alternative uses less storage at the expense of extra computations that have to be done when the quadtree is queried.

The recursive definition of a quadtree immediately translates into a recursive algorithm: split the current square into four quadrants, partition the point set accordingly, and recursively construct quadtrees for each quadrant with its associated point set. The recursion stops when the point set contains less than two points. The only detail that does not follow from the recursive definition is how to find the square with which to start the construction. Sometimes this square will be given as a part of the input. If this is not the case, then we compute a smallest enclosing square for the set of points. This can be done in linear time by computing the extreme points in the x- and y-directions.

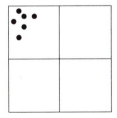

At every step of the quadtree construction the square containing the points is split into four smaller squares. This does not mean that the point set is split as well: it can happen that all the points lie in the same quadrant. Therefore a quadtree can be quite unbalanced, and it is not possible to express the size and depth of a quadtree as a function of the number of points it stores. However, the depth of a quadtree is related to the distance between the points and the size of the initial square. This is made precise in the following lemma.

Lemma 14.1 *The depth of a quadtree for a set P of points in the plane is at most $\log(s/c) + \frac{3}{2}$, where c is the smallest distance between any two points in P and s is the side length of the initial square that contains P.*

Proof. When we descend from a node to one of its children, the size of the corresponding square halves. Hence, the side length of the square of a node at depth i is $s/2^i$. The maximum distance between two points inside a square is given by the length of its diagonal, which is $s\sqrt{2}/2^i$ for the square of a node at depth i. Since an internal node of a quadtree has at least two points in its associated square, and the minimum distance between two points is c, the depth

i of an internal node must satisfy

$$s\sqrt{2}/2^i \geqslant c,$$

which implies

$$i \leqslant \log \frac{s\sqrt{2}}{c} = \log(s/c) + \frac{1}{2}.$$

The lemma now follows from the fact that the depth of a quadtree is exactly one more than the maximum depth of any internal node. ☐

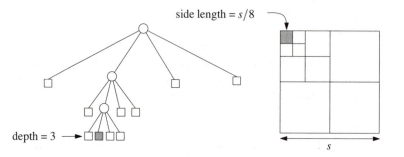

Figure 14.4
A node at depth i corresponds to a square of side length $s/2^i$.

The size of a quadtree and the construction time are a function of the depth of the quadtree and the number of points in P.

Theorem 14.2 *A quadtree of depth d storing a set of n points has $O((d+1)n)$ nodes and can be constructed in $O((d+1)n)$ time.*

Proof. Every internal node in a quadtree has four children, so the total number of leaves is one plus three times the number of internal nodes. Hence, it suffices to bound the number of internal nodes.

Any internal node has one or more points inside its associated square. Moreover, the squares of the nodes at the same depth in the quadtree are disjoint and exactly cover the initial square. This means that the total number of internal nodes at any given depth is at most n. The bound on the size of the quadtree follows.

The most time-consuming task in one step of the recursive construction algorithm is the distribution of points over the quadrants of the current square. Hence, the amount of time spent at an internal node is linear in the number of points that lie in the associated square. We argued above that the total number of points associated with nodes at the same depth in the tree is at most n, from which the time bound follows. ☐

An operation on quadtrees that is often needed is *neighbor finding*: given a node v and a direction—north, east, south, or west—find a node v' such that $\sigma(v')$ is adjacent to $\sigma(v)$ in the given direction. Usually the given node is a leaf, and one wants the reported node to be a leaf as well. This corresponds to finding an adjacent square of a given square in the quadtree subdivision. The algorithm we shall describe is a little bit different: the given node v can also be internal, and the algorithm tries to find the node v' such that σ(v') is adjacent to σ(v) in the given direction and v' and v are at the same depth. If there is no such node, then it finds the deepest node whose square is adjacent. If there is no adjacent square in the given direction—this can happen if σ(v) has an edge contained in an edge of the initial square—then the algorithm reports **nil**.

The neighbor-finding algorithm works as follows. Suppose that we want to find the north-neighbor of v. If v happens to be the SE- or SW-child of its parent, then its north-neighbor is easy to find: it is the NE- or NW-child of the parent, respectively. If v itself is the NE- or NW-child of its parent, then we proceed as follows. We recursively find the north-neighbor, μ, of the parent of v. If μ is an internal node, then the north-neighbor of v is a child of μ; if μ is a leaf, then the north-neighbor we seek is μ itself. The pseudocode for this algorithm is as follows.

north-neighbor of *parent*(v)

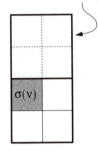

Algorithm NORTHNEIGHBOR(v, $\mathcal{T}$)
Input. A node v in a quadtree $\mathcal{T}$.
Output. The deepest node v' whose depth is at most the depth of v such that σ(v') is a north-neighbor of σ(v), and **nil** if there is no such node.
1. **if** v = *root*($\mathcal{T}$) **then return nil**
2. **if** v = SW-child of *parent*(v) **then return** NW-child of *parent*(v)
3. **if** v = SE-child of *parent*(v) **then return** NE-child of *parent*(v)
4. $\mu \leftarrow$ NORTHNEIGHBOR(*parent*(v), $\mathcal{T}$)
5. **if** μ = **nil or** μ is a leaf
6.   **then return** μ
7.   **else if** v = NW-child of *parent*(v)
8.   **then return** SW-child of μ
9.   **else return** SE-child of μ

This algorithm does not necessarily report a leaf node. If we insist on finding a leaf node, then we have to walk down the quadtree from the node found by our algorithm, always proceeding to a south-child.

The algorithm spends $O(1)$ time at every recursive call. Moreover, at every call the depth of the argument node v decreases by one. Hence, the running time is linear in the depth of the quadtree. We get the following theorem:

Theorem 14.3 *Let $\mathcal{T}$ be a quadtree of depth d. The neighbor of a given node v in $\mathcal{T}$ in a given direction, as defined above, can be found in $O(d+1)$ time.*

We already observed that a quadtree can be quite unbalanced. As a result, large squares can be adjacent to many small squares. In some applications—in particular in the application to meshing—this is unwanted. Therefore we now discuss a variant of quadtrees, the balanced quadtree, that does not have this problem.

A quadtree subdivision is called *balanced* if any two neighboring squares differ at most a factor two in size. A quadtree is called balanced if its subdivision is balanced. So in a balanced quadtree any two leaves whose squares are

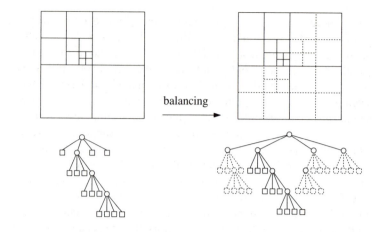

balancing

Figure 14.5
A quadtree and its balanced version

neighbors can differ at most one in depth. Figure 14.5 shows an example of a quadtree subdivision that has been made balanced. The original subdivision is shown solid, and its refinement is dotted.

A quadtree can be made balanced with the following algorithm:

Algorithm BALANCEQUADTREE($\mathcal{T}$)
Input. A quadtree $\mathcal{T}$.
Output. A balanced version of $\mathcal{T}$.
1.   Insert all the leaves of $\mathcal{T}$ into a linear list $\mathcal{L}$.
2.   **while** $\mathcal{L}$ is not empty
3.   **do** Remove a leaf μ from $\mathcal{L}$.
4.   **if** $\sigma(\mu)$ has to be split
5.   **then** Make μ into an internal node with four children, which are leaves that correspond to the four quadrants of $\sigma(\mu)$. If μ stores a point, then store the point in the correct new leaf instead.
6.   Insert the four new leaves into $\mathcal{L}$.
7.   Check if $\sigma(\mu)$ had neighbors that now need to be split and, if so, insert them into $\mathcal{L}$.
8.   **return** $\mathcal{T}$

There are two steps in the algorithm that require some explanation.

First, we have to check whether a given square $\sigma(\mu)$ needs to be split. This means that we have to check whether $\sigma(\mu)$ is adjacent to a square of less than half its size. This can be done using the neighbor-finding algorithm described earlier, as follows. Suppose that we are looking for a north-neighbor of $\sigma(\mu)$ that is less than half the size of $\sigma(\mu)$. There is such a square if and only if NORTHNEIGHBOR($\mu, \mathcal{T}$) reports a node that has a SW-child or a SE-child that is not a leaf.

Second, we have to check if $\sigma(\mu)$ had neighbors that now need to be split. Again, this can be done using the neighbor-finding algorithm: for example, $\sigma(\mu)$ has such a neighbor to the north if and only if NORTHNEIGHBOR($\mu, \mathcal{T}$) reports a node whose square is larger than $\sigma(\mu)$.

So now we have an algorithm to make a quadtree balanced. But before we can analyze the running time of the balancing algorithm we must answer the following question: what happens to the size of the quadtree when we make it balanced? From Figure 14.5 we may get the idea that the complexity of a balanced quadtree subdivision can be quite a lot higher than that of its unbalanced version. First of all, large squares adjacent to very small ones get split up many times. Secondly, the splitting may propagate: sometimes it seems that a square σ need not be split because its neighbors initially have the right size, but these neighbors may have to be split so that σ must be split after all. The next theorem shows that things are not as bad as appears at first sight, and that the balancing can be done efficiently.

Theorem 14.4 *Let T be a quadtree with m nodes. Then the balanced version of T has $O(m)$ nodes and it can be constructed in $O((d+1)m)$ time.*

Proof. We first prove the bound on the number of nodes. Denote the balanced version of T by $\widetilde{T}$. The tree $\widetilde{T}$ is obtained from T by a number of splitting operations, which replace a leaf by one internal node with four leaves. We shall prove that only $O(m)$ splitting operations are performed. Since a single splitting operation increases the total number of (internal and leaf) nodes by four, this proves that the number of nodes in $\widetilde{T}$ is $O(m)$, as claimed.

The difficulty in bounding the number of splits lies in the propagation of splits: a square that is split because it has a small neighbor may in turn cause another neighbor to be split, which causes yet another square to be split, and so on. Therefore we first study the propagation process more carefully.

Let v_1, v_2, and v_3 be three nodes at the same depth such that $\sigma(v_1)$, $\sigma(v_2)$, and $\sigma(v_3)$ are neighbors as in Figure 14.6. Suppose that v_2 and v_3 are leaves of

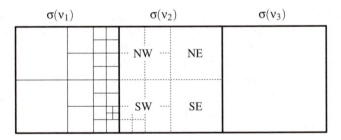

Figure 14.6
The propagation of splits

T, so $\sigma(v_2)$, and $\sigma(v_3)$ are not subdivided before balancing starts. We claim that, no matter how far $\sigma(v_1)$ is subdivided, the splitting that it causes inside $\sigma(v_2)$ does not propagate to $\sigma(v_3)$. We prove this claim by induction on the depth of the subtree rooted at v_1 (which corresponds to the number of times $\sigma(v_1)$ is recursively subdivided). If the subtree has depth one, then $\sigma(v_1)$ is subdivided only once, so $\sigma(v_2)$ need not be split and, trivially, the splitting does not propagate to $\sigma(v_3)$. Now suppose the depth is d for some $d > 1$. Then $\sigma(v_2)$ is split into four quadrants, which may in turn be split further. But by induction the splitting in the NW-quadrant of $\sigma(v_2)$ caused by the small squares in the NE-quadrant of $\sigma(v_1)$ does not propagate to the NE-quadrant of

$\sigma(v_2)$. Similarly, the splitting in the SW-quadrant of $\sigma(v_2)$ caused by the small squares in the SE-quadrant of $\sigma(v_1)$ does not propagate to the SE-quadrant of $\sigma(v_2)$. Hence, the NE- and SE-quadrants of $\sigma(v_2)$ are not split, so $\sigma(v_3)$ need not be split at all, as claimed. We conclude that the subdivision of a square into subsquares can only cause a square of the same size to be split if that square is one of the eight surrounding ones.

Using this observation we can prove the theorem. Call the squares of nodes in $\mathcal{T}$ old squares, and the squares of nodes that are in $\widetilde{\mathcal{T}}$ but not in $\mathcal{T}$ new squares. Suppose that we have to split an—old or new—square σ in the balancing process. (The quadrants of σ may have to be split further, but this will be accounted for separately. Here we only need to account for the increase in the number of nodes by four due to the splitting of σ.) From the observation above we conclude that one of the eight squares of the same size that surround σ must have been an old square that was already subdivided. We charge the splitting of σ to this old square. Every old square—equivalently, every node of $\mathcal{T}$—gets charged at most eight splittings in this manner, from which the theorem follows.

What remains is to prove that BALANCEQUADTREE takes $O((d+1)m)$ time. The time needed to handle a node μ is $O(d+1)$, because a constant number of neighbor-finding operations is performed. Since any node is handled at most once and the total number of nodes is $O(m)$, the total time is $O((d+1)m)$. $\quad\square$

14.3 From Quadtrees to Meshes

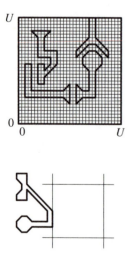

We now return to the mesh generation problem. Recall that the input is a square $[0:U] \times [0:U]$, where $U = 2^j$ for some positive integer j, with a number of disjoint polygonal components inside it. The polygon vertices have integer coordinates, and the polygon edges have one of four possible orientations: the angle that any edge makes with the positive x-axis is $0°$, $45°$, $90°$, or $135°$. The goal is to compute a triangular mesh of the square (both outside and inside the components) that is conforming, respects the input, has well-shaped triangles, and is non-uniform.

The idea is to use a quadtree subdivision as the first step towards a mesh. When we construct a quadtree on a point set, then we stop the recursive construction when a square contains less than two points. Since we are now dealing with polygonal input, we have to reformulate the stopping criterion. Because we want the mesh triangles to be fine near the edges of the components, we keep on splitting as long as a square is intersected by an edge. More precisely, the new stopping criterion is this: we stop splitting when the square is no longer intersected by any edge of any component, or when it has unit size. We consider the square and the edges to be closed, so that for instance a square that has an edge contained in one of its sides is intersected by that edge. This stopping criterion guarantees that the quadtree subdivision will be non-uniform:

the edges of the components will be surrounded by unit size squares, and the squares will be bigger farther away from the edges.

We claim that the interior of any square in the resulting quadtree subdivision can be intersected by an edge of a component in only one way: the intersection must be a diagonal of the square. Indeed, squares whose closure is intersected have unit size, and the vertices of the components have integers coordinates. Hence, the interior of a square cannot be intersected by a horizontal or vertical edge, and the intersection with an edge of orientation 45° or 135° must be a diagonal. It seems that we only have to add diagonals to the squares whose interior is not intersected to obtain a good mesh. The resulting triangles will respect the input, they will be well-shaped, and the mesh will be non-uniform. Unfortunately, the mesh is not conforming. We can remedy this by taking the subdivision vertices on the sides of the square into account when we triangulate it, but this leads to another problem: if a square has many vertices on a side, then not all triangles we get are well-shaped.

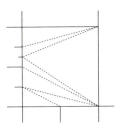

To avoid these problems we make the quadtree subdivision balanced before we triangulate it. Once we have a balanced quadtree subdivision, we can easily generate a mesh with well-shaped triangles as follows. Squares that do not have a vertex in the interior of one of their sides (and that are not already triangulated by an edge of one of the components) are triangulated by adding a diagonal. Because the subdivision is balanced, the remaining squares have at most one vertex in the interior of each side. Moreover, this vertex must be in the middle of the side. Hence, if we add one Steiner point in the center of such a square and connect it to all vertices on the boundary, then we get only triangles with two 45° angles and one 90° angle.

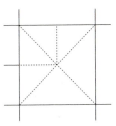

Summarizing, the mesh generation algorithm is as follows:

Algorithm GENERATEMESH(S)
Input. A set S of components inside the square $[0:U] \times [0:U]$ with the properties stated at the beginning of this section.
Output. A triangular mesh $\mathcal{M}$ that is conforming, respects the input, consists of well-shaped triangles, and is non-uniform.
1. Construct a quadtree $\mathcal{T}$ on the set S inside the square $[0:U] \times [0:U]$ with the following stopping criterion: a square is split as long as it is larger than unit size and its closure intersects the boundary of some component.
2. $\mathcal{T} \leftarrow$ BALANCEQUADTREE($\mathcal{T}$)
3. Construct the doubly-connected edge list for the quadtree subdivision $\mathcal{M}$ corresponding to $\mathcal{T}$.
4. **for** each face σ of $\mathcal{M}$
5.   **do if** the interior of σ is intersected by an edge of a component
6.   **then** Add the intersection (which is a diagonal) as an edge to $\mathcal{M}$.
7.   **else if** σ has only vertices at its corners
8.   **then** Add a diagonal of σ as an edge to $\mathcal{M}$.
9.   **else** Add a Steiner point in the center of σ, connect it to all vertices on the boundary of σ, and change $\mathcal{M}$ accordingly.
10. **return** $\mathcal{M}$

GENERATEMESH constructs the mesh in the form of a doubly-connected edge list. We omit the details related to handling the doubly-connected edge list—in particular, how to construct the doubly-connected edge list corresponding to a given quadtree.

The following theorem summarizes the properties of the mesh that is constructed by our algorithm.

Theorem 14.5 *Let S be a set of disjoint polygonal components inside the square* $[0:U] \times [0:U]$ *with the properties stated in the beginning of this section. Then there exists an non-uniform triangular mesh for this input that is conforming, respects the input, and has only well-shaped triangles. The number of triangles is* $O(p(S)\log U)$, *where* $p(S)$ *is the sum of the perimeters of the components in S, and the mesh can be constructed in* $O(p(S)\log^2 U)$ *time.*

Proof. The properties of the mesh—that it is non-uniform, conforming, that it respects the input, and consists of well-shaped triangles—follow from the discussion above. What remains is to bound the size and preprocessing time of the mesh.

The mesh construction has three phases: first we construct a quadtree subdivision, then we make it balanced, and finally we triangulate it.

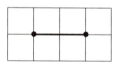

To bound the size of the quadtree subdivision resulting from the first phase we use the following observation. Consider the grid whose cells have unit size. Then the number of cells whose closure is intersected by a line segment of length l is at most $2(l+2)$. It follows that the number of cells whose closure is intersected by an edge of any of the components is $O(p(S))$. Obviously the same is true for any grid with larger cells. Hence, the number of internal nodes of the quadtree at a fixed depth is $O(p(S))$. The depth of the quadtree is $O(\log U)$, since we stop splitting once a cell has unit size. The total number of nodes of the quadtree and, hence, the complexity of the corresponding subdivision, is therefore $O(p(S)\log U)$.

We know from Theorem 14.4 that making a quadtree subdivision balanced does not increase its complexity asymptotically. The same is true for the third phase of the mesh generation, the triangulation of the squares of the balanced subdivision. (This follows for instance from Euler's formula, which implies that the complexity of any subdivision is linear in the number of its vertices.) We conclude that the number of triangles of the final mesh is linear in the complexity of the quadtree subdivision resulting from the first phase, which we just proved to be $O(p(S)\log U)$.

What remains is to prove the bound on the construction time. The first phase is dominated by the time needed to do the splitting steps in the recursive quadtree construction algorithm. For a given node this is linear in the number of component edges intersecting its closure. We argued above that the sum of the number of intersecting edges over all nodes at a fixed depth is $O(p(S))$. Hence, the total time needed in the first phase is $O(p(S)\log U)$. By Theorem 14.4, the time needed to do the balancing introduces an extra $O(\log U)$ factor. Constructing a doubly-connected edge list from a given quadtree, as well as triangulating the balanced subdivision, can be done within the same amount of time. The bound on the preprocessing time follows. ⊟

14.4 Notes and Comments

Quadtrees are one of the first data structures for higher-dimensional data. They were developed by Finkel and Bentley in 1974 [146]. Since then, there have been hundreds of papers dealing with quadtrees. The surveys and two books by Samet [294, 295, 297, 296] give an extensive overview of the various types of quadtrees and their applications.

Mesh generation is but one application of quadtrees. They are also used in computer graphics, image analysis, geographic information systems, and many other areas. Typically they are used to answer range queries, but they can be used for other operations as well. From a theoretical point of view quadtrees do not form the best solution to range searching problems, because usually no sublinear bounds on the query time can be proved. Other solutions to various range searching problems are described in Chapters 5, 10, and 16. In practice, quadtrees often seem to perform well. Quadtrees have also been applied to hidden surface removal, ray tracing, medial axis transforms, map overlay of raster maps, and nearest neighbor finding.

Quadtrees can easily be generalized to higher dimensions, where they are usually called *octrees*.

The mesh generation problem is important in many areas and it has been studied extensively, both in the plane and in 3-dimensional space. The surveys by Bern and Eppstein [44] and by Ho-Le [182] are good starting points to look for results on a specific setting of the mesh generation problem. We shall briefly describe a few of the results.

One can distinguish so-called *structured* and *unstructured* meshes. Structured meshes are usually (deformed) grids; unstructured meshes are often triangulations. We shall restrict our discussion to unstructured meshes. Furthermore, we mainly concentrate on the case where the domain to be meshed is 2-dimensional and polygonal. In most applications there are requirements on the mesh that are similar to the ones imposed in this chapter. In particular, one usually requires the mesh to be conforming—sometimes the term 'consistent' is used—and to respect the input. Furthermore, some sort of well-shapedness is important. This usually means that the triangles have to satisfy one or both of the following criteria: (i) There should be no small angles, that is, every angle should be at least some fixed (not too small) constant θ. This constant should not be larger than the smallest angle of the input domain, because we cannot avoid the input angles in the mesh. (ii) There should be *no obtuse angles*, that is, no angles larger than 90°. In the example studied in this chapter we wanted the triangles to satisfy both criteria, and we then took $\theta = 45°$. Often the goal is to minimize the number of mesh elements under the given conditions. This implies some sort of non-uniformity: small triangles should be used only where needed.

We first consider minimizing the number of triangles in the mesh under the condition that there be no small angles. The number of triangles that we need to mesh a polygonal domain under this condition not only depends on the number of vertices of the domain; it also depends on the shape of the domain. To see this we can introduce a parameter that is closely related to the minimum angle of a triangle, namely the *aspect ratio* of the triangle. This is the ratio of the length of the longest side of the triangle to the height of the triangle, where the height of a triangle is the Euclidean distance of the longest edge to its opposite vertex. If the smallest angle of a triangle is θ then the aspect ratio is between $1/\sin\theta$ and $2/\sin\theta$. Now consider a rectangular domain whose shorter sides have length 1 and whose longer sides have length A. Suppose we require that the minimum angle be, say, $30°$. This implies that the aspect ratio of any triangle in the mesh must be less than or equal to $2/\sin 30° = 4$. Furthermore, the height of any triangle in the domain is at most one. Hence, the area of any triangle is $O(1)$. Because the total area of the rectangular domain is A, this implies that we need at least $\Omega(A)$ triangles in the mesh. Bern et al. [46] describe a method based on quadtrees that produces an asymptotically optimal number of triangles. The method described in this chapter is based on their technique.

If the only requirement is that the triangles in the mesh are non-obtuse then it turns out to be possible to construct a mesh for a given polygonal domain whose number of triangles only depends on the number of vertices of the domain. More precisely, Bern and Eppstein [45] have shown that for any polygonal domain with n vertices there is a mesh consisting of $O(n^2)$ non-obtuse triangles. Quite recently, Bern et al. [47] improved this bound to $O(n)$.

Melissaratos and Souvaine [244] extended the approach of Bern et al. [46] for computing a mesh without small angles so that it also avoids obtuse triangles. The number of triangles in the mesh is still at most a constant factor from optimal.

Minimizing the number of triangles is not always the goal of meshing algorithms. It can also be important to be able to control the mesh density, so that one can have a dense mesh in interesting areas and a coarse mesh in uninteresting areas. This is the setting studied by Chew [95]. He describes a meshing algorithm that allows the user to define a function that determines whether a triangle of the mesh is fine enough. The angles of the triangles produced by his algorithm are between $30°$ and $120°$. Another nice aspect of his work is that the algorithm not only deals with planar regions, but also with regions on surface patches.

14.5 Exercises

14.1 In Figure 14.2 a uniform and a non-uniform mesh are shown in a square with integer coordinates $(0,0)$, $(16,0)$, $(0,16)$ and $(16,16)$. Consider similar meshes in squares of larger sizes $U = 2^j$ for an integer j. Express the number of triangles in the mesh for both meshes in terms of j.

14.2 Suppose a triangular mesh is needed inside a rectangle whose sides have length 1 and length $k > 1$. Steiner points may not be used on the sides, but they may be used inside the rectangle. Also assume that all triangles must have angles between 30° and 90°. Is it always possible to create a triangular mesh with these properties? Suppose it is possible to create a mesh for a particular input, what is the minimum number of Steiner points needed?

14.3 All triangles produced by the meshing algorithm of this chapter are non-obtuse, that is, they do not have angles larger than 90°. Prove that if a triangulation of a set P of points in the plane contains only non-obtuse triangles, then it must be the Delaunay triangulation of P.

14.4 Let P be a set of point in 3-dimensional space. Describe an algorithm to construct an octree on P. (An octree is the 3-dimensional variant of the quadtree.)

14.5 It is possible to reduce the size of a quadtree of depth d for a set of points (with real coordinates) inside a square from $O((d+1)n)$ to $O(n)$. The idea is to discard any node v that has only one child under which points are stored. The node is discarded by replacing the pointer from the parent of v to v with the pointer from the parent to the only interesting child of v. Prove that the resulting tree has linear size. Can you also improve upon the $O((d+1)n)$ construction time?

14.6 In this chapter we called a quadtree balanced if two adjacent squares of the quadtree subdivision differ by no more than a factor two in size. To save a constant factor in the number of extra nodes needed to balance a quadtree, we could weaken the balance condition by allowing adjacent squares to differ by a factor of four in size. Can you still complete such a weakly balanced quadtree subdivision to a mesh such that all angles are between 45° and 90° by using only $O(1)$ triangles per square?

14.7 Suppose we make the balancing condition for quadtrees more severe: we no longer allow adjacent squares to differ by a factor two in size, but we require them to have exactly the same size. Is the number of nodes in the new balanced version still linear in the number of nodes of the original quadtree? If not, can you say anything about this number?

14.8 The algorithm to construct a balanced quadtree had two phases: first, a normal quadtree was constructed, which was then balanced in a postprocessing step. It is also possible to construct a balanced quadtree without first constructing the unbalanced version. To this end we maintain the current quadtree subdivision in a doubly-connected edge list during the quadtree construction, and whenever we split a square we check whether any neighbors have to be split. Describe the algorithm in detail and analyze its running time.

14.9 One of the steps in Algorithm GENERATEMESH is to construct a doubly-connected edge list for the quadtree subdivision of a given quadtree. Describe an algorithm for this step, and analyze its running time.

14.10 A quadtree can also be used to store a subdivision for efficient point location. The idea is to keep splitting a bounding square of the subdivision until all leaf nodes correspond to squares that contain at most one vertex and only edges incident to that vertex, or no vertex and at most one edge.

   a.   Since a vertex can be incident to many edges, we need an additional data structure at the quadtree leaves storing vertices. Which data structure would you use?

   b.   Describe the algorithm for constructing the point location data structure in detail, and analyze its running time.

   c.   Describe the query algorithm in detail, and analyze its running time.

14.11 Quadtrees are often used to store pixel images. In this case the initial square is exactly the size of the image (which is assumed to be a $2^k \times 2^k$ grid for some integer k). A square is split into subsquares if not all pixels inside have the same intensity.

   Prove a bound on the complexity of the quadtree subdivision. *Hint:* This is similar to the bound we proved on the size of the quadtree mesh.

14.12 Suppose we have quadtrees on pixel images I_1 and I_2 (see the previous exercise). Both images have size $2^k \times 2^k$, and contain only two intensities, 0 and 1. Give algorithms for Boolean operations on these images, that is, give algorithms to compute a quadtree for $I_1 \vee I_2$ and $I_1 \wedge I_2$. (Here $I_1 \vee I_2$ is the $2^k \times 2^k$ image where pixel (i, j) has intensity 1 if and only if (i, j) has intensity 1 in image I_1 or in image I_2. The image $I_1 \wedge I_2$ is defined similarly.)

14.13 Quadtrees can be used to perform range queries. Describe an algorithm for querying a quadtree on a set P of points with a query region R. Analyze the worst-case query time for the case where R is a rectangle, and for the case where R is a half-plane bounded by a vertical line.

14.14 In this chapter we studied quadtrees that store a set of point in the plane. In Chapter 5 we studied two other data structures for storing sets of points in the plane, the kd-tree and the range tree. Discuss the advantages and disadvantages of each of the three structures.

15 Visibility Graphs

Finding the Shortest Route

In Chapter 13 we saw how to plan a path for a robot from a given start position to a given goal position. The algorithm we gave always finds a path if it exists, but we made no claims about the quality of the path: it could make a large detour, or make lots of unnecessary turns. In practical situations we would prefer to find not just any path, but a good path.

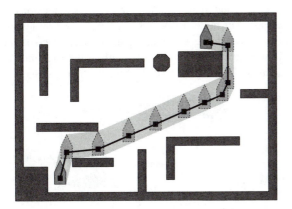

Figure 15.1
A shortest path

What constitutes a good path depends on the robot. In general, the longer a path, the more time it will take the robot to reach its goal position. For a mobile robot on a factory floor this means it can transport less goods per time unit, resulting in a loss of productivity. Therefore we would prefer a short path. Often there are other issues that play a role as well. For example, some robots can only move in a straight line; they have to slow down, stop, and rotate, before they can start moving into a different direction, so any turn along the path causes some delay. For this type of robot not only the path length but also the number of turns on the path has to be taken into account. In this chapter we ignore this aspect; we only show how to compute the Euclidean shortest path for a translating planar robot.

15.1 Shortest Paths for a Point Robot

As in Chapter 13 we first consider the case of a point robot moving among a set S of disjoint simple polygons in the plane. The polygons in S are called *obstacles*, and their total number of edges is denoted by n. Obstacles are open sets, so the robot is allowed to touch them. We are given a start position p_{start} and a goal position p_{goal}, which we assume are in the free space. Our goal is to compute a shortest collision-free path from p_{start} to p_{goal}, that is, a shortest path that does not intersect the interior of any of the obstacles. Notice that we cannot say *the* shortest path, because it need not be unique. For a shortest path to exist, it is important that obstacles are open sets; if they were closed, then (unless the robot can move to its goal in a straight line) a shortest path would not exist, as it would always be possible to shorten a path by moving closer to an obstacle.

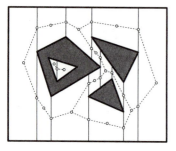

Let's quickly review the method from Chapter 13. We computed a trapezoidal map $\mathcal{T}(C_{free})$ of the free configuration space C_{free}. For a point robot, C_{free} was simply the empty space between the obstacles, so this was rather easy. The key idea was then to replace the continuous work space, where there are infinitely many paths, by a discrete road map G_{road}, where there are only finitely many paths. The road map we used was a plane graph with nodes in the centers of the trapezoids of $\mathcal{T}(C_{free})$ and in the middle of the vertical extensions that separate adjacent trapezoids. The nodes in the center of each trapezoid were connected to the nodes on its boundary. After finding the trapezoids containing the start and goal position of the robot, we found a path in the road map between the nodes in the centers of these trapezoids by breadth-first search.

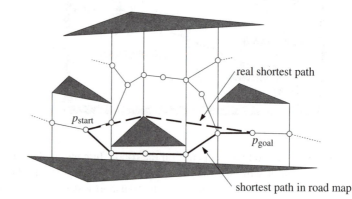

Figure 15.2
The shortest path does not follow the road map.

Because we used breadth-first search, the path that is found uses a minimum number of arcs in G_{road}. This is not necessarily a short path, because some arcs are between nodes that are far apart, whereas others are between nodes that are close to each other. An obvious improvement would be to give each arc a weight corresponding to the Euclidean length of the segment connecting its incident nodes, and to use a graph search algorithm that finds a shortest path in a weighted graph, such as Dijkstra's algorithm. Although this may improve the path length, we still do not get the shortest path. This is illustrated in Figure 15.2: the shortest path from p_{start} to p_{goal} following the road map passes below the triangle, but the real shortest path passes above it. What we need is a different road map, one which guarantees that the shortest path following the road map is the real shortest path.

Let's see what we can say about the shape of a shortest path. Consider some path from p_{start} to p_{goal}. Think of this path as an elastic rubber band, whose endpoints we fix at the start and goal position and which we force to take the shape of the path. At the moment we release the rubber band, it will try to contract and become as short as possible, but it will be stopped by the obstacles. The new path will follow parts of the obstacle boundaries and straight line segments through open space. The next lemma formulates this observation more precisely. It uses the notion of an *inner vertex* of a polygonal path, which is a vertex that is not the begin- or endpoint of the path.

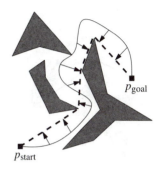

Lemma 15.1 *Any shortest path between p_{start} and p_{goal} among a set S of disjoint polygonal obstacles is a polygonal path whose inner vertices are vertices of S.*

Proof. Suppose for a contradiction that a shortest path τ is not polygonal. Since the obstacles are polygonal, this means there is a point p on τ that lies in the interior of the free space with the property that no line segment containing p is contained in τ. Since p is in the interior of the free space, there is a disc of positive radius centered at p that is completely contained in the free space. But then the part of τ inside the disc, which is not a straight line segment, can be shortened by replacing it with the segment connecting the point where it enters the disc to the point where it leaves the disc. This contradicts the optimality of τ, since any shortest path must be *locally shortest*, that is, any subpath connecting points q and r on the path must be the shortest path from q to r.

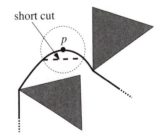

Now consider a vertex v of τ. It cannot lie in the interior of the free space: then there would be a disc centered at p that is completely in the free space, and we could replace the subpath of τ inside the disc— which turns at v—by a straight line segment which is shorter. Similarly, v cannot lie in the relative interior of an obstacle edge: then there would be a disc centered at v such that half of the disc is contained in the free space, which again implies that we can replace the subpath inside the disc with a straight line segment. The only possibility left is that v is an obstacle vertex. $\quad\square$

With this characterization of the shortest path, we can construct a road map that allows us to find the shortest path. This road map is the *visibility graph* of S, which we denote by $G_{\text{vis}}(S)$. Its nodes are the vertices of S, and there is an arc between vertices v and w if they can *see* each other, that is, if the segment $\overline{vw}$ does not intersect the interior of any obstacle in S. Two vertices that can see each other are also called *(mutually) visible*, and the segment connecting them is called a *visibility edge*. Note that endpoints of the same obstacle edge always see each other. Hence, the obstacle edges form a subset of the arcs of $G_{vis}(S)$.

By Lemma 15.1 the segments on a shortest path are visibility edges, except for the first and last segment. To make them visibility edges as well, we add the start and goal position as vertices to S, that is, we consider the visibility graph of the set $S^* := S \cup \{p_{start}, p_{goal}\}$. By definition, the arcs of $G_{vis}(S^*)$ are between vertices—which now include p_{start} and p_{goal}—that can see each other. We get the following corollary.

Corollary 15.2 *The shortest path between p_{start} and p_{goal} among a set S of disjoint polygonal obstacles consists of arcs of the visibility graph $G_{vis}(S^*)$, where $S^* := S \cup \{p_{start}, p_{goal}\}$.*

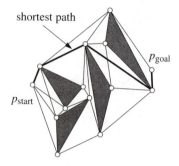

shortest path

p_{goal}

p_{start}

Corollary 15.2 implies that we can compute a shortest path from p_{start} to p_{goal} as follows.

Algorithm SHORTESTPATH(S, p_{start}, p_{goal})
Input. A set S of disjoint polygonal obstacles, and two points p_{start} and p_{goal} in the free space.
Output. The shortest collision-free path connecting p_{start} and p_{goal}.
1. $G_{vis} \leftarrow$ VISIBILITYGRAPH($S \cup \{p_{start}, p_{goal}\}$)
2. Assign each arc (v, w) in G_{vis} a weight, which is the Euclidean length of the segment $\overline{vw}$.
3. Use Dijkstra's algorithm to compute a shortest path between p_{start} and p_{goal} in G_{vis}.

In the next section we show how to compute the visibility graph in $O(n^2 \log n)$ time, where n is the total number of obstacle edges. The number of arcs of G_{vis} is of course bounded by $\binom{n+2}{2}$. Hence, line 2 of the algorithm takes $O(n^2)$ time. Dijkstra's algorithm computes the shortest path between two nodes in graph with k arcs, each having a non-negative weight, in $O(n \log n + k)$ time. Since $k = O(n^2)$, we conclude that the total running time of SHORTESTPATH is $O(n^2 \log n)$, leading to the following theorem.

Theorem 15.3 *A shortest path between two points among a set of polygonal obstacles with n edges in total can be computed in $O(n^2 \log n)$ time.*

15.2 Computing the Visibility Graph

Let S be a set of disjoint polygonal obstacles in the plane with n edges in total. To compute the visibility graph of S, we have to find the pairs of vertices that can see each other. This means that for every pair we have to test whether the line segment connecting them intersects any obstacle. Such a test would cost $O(n)$ time when done naively, leading to an $O(n^3)$ running time. We will see shortly that the test can be done more efficiently if we don't consider the pairs in arbitrary order, but concentrate on one vertex at a time and identify all vertices visible from it, as in the following algorithm.

Algorithm VISIBILITYGRAPH(S)

Input. A set S of disjoint polygonal obstacles.

Output. The visibility graph $\mathcal{G}_{vis}(S)$.

1. Initialize a graph $G = (V, E)$ where V is the set of all vertices of the polygons in S and $E = \emptyset$.
2. **for** all vertices $v \in V$
3.   **do** $W \leftarrow$ VISIBLEVERTICES(v, S)
4.   For every vertex $w \in W$, add the arc (v, w) to E.
5. **return** $\mathcal{G}$

The procedure VISIBLEVERTICES has as input a set S of polygonal obstacles and a point p in the plane; in our case p is a vertex of S, but that is not required. It should return all obstacle vertices visible from p.

If we just want to test whether one specific vertex w is visible from p, there is not much we can do: we have to check the segment $\overline{pw}$ against all obstacles. But there is hope if we want to test all vertices of S: we might be able to use the information we get when we test one vertex to speed up the test for other vertices. Now consider the set of all segments $\overline{pw}$. What would be a good order to treat them, so that we can use the information from one vertex when we treat the next one? The logical choice is the cyclic order around p. So what we will do is treat the vertices in cyclic order, meanwhile maintaining information that will help us to decide on the visibility of the next vertex to be treated.

A vertex w is visible from p if the segment $\overline{pw}$ does not intersect the interior of any obstacle. Consider the half-line ρ starting at p and passing through w. If w is not visible, then ρ must hit an obstacle edge before it reaches w. To check this we perform a binary search with the vertex w on the obstacle edges intersected by ρ. This way we can find out whether w lies behind any of these edges, as seen from p. (If p itself is also an obstacle vertex, then there is another case where w is not visible, namely when p and w are vertices of the same obstacle and $\overline{pw}$ is contained in the interior of that obstacle. This case can be checked by looking at the edges incident to w, to see whether ρ is in the interior of the obstacle before it reaches w. For the moment we ignore degenerate cases, where one of the incident edges of w is contained in $\overline{pw}$.)

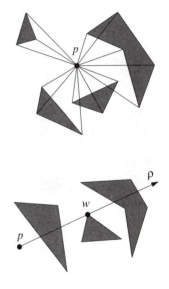

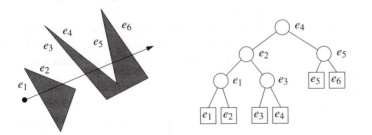

Figure 15.3
The search tree on the intersected edges

While treating the vertices in cyclic order around p we therefore maintain the obstacle edges intersected by ρ in a balanced search tree $\mathcal{T}$. (As we will see later, edges that are contained in ρ need not be stored in $\mathcal{T}$.) The leaves of

$\mathcal{T}$ store the intersected edges in order: the leftmost leaf stores the first segment intersected by ρ, the next leaf stores the segment that is intersected next, and so on. The interior nodes, which guide the search in $\mathcal{T}$, also store edges. More precisely, an interior node ν stores the rightmost edge in its left subtree, so that all edges in its right subtree are greater (with respect to the order along ρ) than this segment e_ν, and all segments in its left subtree are smaller than or equal to e_ν (with respect to the order along ρ). Figure 15.3 shows an example.

Treating the vertices in cyclic order effectively means that we rotate the half-line ρ around p. So our approach is similar to the plane sweep paradigm we used at various other places; the difference is that instead of using a horizontal line moving downward to sweep the plane, we use a rotating half-line.

The status of our *rotational plane sweep* is the ordered sequence of obstacle edges intersected by ρ. It is maintained in $\mathcal{T}$. The events in the sweep are the vertices of S. To deal with a vertex w we have to decide whether w is visible from p by searching in the status structure $\mathcal{T}$, and we have to update $\mathcal{T}$ by inserting and/or deleting the obstacle edges incident to w.

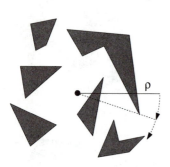

Algorithm VISIBLEVERTICES summarizes our rotational plane sweep. The sweep is started with the half-line ρ pointing into the positive x-direction and proceeds in clockwise direction. So the algorithm first sorts the vertices by the clockwise angle that the segment from p to each vertex makes with the positive x-axis. What do we do if this angle is equal for two or more vertices? To be able to decide on the visibility of a vertex w, we need to know whether $\overline{pw}$ intersects the interior of any obstacle. Hence, the obvious choice is to treat any vertices that may lie in the interior of $\overline{pw}$ before we treat w. In other words, vertices for which the angle is the same are treated in order of increasing distance to p. The algorithm now becomes as follows:

Algorithm VISIBLEVERTICES(p,S)
Input. A set S of polygonal obstacles and a point p that does not lie in the interior of any obstacle.
Output. The set of all obstacle vertices visible from p.
1. Sort the obstacle vertices according to the clockwise angle that the half-line from p to each vertex makes with the positive x-axis. In case of ties, vertices closer to p should come before vertices farther from p. Let $w_1, \ldots, w_n$ be the sorted list.
2. Let ρ be the half-line parallel to the positive x-axis starting at p. Find the obstacle edges that are properly intersected by ρ, and store them in a balanced search tree $\mathcal{T}$ in the order in which they are intersected by ρ.
3. $W \leftarrow \emptyset$
4. **for** $i \leftarrow 1$ **to** n
5. **do if** VISIBLE(w_i) **then** Add w_i to W.
6. Insert into $\mathcal{T}$ the obstacle edges incident to w_i that lie on the clockwise side of the half-line from p to w_i.
7. Delete from $\mathcal{T}$ the obstacle edges incident to w_i that lie on the counterclockwise side of the half-line from p to w_i.
8. **return** W

The subroutine VISIBLE must decide whether a vertex w_i is visible. Normally, this only involves searching in $\mathcal{T}$ to see if the edge closest to p, which is stored in the leftmost leaf, intersects $\overline{pw_i}$. But we have to be careful when $\overline{pw_i}$ contains other vertices. Is w_i visible or not in such a case? That depends. See Figure 15.4 for some of the cases that can occur. $\overline{pw_i}$ may or may not intersect the interior of the obstacles incident to these vertices. It seems that we have to inspect all edges with a vertex on $\overline{pw_i}$ to decide if w_i is visible. Fortunately we have already inspected them while treating the preceding vertices that lie on $\overline{pw_i}$. We can therefore decide on the visibility of w_i as follows. If w_{i-1} is not visible then w_i is not visible either. If w_{i-1} is visible then there are two ways in which w_i can be invisible. Either the whole segment $\overline{w_{i-1}w_i}$ lies in an obstacle of which both w_{i-1} and w_i are vertices, or the segment $\overline{w_{i-1}w_i}$ is intersected by an edge in $\mathcal{T}$. (Because in the latter case this edge lies between w_{i-1} and w_i is must properly intersect $\overline{w_{i-1}w_i}$.) This test is correct because $\overline{pw_i} = \overline{pw_{i-1}} \cup \overline{w_{i-1}w_i}$. (If $i = 1$, then there is no vertex in between p and w_i, so we only have to look at the segment $\overline{pw_i}$.) We get the following subroutine:

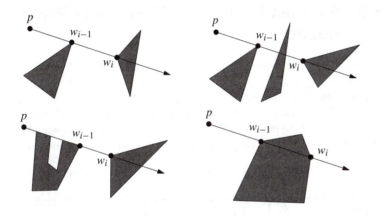

Figure 15.4
Some examples where p contains multiple vertices. In all these cases w_{i-1} is visible. In the left two cases w_i is also visible and in the right two cases w_i is not visible.

VISIBLE(w_i)
1. **if** $\overline{pw_i}$ intersects the interior of the obstacle of which w_i is a vertex, locally at w_i
2. **then return false**
3. **else if** $i = 1$ **or** w_{i-1} is not on the segment $\overline{pw_i}$
4. **then** Search in $\mathcal{T}$ for the edge e in the leftmost leaf.
5. **if** e exists and $\overline{pw_i}$ intersects e
6. **then return false**
7. **else return true**
8. **else if** w_{i-1} is not visible
9. **then return false**
10. **else** Search in $\mathcal{T}$ for an edge e that intersects $\overline{w_{i-1}w_i}$.
11. **if** e exists
12. **then return false**
13. **else return true**

This finishes the description of the algorithm VISIBLEVERTICES to compute the vertices visible from a given point p.

What is the running time of VISIBLEVERTICES? The time we spent before line 4 is dominated by the time to sort the vertices in cyclic order around p, which is $O(n \log n)$. Each execution of the loop involves a constant number of operations on the balanced search tree $\mathcal{T}$, which take $O(\log n)$ time, plus a constant number of geometric tests that take constant time. Hence, one execution takes $O(\log n)$ time, leading to an overall running time of $O(n \log n)$.

Recall that we have to apply VISIBLEVERTICES to each of the n vertices of S in order to compute the entire visibility graph. We get the following theorem:

Theorem 15.4 *The visibility graph of a set S of disjoint polygonal obstacles with n edges in total can be computed in $O(n^2 \log n)$ time.*

15.3 Shortest Paths for a Translating Polygonal Robot

In Chapter 13 we have seen that we can reduce the motion planning problem for a translating, convex, polygonal robot $\mathcal{R}$ to the case of a point robot by computing the free configuration space $\mathcal{C}_{\text{free}}$. The reduction involves computing the Minkowski sum of $-\mathcal{R}$, a reflected copy of $\mathcal{R}$, with each of the obstacles, and taking the union of the resulting configuration-space obstacles. This gives us a set of disjoint polygons, whose union is the forbidden configu-

work space   configuration space   visibility graph

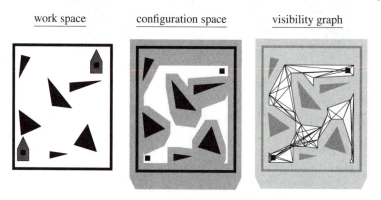

Figure 15.5
Computing a shortest path for a
polygonal robot

ration space. We can then compute a shortest path with the method we used for a point robot: we extend the set of polygons with the points in configuration space that correspond to the start and goal placement, compute the visibility graph of the polygons, assign each arc a weight which is the Euclidean length of the corresponding visibility edge, and find a shortest path in the visibility graph using Dijkstra's algorithm.

To what running time does this approach lead? Lemma 13.13 states that the forbidden space can be computed in $O(n \log^2 n)$ time. Furthermore, the complexity of the forbidden space is $O(n)$ by Theorem 13.12, so from the previous section we know that the visibility graph of the forbidden space can be computed in $O(n^2 \log n)$ time.

This leads to the following result:

Theorem 15.5 *Let $\mathcal{R}$ be a convex, constant-complexity robot that can translate among a set of polygonal obstacles with n edges in total. A shortest collision-free path for $\mathcal{R}$ from a given start placement to a given goal placement can be computed in $O(n^2 \log n)$ time.*

15.4 Notes and Comments

The problem of computing the shortest path in a weighted graph has been studied extensively. Dijkstra's algorithm, and other solutions, are described in most books on graph algorithms and in many books on algorithms and data structures. In Section 15.1 we stated that Dijkstra's algorithm runs in $O(n \log n + k)$ time. To achieve this time bound, one has to use Fibonacci heaps in the implementation. In our application an $O((n + k) \log n)$ algorithm would also do fine, since the rest of the algorithm needs that much time anyway.

The geometric version of the shortest path problem has also received considerable attention. The algorithm given here is due to Lee [212]. More efficient algorithms based on arrangements have been proposed; they run in $O(n^2)$ time [11, 129, 331].

Any algorithm that computes a shortest path by first constructing the entire visibility graph is doomed to have at least quadratic running time in the worst case, because the visibility graph can have a quadratic number of edges. For a long time no approaches were known with a subquadratic worst-case running time. Mitchell [247] was the first to break the quadratic barrier: he showed that the shortest path for a point robot can be computed in $O(n^{5/3+\varepsilon})$ time. Later he improved the running time of his algorithm to $O(n^{3/2+\varepsilon})$ [248]. In the mean time, however, Hershberger and Suri [178, 179] succeeded in developing an optimal $O(n \log n)$ time algorithm.

In the special case where the free space of the robot is a polygon without holes, a shortest path can be computed in linear time by combining the linear-time triangulation algorithm of Chazelle [73] with the shortest path method of Guibas et al. [163].

The 3-dimensional version of the Euclidean shortest path problem is much harder. This is due to the fact that there is no easy way to discretize the problem: the inflection points of the shortest path are not restricted to a finite set of points, but they can lie anywhere on the obstacle edges. Canny [60] proved that the problem of computing a shortest path connecting two points among polyhedral obstacles in 3-dimensional space is NP-hard. Reif and Storer [289] gave a single-exponential algorithm for the problem, by reducing it to a decision problem in the theory of real numbers. There are also several papers that approximate the shortest path in polynomial time, for example, by adding points on obstacle edges and searching a graph with these points as nodes [6, 103, 104, 223, 279].

In this chapter we concentrated on the Euclidean metric. Various papers deal with shortest paths under a different metric. Because the number of settings is quite large, we mention only a few, and we give only a few references for each setting. An interesting metric that has been studied is the *link metric*, where the length of a polygonal path is the number of links it consists of [320, 250, 9, 100]. Another case that has been studied extensively is that of rectilinear paths. Such paths play an important role in VLSI design, for instance. Lee et al. [218] give a survey of rectilinear path problems. An interesting metric that has been studied for rectilinear paths is the *combined metric*, which is a linear combination of the Euclidean metric and the link metric [40]. Finally, there are papers that consider paths in a subdivision where each region has a weight associated with it. The cost of a path through a region is then its Euclidean length times the weight of the region. Obstacles can be modeled by regions with infinite weight [92, 249].

While there are many obvious metrics for translating robots—in particular, the Euclidean metric immediately comes to mind—it is not so easy to give a good definition of a shortest path for a robot that can rotate as well as translate. Some results have been obtained for the case where the robot is a line segment [12, 93, 185].

The visibility graph was introduced for planning motions by Nilson [259]. The $O(n^2 \log n)$ time algorithm that we described to compute it is due to Lee [212]. A number of faster algorithms are known [11, 331], including an optimal algorithm by Gosh and Mount [157], which runs in $O(n \log n + k)$ time, where k is the number of arcs of the visibility graph.

To compute a shortest path for a point robot among a set of convex polygonal obstacles, one does not need all the visibility edges. One only needs the visibility edges that define a common tangent. Rohnert [291] gave an algorithm that computes this reduced visibility graph in time $O(n + c^2 \log n)$, where c is the number of obstacles, and n is their total number of edges.

The *visibility complex*, introduced by Vegter and Pocchiola [282, 283, 326] is a structure that has the same complexity as the visibility graph, but contains more information. It is defined on a set of convex (not necessarily polygonal) objects in the plane, and can be used for shortest path problems and ray shooting. It can be computed in $O(n \log n + k)$ time.

15.5 Exercises

15.1 Prove that the number of segments on the shortest path is bounded by $O(n)$. Give an example where it is $\Theta(n)$.

15.2 Algorithm VISIBILITYGRAPH calls algorithm VISIBLEVERTICES with each obstacle vertex. VISIBLEVERTICES sorts all vertices around its input point. This means that n cyclic sortings are done, one around each obstacle vertex. In this chapter we simply did every sort in $O(n\log n)$ time, leading to $O(n^2\log n)$ time for all sortings. Show that this can be improved to $O(n^2)$ time using dualization (see Chapter 8). Does this improve the running time of VISIBILITYGRAPH?

15.3 The algorithm for finding the shortest path can be extended to objects other than polygons. Let S be a set of n disjoint disc obstacles (not necessarily of equal radius).

    a.   Prove that in this case the shortest path between two points consists of parts of boundaries of the disks and common tangents of disks.

    b.   Adapt the notion of a visibility graph for this situation.

    c.   Adapt the shortest path algorithm to find the shortest path between two points among the disks in S.

15.4 What is the maximal number of shortest paths connecting two fixed points among a set of n triangles in the plane?

15.5 Let S be a set of disjoint polygons and let a starting point p_{start} be given. We want to preprocess the set S (and p_{start}) such that for different goal points we can efficiently find the shortest path from p_{start} to the goal. Describe how the preprocessing can be done in time $O(n^2\log n)$ such that for any given goal point p_{goal} we can compute the shortest path from p_{start} to p_{goal} in time $O(n\log n)$.

15.6 Design an algorithm to find a shortest paths between two points inside a simply polygon. Your algorithm should run in subquadratic time.

15.7 When all obstacles are convex polygons we can improve the shortest path algorithm by only considering common tangents rather than all visibility edges.

    a.   Prove that that the only visibility edges that are required in the shortest path algorithm are the common tangents of the polygons.

    b.   Give a fast algorithm to find the common tangents of two disjoint convex polygons.

    c.   Give an algorithm to compute those common tangents that are also visibility edges among a set of convex polygons.

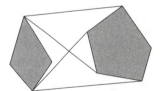

15.8* If you are familiar with homogeneous coordinates, it is interesting to see that the rotational sweep that we used in this chapter can be transformed into a normal plane sweep using a horizontal line that translates over the plane. Show that this is the case using a projective transformation that moves the center of the sweep to a point at infinity.

16 Simplex Range Searching

Windowing Revisited

In Chapter 2 we saw that geographic information systems often store a map in a number of separate layers. Each layer represents a *theme* from the map, that is, a specific type of feature such as roads or cities. Distinguishing layers makes it easy for the user to concentrate her attention on a specific feature. Sometimes one is not interested in all the features of a given type, but only in the ones lying inside a certain region. Chapter 10 contains an example of this: from a road map of the whole of Europe we wanted to select the part lying inside a much smaller region. There the query region, or *window*, was rectangular, but it is easy to imagine situations where the region has a different shape. Suppose that

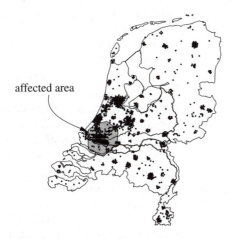

affected area

Figure 16.1
Population density of the Netherlands

we have a map layer whose theme is population density. The density is shown on the map by plotting a point for every 5,000 people, say. An example of such a map is given in Figure 16.1. If we want to estimate the impact of building, say, a new airport at a given location, it is useful to know how many people live in the affected area. In geometric terms we have a set of points in the plane and we want to count the points lying inside a query region (for instance, the region within which the noise of planes exceeds a certain level).

In Chapter 5, where we studied database queries, we developed a data structure to report the points inside an axis-parallel query rectangle. The area affected by the airport, however, is determined by the dominating direction of the wind, and unlikely to be rectangular, so the data structures from Chapter 5 are of little help here. We need to develop a data structure that can handle more general query ranges.

16.1 Partition Trees

Given a set of points in the plane, we want to count the points lying inside a query region. (Here and in the following we follow the convention of using the expression "to count the points" in the sense of "to report the number of points", not in the sense of enumerating the points.) Let's assume that the query region is a simple polygon; if it is not, we can always approximate it. To simplify the query answering algorithm we first triangulate the query region, that is, we decompose it into to triangles. Chapter 3 describes how to do this. After we have triangulated the region, we query with each of the resulting triangles. The set of points lying inside the region is just the union of the sets of points inside the triangles. When counting the points we have to be a bit careful with points lying on the common boundary of two triangles, but that is not difficult to take care of.

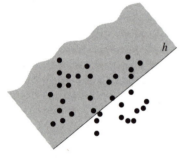

We have arrived at the *triangular range searching problem*: given a set S of n points in the plane, count the points from S lying in a query triangle t. Let's first study a slightly simpler version of this problem, where the query triangle degenerates into a half-plane. What should a data structure for half-plane range queries look like? As a warm-up, let's look at the one-dimensional problem first: in the one-dimensional version of the problem we are given a set of n points on the real line, and we want to count the points in a query half-line (that is, the points lying on a specified side of a query point). Using a balanced binary search tree where every node also stores the number of points in its subtree, we can answer such queries in $O(\log n)$ time. How can we generalize this to a 2-dimensional setting? To answer this question we must first interpret a balanced search tree geometrically. Each node of the tree contains a key— the coordinate of a point—that is used to split the point set into the sets that are stored in the left and right subtrees. Similarly, we can consider this key value to split the real line into two pieces. In this way, every node of the tree corresponds to a region on the line—the root to the whole line, the two children of the root to two half-lines, and so on. For any query half-line and any node, the region of one child of the node is either completely contained in the half-line or completely disjoint from it. All points in that region are in the half-line, or none of them is. Hence, we only have to search recursively in the other subtree of the node. This is shown in Figure 16.2. The points are drawn below the tree as black dots; the two regions of the real line corresponding to the two subtrees are also indicated. The query half-line is the grey region. The region corresponding to the subtree rooted at the black node is completely inside the query half-line. Hence, we only have to recurse on the right subtree.

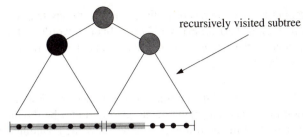

recursively visited subtree

Figure 16.2
Answering a half-line query with a
binary tree

To generalize this to two dimensions we could try to partition the plane into
two regions, such that for any query half-plane there is one region that is either
completely contained in the half-plane or completely disjoint from it. Unfortunately, such a partitioning does not exist, so we need a further generalization:
instead of partitioning into two regions, we must partition into more regions.
The partitioning should be such that for any query half-plane we have to search
recursively in only few of the regions.

We now give a formal definition of the type of partitioning we need. A *simplicial partition* for a set S of n points in the plane is a collection $\Psi(S) :=
\{(S_1, t_1), \ldots, (S_r, t_r)\}$, where the S_i's are disjoint subsets of S whose union is
S, and t_i is a triangle containing S_i. The subsets S_i are called *classes*. We do
not require the triangles to be disjoint, so a point of S may lie in more than one
triangle. Still, such a point is a member of only one class. We call r, the number of triangles, the *size* of $\Psi(S)$. Figure 16.3 gives an example of a simplicial
partition of size five; different shades of grey are used to indicate the different
classes. We say that a line ℓ crosses a triangle t_i if ℓ intersects the interior of t_i.

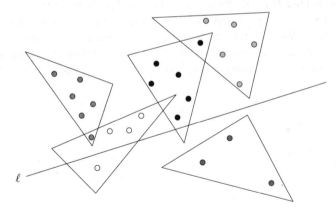

Figure 16.3
A fine simplicial partition

When the point set S is not in general position we sometimes need to use (relatively open) segments as well as triangles in the simplicial partition. A line is
defined to cross a segment when it intersects its relative interior but does not
contain it. The *crossing number* of a line ℓ with respect to $\Psi(S)$ is the number
of triangles of $\Psi(S)$ crossed by ℓ. Thus the crossing number of the line ℓ in
Figure 16.3 is two. The crossing number of $\Psi(S)$ is the maximum crossing
number over all possible lines ℓ. In Figure 16.3 you can find lines that intersect

four triangles, but no line intersects all five. Finally, we say that a simplicial partition is *fine* if $|S_i| \leqslant 2n/r$ for every $1 \leqslant i \leqslant r$. In other words, in fine simplicial partitions none of the classes contains more than twice the average number of points of the classes.

Now that we have formalized the notion of partition, let's see how we can use such a partition to answer half-plane range queries. Let h be the query half-plane. If a triangle t_i of the partition is not crossed by the bounding line of h, then its class S_i either lies completely in h, or it is completely disjoint from h. This means we only have to recurse on the classes S_i for which t_i is crossed by the bounding line of h. For example, if in Figure 16.3 we queried with ℓ^+, the half-plane lying above ℓ, we would have to recurse on two of the five classes. The efficiency of the query answering process therefore depends on the crossing number of the simplicial partition: the lower the crossing number, the better the query time. The following theorem states that it is always possible to find a simplicial partition with crossing number $O(\sqrt{r})$; later we shall see what this implies for the query time.

Theorem 16.1 *For any set S of n points in the plane, and any parameter r with $1 \leqslant r \leqslant n$, a fine simplicial partition of size r and crossing number $O(\sqrt{r})$ exists. Moreover, for any $\varepsilon > 0$ such a simplicial partition can be constructed in time $O(n^{1+\varepsilon})$.*

It seems a bit strange to claim a construction time of $O(n^{1.1})$ or $O(n^{1.01})$ or with the exponent even closer to 1. Still, no matter how small ε is, as long as it is a positive constant, the bound in the theorem can be attained. But a better upper bound like $O(n)$ or $O(n \log n)$ isn't claimed in the theorem.

Section 16.4 gives pointers to the literature where a proof of this theorem can be found. We shall take the theorem for granted, and concentrate on how to use it in the design of an efficient data structure for half-plane range queries. The data structure we'll obtain is called a *partition tree*. Probably you can al-

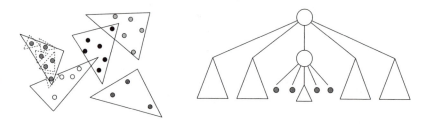

Figure 16.4
A simplicial partition and the
corresponding tree

ready guess what such a partition tree looks like: it's a tree whose root has r children, each being the root of a recursively defined partition tree for one of the classes in a simplicial partition. There is no specific order on the children; it happens to be irrelevant. Figure 16.4 shows a simplicial partition and the corresponding tree. The dotted triangles form the partition computed recursively for the class corresponding to the middle child of the root; the five "subclasses" are stored in five subtrees below the middle child. Depending on the application, we also store some extra information about the classes. The basic structure of a partition tree is thus as follows:

- If S contains only one point, p, the partition tree consists of a single leaf where p is stored explicitly. The set S is the *canonical subset* of the leaf.

- Otherwise, the structure is a tree $\mathcal{T}$ of branching degree r, where r is a sufficiently large constant. (Below we shall see how r should be chosen.) The children of the root of the tree are in one-to-one correspondence with the triangles of a fine simplicial partition of size r for the set S. The triangle of the partition corresponding to child v is denoted by $t(v)$. The corresponding class in S is called the *canonical subset* of v; it is denoted $S(v)$. The child v is the root of a recursively defined partition tree $\mathcal{T}_v$ on the set $S(v)$.

- With each child v we store the triangle $t(v)$. We also store information about the subset $S(v)$; for half-plane range counting this information is the cardinality of $S(v)$, but for other applications we may want to store other information.

We can now describe the query algorithm for counting the number of points from S in a query half-plane h. The algorithm returns a set Υ of nodes from the partition tree $\mathcal{T}$, called the *selected nodes*, such that the subset of points from S lying in h is the disjoint union of the canonical subsets of the nodes in Υ. In other words, Υ is a set of nodes whose canonical subsets are disjoint, and such that

$$S \cap h = \bigcup_{v \in \Upsilon} S(v).$$

The selected nodes are exactly the nodes v with the property: $t(v) \subset h$ (or, in case v is a leaf, the point stored at v lies in h) and there is no ancestor μ of v such that $t(\mu) \subset h$. The number of points in h can be computed by summing the cardinalities of the selected canonical subsets.

Algorithm SELECTINHALFPLANE($h, \mathcal{T}$)
Input. A query half-plane h and a partition tree or subtree of it.
Output. A set of canonical nodes for all points in the tree that lie in h.
1. $\Upsilon \leftarrow \emptyset$
2.   **if** $\mathcal{T}$ consists of a single leaf μ
3.   **then if** the point stored at μ lies in h **then** $\Upsilon \leftarrow \{\mu\}$
4.   **else for** each child v of the root of $\mathcal{T}$
5.   **do if** $t(v) \subset h$
6.   **then** $\Upsilon \leftarrow \Upsilon \cup \{v\}$
7.   **else if** $t(v) \cap h \neq \emptyset$
8.   **then** $\Upsilon \leftarrow \Upsilon \cup$ SELECTINHALFPLANE($h, \mathcal{T}_v$)
9.   **return** Υ

Figure 16.5 illustrates the operation of the query algorithm. The selected children of the root are shown in black. The children that are visited recursively (as well as the root itself, since it has also been visited) are grey. As said before, a

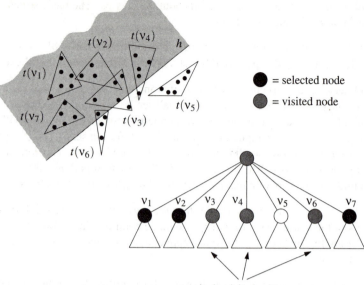

Figure 16.5
Answering a half-plane range query
using a partition tree

recursively visited subtrees

half-plane range counting query can be answered by calling SELECTINHALF-PLANE and summing the cardinalities of the selected nodes, which are stored at the nodes. In practice, one would probably not keep track of the set Υ, but one would maintain a counter; when a node is selected, the cardinality of its canonical subset is added the counter.

We have described the partition tree, a data structure for half-plane range counting, and its query algorithm. Now it's time to analyze our structure. We start with the amount of storage.

Lemma 16.2 *Let S be a set of n points in the plane. A partition tree on S uses* $O(n)$ *storage.*

Proof. Let $M(n)$ be the maximum number of nodes that a partition tree on a set of n points can have, and let n_v denote the cardinality of the canonical subset $S(v)$. Then $M(n)$ satisfies the following recurrence:

$$M(n) \leqslant \begin{cases} 1 & \text{if } n = 1, \\ 1 + \sum_v M(n_v) & \text{if } n > 1, \end{cases}$$

where we sum over all children v of the root of the tree. Because the classes in a simplicial partition are disjoint, we have $\sum_v n_v = n$. Furthermore, $n_v \leqslant 2n/r$ for all v. Hence, for any constant $r > 2$ the recurrence solves to $M(n) = O(n)$.

The storage needed for a single node of the tree is $O(r)$. Since r is a constant, the lemma follows. ◻

Linear storage is the best one could hope for, but what about the query time? Here it becomes important what the exact value of r is. When we perform a query we have to recurse in at most $c\sqrt{r}$ subtrees, for some constant c that does not depend on r or n. It turns out that this constant has an influence on the exponent of n in the query time. To decrease this influence, we need to make r large enough, and then, as we will see, we can get a query time close to $O(\sqrt{n})$.

Lemma 16.3 *Let S be a set of n points in the plane. For any $\varepsilon > 0$, there is a partition tree for S such that for a query half-plane h we can select $O(n^{1/2+\varepsilon})$ nodes from the tree with the property that the subset of points from S in h is the disjoint union of the canonical subsets of the selected nodes. The selection of these nodes takes $O(n^{1/2+\varepsilon})$ time. As a consequence, half-plane range counting queries can be answered in $O(n^{1/2+\varepsilon})$ time.*

Proof. Let $\varepsilon > 0$ be given. According to Theorem 16.1 there is a constant c such that for any parameter r we can construct a simplicial partition of size r with crossing number at most $c\sqrt{r}$. We base the partition tree on simplicial partitions of size $r := \lceil 2(c\sqrt{2})^{1/\varepsilon} \rceil$. Let $Q(n)$ denote the maximum query time for any query in a tree for a set of n points. Let h be a query half-plane, and let n_v denote the cardinality of the canonical subset $S(v)$. Then $Q(n)$ satisfies the following recurrence:

$$Q(n) \leqslant \begin{cases} 1 & \text{if } n = 1, \\ r + \sum_{v \in C(h)} Q(n_v) & \text{if } n > 1, \end{cases}$$

where we sum over the set $C(h)$ of all children v of the root such that the boundary of h crosses $t(v)$. Because the simplicial partition underlying the data structure has crossing number $c\sqrt{r}$, we know that the number of nodes in the set $C(h)$ is at most $c\sqrt{r}$. We also know that $n_v \leqslant 2n/r$ for each v, because the simplicial partition is fine. Using these two facts one can show that with our choice of r the recurrence for $Q(n)$ solves to $O(n^{1/2+\varepsilon})$. $\qquad\qquad\square$

You may be a bit disappointed by the query time: the query time of most geometric data structures we have seen up to now is $O(\log n)$ or a polynomial in $\log n$, whereas the query time for the partition tree is around $O(\sqrt{n})$. Apparently this is the price we have to pay if we want to solve truly 2-dimensional query problems, such as half-plane range counting. Is it impossible to answer such queries in logarithmic time? No: later in this chapter we shall design a data structure for half-plane range queries with logarithmic query time. But the improvement in query time will not come for free, as that data structure will need quadratic storage.

It is helpful to compare the approach we have taken here with the range trees from Chapter 5 and the segment trees from Chapter 10. In these data structures, we would like to return information about a subset of a given set of geometric objects (points in range trees and partition trees, intervals in segment trees), or to report the subset itself. If we could precompute the requested information for every possible subset that can appear in a query, queries could be answered very fast. However, the number of possible different answers often prohibits such an approach. Instead, we identified what we have called *canonical subsets*, and we precomputed the required information for these subsets only. A query is then solved by expressing the answer to the query as the disjoint union of some of these canonical subsets. The query time is roughly linear in the number of canonical subsets that are required to express any possible query subset. The storage is proportional to the total number of precomputed canonical subsets for range trees and partition trees, and proportional to the sum of the sizes of the precomputed canonical subsets for segment trees. There is a trade-off between query time and storage: to make sure that every possible query can be expressed as the union of only a few canonical subsets, we need to provide a large repertoire of such subsets, and need a lot of storage. To decrease the storage, we need to decrease the number of precomputed canonical subsets— but that may mean that the number of canonical subsets needed to express a given query will be larger, and the query time increases.

This phenomenon can be observed clearly for 2-dimensional range searching: the partition tree we constructed in this section provides a repertoire of only $O(n)$ canonical subsets, and therefore needs only linear storage, but in general one needs $\Omega(\sqrt{n})$ canonical subsets to express the set of points lying in a half-plane. Only by providing roughly a quadratic number of canonical subsets can one achieve logarithmic query time.

Now let's go back to the problem that we wanted to solve, namely triangular range queries. Which modifications do we need if we want to use partition trees for triangles instead of half-planes as query regions? The answer is simple: none. We can use exactly the same data structure and query algorithm, with the query half-plane replaced by a query triangle. In fact, the solution works for any query range γ. The only question is what happens to the query time.

When the query algorithm visits a node, there are three types of children: the children v for which $t(v)$ lies completely inside the query range, the children for which $t(v)$ lies outside the range, and the children for which $t(v)$ lies partially inside the query range. Only the children of the third type have to be visited recursively. The query time therefore depends on the number of triangles in the partition that are crossed by the boundary of the query range γ. In other words, we have to see what the crossing number of γ is with respect to the simplicial partition. For a triangular query region, this is easy: a triangle in the partition is crossed by the boundary of γ only if it is crossed by one of the three lines through the edges of γ. Since each one of these lines intersects at most $c\sqrt{r}$ triangles, the crossing number of γ is at most $3c\sqrt{r}$.

The recursion for the query time therefore remains nearly the same, only the constant c changes to $3c$. As a result, we will need to choose r larger, but in the end the query time remains asymptotically the same. We get the following theorem:

Theorem 16.4 *Let S be a set of n points in the plane. For any $\varepsilon > 0$, there is a data structure for S, called a partition tree, that uses $O(n)$ storage, such that the points from S lying inside a query triangle can be counted in $O(n^{1/2+\varepsilon})$ time. The points can be reported in $O(k)$ additional time, where k is the number of reported points. The structure can be constructed in $O(n^{1+\varepsilon})$ time.*

Proof. The only two issues that have not been discussed yet are the construction time and the reporting of points.

Constructing a partition tree is easy: the recursive definition given before immediately implies a recursive construction algorithm. We denote the time this algorithm needs to construct a partition tree for a set of n points by $T(n)$. Let $\varepsilon > 0$ be given. According to Theorem 16.1 we can construct a fine simplicial partition for S of size r with crossing number $O(\sqrt{r})$ in time $O(n^{1+\varepsilon'})$, for any $\varepsilon' > 0$. We let $\varepsilon' = \varepsilon/2$. Hence, $T(n)$ satisfies the recurrence

$$T(n) = \begin{cases} O(1) & \text{if } n = 1, \\ O(n^{1+\varepsilon/2}) + \sum_v T(n_v) & \text{if } n > 1, \end{cases}$$

where we sum over all children v of the root of the tree. Because the classes in a simplicial partition are disjoint, we have $\sum_v n_v = n$, and the recurrence solves to $T(n) = O(n^{1+\varepsilon})$.

It remains to show that the k points in a query triangle can be reported in $O(k)$ additional time. These points are stored in the leaves below the selected nodes. Hence, they can be reported by traversing the subtrees rooted at the selected nodes. Because the number of internal nodes of a tree is linear in the number of leaves of that tree this takes time linear in the number of reported points. ◻

16.2 Multi-Level Partition Trees

Partition trees are powerful data structures. Their strength is that the points lying in a query half-plane can be selected in a small number of groups, namely, the canonical subsets of the nodes selected by the query algorithm. In the example above we used a partition tree for half-plane range counting, so the only information that we needed about the selected canonical subsets was their cardinality. In other query applications we will need other information about the canonical subsets, so we have to precompute and store that data. The information that we store about a canonical subset does not have to be a single number, like its cardinality. We can also store the elements of the canonical subset in a list, or a tree, or any kind of data structure we like. This way we get a *multi-level data structure*. The concept of multi-level data structures is not new: we already used it in Chapter 5 to answer multi-dimensional rectangular range queries and in Chapter 10 for windowing queries.

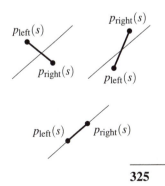

We now give an example of a multi-level data structure based on partition trees. Let S be a set of n line segments in the plane. We want to count the number of segments intersected by a query line ℓ. Let $p_{\text{right}}(s)$ and $p_{\text{left}}(s)$ denote the right and left endpoint of a segment s, respectively. A line ℓ intersects s if and only if either the endpoints of s lie to distinct sides of ℓ, or s has an endpoint on ℓ. We show how to count the number of segments $s \in S$ with $p_{\text{right}}(s)$ lying above ℓ and $p_{\text{left}}(s)$ lying below ℓ. The segments with an endpoint on ℓ, and the ones with $p_{\text{right}}(s)$ lying below ℓ and $p_{\text{left}}(s)$ above ℓ, can be counted with a similar data structure. Here we choose—for a vertical line ℓ—the left side to be below ℓ and the right side above ℓ.

The idea of the data structure is simple. We first find all the segments $s \in S$ such that $p_{\text{right}}(s)$ lies above ℓ. In the previous section we saw how to use a partition tree to select these segments in a number of canonical subsets. For each of the selected canonical subsets we are interested in the number of segments s with $p_{\text{left}}(s)$ below ℓ. This is a half-plane range counting query, which can be answered if we store each canonical subset in a partition tree. Let's describe this solution in a little more detail. The data structure is defined as follows. For a set S' of segments, let $P_{\text{right}}(S') := \{p_{\text{right}}(s) : s \in S'\}$ be the set of right endpoints of the segments in S', and let $P_{\text{left}}(S') := \{p_{\text{left}}(s) : s \in S'\}$ be the set of left endpoints of the segments in S'.

- The set $P_{\text{right}}(S)$ is stored in a partition tree $\mathcal{T}$. The canonical subset of a node v of $\mathcal{T}$ is denoted $P_{\text{right}}(v)$. The set of segments corresponding to the left endpoints in $P_{\text{right}}(v)$ is denoted $S(v)$, that is, $S(v) = \{s : p_{\text{right}}(s) \in P_{\text{right}}(v)\}$. (Abusing the terminology slightly, we sometimes call $S(v)$ the canonical subset of v.)

- With each node v of the first-level tree $\mathcal{T}$, we store the set $P_{\text{left}}(S(v))$ in a second-level partition tree $\mathcal{T}_v^{\text{assoc}}$ for half-plane range counting. This partition tree is the *associated structure* of v.

With this data structure we can select the segments $s \in S$ with $p_{\text{right}}(s)$ above ℓ and $p_{\text{left}}(s)$ below ℓ in a number of canonical subsets. The query algorithm for this is described below. To count the number of such segments, all we have to do is sum the cardinalities of the selected subsets. Let $\mathcal{T}_v$ denote the subtree of $\mathcal{T}$ rooted at v.

Algorithm SELECTINTSEGMENTS($\ell, \mathcal{T}$)
Input. A query line ℓ and a partition tree or subtree of it.
Output. A set of canonical nodes for all segments in the tree that are intersected by ℓ.
1. $\Upsilon \leftarrow \emptyset$
2. **if** $\mathcal{T}$ consists of a single leaf μ
3.   **then if** the segment stored at μ intersects ℓ **then** $\Upsilon \leftarrow \{\mu\}$
4.   **else for** each child v of the root of $\mathcal{T}$
5.   **do if** $t(v) \subset \ell^+$
6.   **then** $\Upsilon \leftarrow \Upsilon \cup$ SELECTINHALFPLANE($\ell^-, \mathcal{T}_v^{\text{assoc}}$)
7.   **else if** $t(v) \cap h \neq \emptyset$
8.   **then** $\Upsilon \leftarrow \Upsilon \cup$ SELECTINTSEGMENTS($\ell, \mathcal{T}_v$)
9. **return** Υ

The query algorithm just given can find the segments with the right endpoint above the query line and the left endpoint below it. Interestingly, the same partition tree can be used to find the segments with the left endpoint above the query line and the right endpoint below it. Only the query algorithm has to be changed: exchange the "ℓ^+" and the "ℓ^-" and we are done.

Let's analyze our multi-level partition tree for segment intersection selection. We start with the amount of storage.

Lemma 16.5 *Let S be a set of n segments in the plane. A two-level partition tree for segment intersection selection queries in S uses $O(n \log n)$ storage.*

Proof. Let n_v denote the cardinality of the canonical subset $S(v)$ in the first-level partition tree. The storage for this node consists of a partition tree for S_v, and as we know from the previous section, it needs linear storage. Hence, the storage $M(n)$ for a two-level partition tree on n segments satisfies the recurrence

$$M(n) = \begin{cases} O(1) & \text{if } n = 1, \\ \sum_v [O(n_v) + M(n_v)] & \text{if } n > 1, \end{cases}$$

where we sum over all children v of the root of the tree. We know that $\sum_v n_v = n$ and $n_v \leqslant 2n/r$. Since $r > 2$ is a constant the recurrence for $M(n)$ solves to $M(n) = O(n \log n)$. $\quad\square$

Adding a second level to the partition tree has increased the amount of storage by a logarithmic factor. What about the query time? Surprisingly, the asymptotic query time does not change at all.

Lemma 16.6 *Let S be a set of n segments in the plane. For any $\varepsilon > 0$, there is a two-level partition tree for S such that for a query line ℓ we can select $O(n^{1/2+\varepsilon})$ nodes from the tree with the property that the subset of segments from S intersected by ℓ is the disjoint union of the canonical subsets of the selected nodes. The selection of these nodes takes $O(n^{1/2+\varepsilon})$ time. As a consequence, the number of intersected segments can be counted in $O(n^{1/2+\varepsilon})$ time.*

Proof. Again we use a recurrence to analyze the query time. Let $\varepsilon > 0$ be given. Let n_v denote the cardinality of the canonical subset $S(v)$. Lemma 16.3 tells us that we can construct the associated structure T_v^{assoc} of node v in such a way that the query time in T_v^{assoc} is $O(n_v^{1/2+\varepsilon})$. Now consider the full two-level tree $\mathcal{T}$ on S. We base this tree on a fine simplicial partition of size r with crossing number at most $c\sqrt{r}$, for $r := \lceil 2(c\sqrt{2})^{1/\varepsilon} \rceil$; such a partition exists by Theorem 16.1. Let $Q(n)$ denote the query time in the two-level tree for a set of n segments. Then $Q(n)$ satisfies the recurrence:

$$Q(n) = \begin{cases} O(1) & \text{if } n = 1, \\ O(rn^{1/2+\varepsilon}) + \sum_{i=1}^{c\sqrt{r}} Q(2n/r) & \text{if } n > 1. \end{cases}$$

With our choice of r the recurrence for $Q(n)$ solves to $O(n^{1/2+\varepsilon})$. This bound on the query time immediately implies the bound on the number of selected canonical subsets. $\quad\square$

16.3 Cutting Trees

In the previous sections we have solved planar range searching problems with partition trees. The storage requirements of partition trees are good: they use roughly linear storage. The query time, however, is $O(n^{1/2+\varepsilon})$, and this is rather high. Can we achieve a better query time, for example $O(\log n)$, if we are willing to spend more than linear storage? To have any hope of success, we must abandon the approach of using simplicial partitions: it is not possible to construct simplicial partitions with less than $O(\sqrt{r})$ crossing number, which would be needed to achieve a query time faster than $O(\sqrt{n})$.

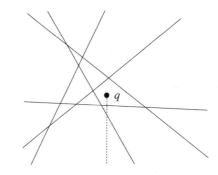

Figure 16.6
Half-plane range counting in the dual plane: how may lines are below a query point?

To come up with a new approach to the problem, we need to look at it in a different light. We apply the duality transform from Chapter 8. The first problem we solved in Section 16.1 was the half-plane range counting problem: given a set of points, count the number of points lying in a query half-plane. Let's see what we get when we look at this problem in the dual plane. Assume that the query half-plane is positive, that is, we want to count the points above the query line. In the dual plane we then have the following setting: given a set L of n lines in the plane, count the number of lines below a query point q. With the tools we constructed in the previous chapters it is easy to design a data structure with logarithmic query time for this problem: the key observation is that the number of lines below the query point q is uniquely determined by the face of the arrangement $\mathcal{A}(L)$ that contains q. Hence, we can construct $\mathcal{A}(L)$ and preprocess it for point location queries, as described in Chapter 6, and store with each face the number of lines below it. Counting the number of lines below a query point now boils down to doing point location. This solution uses $O(n^2)$ storage and it has $O(\log n)$ query time.

Note that this was a situation where we could afford to precompute the answer for every possible query—in other words, the set of canonical subsets consists of all possible subsets that can appear. But if we go to triangular range counting, this approach is not so good: there are just too many possible triangles to precompute all possible answers. Instead, we'll try to express the set of lines below a query point by a a small number of canonical subsets in a recursive way. Then we can then use the multi-level approach from the previous section to solve the triangular range searching problem.

We construct the whole collection of canonical subsets using a data structure called a *cutting tree*. The idea behind cutting trees is the same as for partition trees: the plane is partitioned into triangular regions, as depicted in Figure 16.7. This time, however, we require that the triangles be disjoint. How can such

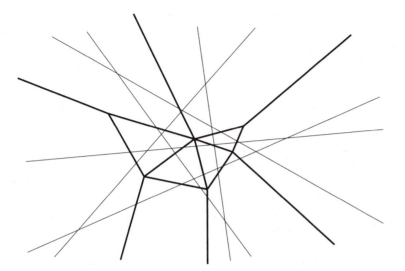

Figure 16.7
A $(1/2)$-cutting of size ten for a set of six lines

a partitioning help to count the number of lines below a query point? Let $L := \{\ell_1, \ell_2, \ldots, \ell_n\}$ be the set of lines that we obtained after dualizing the points to be preprocessed for triangular range queries. Consider a triangle t of the partitioning, and a line ℓ_i that does not intersect t. If ℓ_i lies below t, then ℓ_i lies below any query point inside t. Similarly, if ℓ_i lies above t, it lies above any query point inside t. This means that if our query point q lies in t, then the only lines of which we don't know yet whether they lie above or below q are the ones that intersect t. Our data structure will store each triangle of the partitioning, with a counter indicating the number of lines below it; for each triangle we also have a recursively defined structure on the lines intersecting it. To query in this structure, we first determine in which triangle t the query point q falls. We then compute how many lines from the ones that intersect t are below q, by recursively visiting the subtree corresponding to t. Finally, we add the number we computed in the recursive call to the number of lines below t. The efficiency of this approach depends on the number of lines intersecting a triangle: the smaller this number, the fewer lines on which we have to recurse. We now formally define the kind of partitioning we need.

Let L be a set of n lines in the plane, and let r be a parameter with $1 \leqslant r \leqslant n$. A line is said to cross a triangle if it intersects the interior of the triangle. A $(1/r)$-*cutting* for L is a set $\Xi(L) := \{t_1, t_2, \ldots, t_m\}$ of possibly unbounded triangles with disjoint interiors that together cover the plane, with the property that no triangle of the partitioning is crossed by more than n/r lines from L. The *size* of the cutting $\Xi(L)$ is the number of triangles it consists of. Figure 16.7 gives an example of a cutting.

Theorem 16.7 *For any set L of n lines in the plane, and any parameter r with $1 \leqslant r \leqslant n$, a $(1/r)$-cutting of size $O(r^2)$ exists. Moreover, such a cutting (with for each triangle in the cutting the subset of lines from L that cross it) can be constructed in $O(nr)$ time.*

In Section 16.4 references are given to the papers where this theorem is proved. We shall only concern ourselves with how cuttings can be used to design data structures. The data structure based on cuttings is called a *cutting tree*. The basic structure of a cutting tree for a set L of n lines is as follows.

- If the cardinality of L is one then the cutting tree consists of a single leaf where L is stored explicitly. The set L is the *canonical subset* of the leaf.

- Otherwise, the structure is a tree $\mathcal{T}$. There is a one-to-one correspondence between the children of the root of the tree and the triangles of a $(1/r)$-cutting $\Xi(L)$ for the set L, where r is a sufficiently large constant. (Below we shall see how r should be chosen.) The triangle of the cutting that corresponds to a child v is denoted by $t(v)$. The subset of lines in L that lie below $t(v)$ is called the *lower canonical subset* of v; it is denoted $L^-(v)$. The subset of lines in L that lie above $t(v)$ is called the *upper canonical subset* of v; it is denoted $L^+(v)$. The subset of lines that cross $t(v)$ is called the *crossing subset* of $t(v)$. The child v is the root of a recursively defined partition tree on its crossing subset; this subtree is denoted by $\mathcal{T}_v$.

- With each child v we store the triangle $t(v)$. We also store information about the lower and upper canonical subsets $L^-(v)$ and $L^+(v)$; for counting the number of lines below the query point we only need to store the cardinality of the set $L^-(v)$, but for other applications we may store other information.

Figure 16.8 illustrates the notions of lower canonical subset, upper canonical subset, and crossing subset. We describe an algorithm for selecting the lines

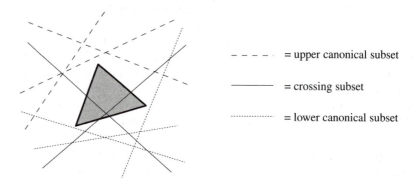

Figure 16.8
The canonical subsets and the crossing subset for a triangle

— — — — = upper canonical subset

———— = crossing subset

·········· = lower canonical subset

from L below a query point in a number of canonical subsets. To count the number of such lines we have to sum the cardinalities of the selected canonical subsets. Let q be the query point. The set of selected nodes will be denoted by Υ.

Algorithm SELECTBELOWPOINT$(q, \mathcal{T})$

Input. A query point q and a cutting tree or subtree of it.

Output. A set of canonical nodes for all lines in the tree that lie below q.

1. $\Upsilon \leftarrow \emptyset$
2.   **if** $\mathcal{T}$ consists of a single leaf μ
3.   **then if** the line stored at μ lies below q **then** $\Upsilon \leftarrow \{\mu\}$
4.   **else for** each child v of the root of $\mathcal{T}$
5.   **do** Check if q lies in $t(v)$.
6.   Let v_q be the child such that $q \in t(v_q)$.
7. $\Upsilon \leftarrow \{v_q\} \cup$ SELECTBELOWPOINT$(q, \mathcal{T}_{v_q})$
8.   **return** Υ

Lemma 16.8 *Let L be a set of n lines in the plane. Using a cutting tree, the lines from L below a query point can be selected in $O(\log n)$ time in $O(\log n)$ canonical subsets. As a consequence, the number of such lines can be counted in $O(\log n)$ time. For any $\varepsilon > 0$, a cutting tree on L can be constructed that uses $O(n^{2+\varepsilon})$ storage.*

Proof. Let $Q(n)$ denote the query time in a cutting tree for a set of n lines. Then $Q(n)$ satisfies the recurrence:

$$Q(n) = \begin{cases} O(1) & \text{if } n = 1, \\ O(r^2) + Q(n/r) & \text{if } n > 1. \end{cases}$$

This recurrence solves to $Q(n) = O(\log n)$ for any constant $r > 1$.

Let $\varepsilon > 0$ be given. According to Theorem 16.7 we can construct a $(1/r)$-cutting for L of size cr^2, where c is a constant. We construct a cutting tree based on $(1/r)$-cuttings for $r = \lceil (2c)^{1/\varepsilon} \rceil$. The amount of storage used by the cutting tree, $M(n)$, satisfies

$$M(n) = \begin{cases} O(1) & \text{if } n = 1, \\ O(r^2) + \sum_v M(n_v) & \text{if } n > 1, \end{cases}$$

where we sum over all children v of the root of the tree. The number of children of the root is cr^2, and $n_v \leqslant n/r$ for each child v. Hence, with our choice of r the recurrence solves to $M(n) = O(n^{2+\varepsilon})$. $\quad\boxed{}$

We conclude that we can count the number of lines below a query point in $O(\log n)$ time with a structure that uses $O(n^{2+\varepsilon})$ storage. By duality, we can do half-plane range counting within the same bounds. Now let's look at triangular range counting again: given a set S of points in the plane, count the number of points inside a query triangle. Following the approach for half-plane queries, we go to the dual plane. What problem do we get in the dual plane? The set of points dualizes to a set of lines, of course, but it is less clear what happens to the query triangle. A triangle is the intersection of three half-planes, so a point p lies in a triangle if and only if it lies in each of the half-planes. In Figure 16.9, for instance, point p lies in the triangle because $p \in \ell_1^+$ and $p \in \cap \ell_2^-$ and $p \in \cap \ell_3^-$. The line dual to p therefore has ℓ_1^* above it, and ℓ_2^* and ℓ_3^*

primal plane   dual plane

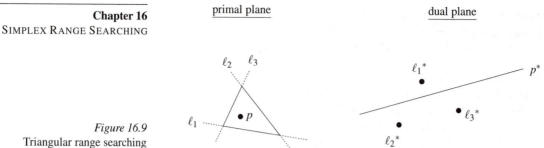

Figure 16.9
Triangular range searching

below it. In general, the dual statement of the triangular range searching problem is: given a set L of lines in the plane, and a triple q_1, q_2, q_3 of query points labeled "above" or "below", count the number of lines from L that lie on the specified sides of the three query points. This problem can be solved with a three-level cutting tree. We now describe a data structure for the following, slightly simpler, problem: given a set L of lines, and a pair q_1, q_2 of query points, select the lines that lie below both query points. After having seen the two-level cutting tree that solves this problem, designing a three-level cutting tree for the dual of the triangular range searching problem should be easy.

A two-level cutting tree on a set L of n lines for selecting the lines below a pair q_1, q_2 of query points is defined as follows.

■ The set L is stored in a cutting tree $\mathcal{T}$.

■ With each node v of the first-level tree $\mathcal{T}$, we store its lower canonical subset in a second-level cutting tree $\mathcal{T}_v^{\text{assoc}}$.

The idea is that the first level of the tree is used to select the lines below q_1 in a number of canonical subsets. The associated structures (or, trees at the second level) storing the selected canonical subsets are then used to select the lines that lie below q_2. Because the associated structures are one-level cuttings trees, we can use algorithm SELECTBELOWPOINT to query them. The total query algorithm is thus as follows.

Algorithm SELECTBELOWPAIR($q_1, q_2, \mathcal{T}$)
Input. Two query points q_1 and q_2 and a cutting tree or subtree of it.
Output. A set of canonical nodes for all lines in the tree that lie below q_1 and q_2.
1. $\Upsilon \leftarrow \emptyset$
2. **if** $\mathcal{T}$ consists of a single leaf μ
3.   **then if** the line stored at μ lies below q_1 and q_2 **then** $\Upsilon \leftarrow \{\mu\}$
4.   **else for** each child v of the root of $\mathcal{T}$
5.   **do** Check if q_1 lies in $t(v)$.
6.   Let v_{q_1} be the child such that $q_1 \in t(v_{q_1})$.
7. $\Upsilon_1 \leftarrow$ SELECTBELOWPOINT($q_2, \mathcal{T}_{v_{q_1}}^{\text{assoc}}$)
8. $\Upsilon_2 \leftarrow$ SELECTBELOWPAIR($q_1, q_2, \mathcal{T}_{v_{q_1}}$)
9. $\Upsilon \leftarrow \Upsilon_1 \cup \Upsilon_2$
10. **return** Υ

Recall that adding an extra level to a partition tree did not increase its query time, whereas the amount of storage it used increased by a logarithmic factor. For cutting trees this is exactly the other way around: adding an extra level increases the query time by a logarithmic factor, whereas the amount of storage stays the same. This is proved in the next two lemmas.

Lemma 16.9 *Let L be a set of n lines in the plane. Using a two-level cutting tree, the lines from L below a pair of query points can be selected in $O(\log^2 n)$ time in $O(\log^2 n)$ canonical subsets. As a consequence, the number of such lines can be counted in $O(\log^2 n)$ time. For any $\varepsilon > 0$, such a two-level cutting tree on L can be constructed that uses $O(n^{2+\varepsilon})$ storage.*

Proof. Let $Q(n)$ denote the query time in a two-level cutting tree for a set of n lines. The associated structures are one-level cutting trees, so the query time for the associated structures is $O(\log n)$ by Lemma 16.8. Hence, $Q(n)$ satisfies the recurrence:

$$Q(n) = \begin{cases} O(1) & \text{if } n = 1, \\ O(r^2) + O(\log n) + Q(n/r) & \text{if } n > 1. \end{cases}$$

This recurrence solves to $Q(n) = O(\log^2 n)$ for any constant $r > 1$.

Let $\varepsilon > 0$ be given. According to Lemma 16.8 we can construct the associated structures of the children of the root such that each of them uses $O(n^{2+\varepsilon})$ storage. Hence, the amount of storage used by the cutting tree, $M(n)$, satisfies

$$M(n) = \begin{cases} O(1) & \text{if } n = 1, \\ \sum_v [O(n^{2+\varepsilon} + M(n_v))] & \text{if } n > 1, \end{cases}$$

where we sum over all children v of the root of the tree. The number of children of the root is $O(r^2)$, and $n_v \leqslant n/r$ for each child v. It follows that, if r is a large enough constant, the recurrence solves to $M(n) = O(n^{2+\varepsilon})$. (If you are a bit bored by now, you are on the right track: cutting trees, partition trees, and their multi-level variants are all analyzed in the same way.) $\quad\square$

We designed and analyzed a two-level cutting tree for selecting (or counting) the number of lines below a pair of query points. For the triangular range searching we need a three-level cutting tree. The design and analysis of three-level cutting trees follows exactly the same pattern as for two-level cutting trees. Therefore you should hopefully have no difficulties in proving the following result.

Theorem 16.10 *Let S be a set of n points in the plane. For any $\varepsilon > 0$, there is a data structure for S, called a cutting tree, that uses $O(n^{2+\varepsilon})$ storage such that the points from S lying inside a query triangle can be counted in $O(\log^3 n)$ time. The points can be reported in $O(k)$ additional time, where k is the number of reported points. The structure can be constructed in $O(n^{2+\varepsilon})$ time.*

One can even do a little bit better than in this theorem. This is discussed in Section 16.4 and in the exercises.

16.4 Notes and Comments

Range searching is one of the best studied problems in computational geometry. We can distinguish between *orthogonal range searching* and *simplex range searching*. Orthogonal range searching was the topic of Chapter 5. In this chapter we discussed the planar variant of simplex range searching, namely triangular range searching. We conclude with a brief overview of the history of simplex range searching, and a discussion of the higher-dimensional variants of the theory we presented.

We begin our discussion with data structures for simplex range searching in the plane that use roughly linear storage. Willard [336] was the first to present such a data structure. His structure is based on the same idea as the partition trees described in this chapter, namely a partition of the plane into regions. His partition, however, did not have such a good crossing number, so the query time of his structure was $O(n^{0.774})$. As better simplicial partitions were developed, more efficient partition trees were possible [90, 140, 176, 341]. Improvements were also obtained using a somewhat different structure than a partition tree, namely a spanning tree with low stabbing number [332, 91]. The best solution for triangular range searching so far is given by Matoušek [228]. Theorem 16.1 is proved in that paper. Matoušek also describes a more complicated data structure with $O(\sqrt{n}2^{O(\log^* n)})$ query time. This structure, however, cannot be used so easily as a basis for multi-level trees.

The simplex range searching problem in $\mathbb{R}^d$ is stated as follows: preprocess a set S of points in $\mathbb{R}^d$ into a data structure, such that the points from S lying in a query simplex can be counted (or reported) efficiently. Matoušek also proved results for simplicial partitions in higher dimensions. The definition of simplicial partitions in $\mathbb{R}^d$ is similar to the definition in the plane; the only difference is that the triangles of the partition are replaces by d-simplices, and that the crossing number is defined with respect to hyperplanes instead of lines. Matoušek proved that any set of points in $\mathbb{R}^d$ admits a simplicial partition of size r with crossing number $O(r^{1-1/d})$. Using such a simplicial partition one can construct, for any $\varepsilon > 0$, a partition tree for simplex range searching in $\mathbb{R}^d$ that uses linear space and has $O(n^{1-1/d+\varepsilon})$ query time. The query time can be improved to $O(n^{1-1/d}(\log n)^{O(1)})$. The query time of Matoušek's structure comes close to the lower bounds proved by Chazelle [68], which state that a data structure for triangular range searching that uses $O(m)$ storage must have $\Omega(n/(m^{1/d}\log n))$ query time. A structure that uses linear space must thus have $\Omega(n^{1-1/d}/\log n)$ query time. (In the plane a slightly sharper lower bound is known, namely $\Omega(n/\sqrt{m})$.)

Data structures for simplex range searching with logarithmic query time have also received a lot of attention. Clarkson [109] was the first to realize that cuttings can be used as the basis for a data structure for range searching. Using a probabilistic argument, he proved the existence of $O(\log r/r)$-cuttings of size $O(r^d)$ for sets of hyperplanes in $\mathbb{R}^d$, and he used this to develop a data structure for half-space range queries. After this, several people worked on improving the results and on developing efficient algorithms for computing cuttings. Currently the best known algorithm is by Chazelle [74]. He has shown that for any parameter r, it is possible to compute a $(1/r)$-cutting of size $O(r^d)$ with a deterministic algorithm that takes $O(nr^{d-1})$ time. These cuttings can be used to design a (multi-level) cutting tree for simplex range searching, as shown in this chapter for the planar case. The resulting data structure has $O(\log^d n)$ query time and uses $O(n^{d+\varepsilon})$ storage. The query time can be reduced to $O(\log n)$. Due to a special property of Chazelle's cuttings, it is also possible to get rid of the $O(n^\varepsilon)$ factor in the storage [230], but for the new structure it is no longer possible to reduce the query time to $O(\log n)$. These bounds are again close to Chazelle's lower bounds.

By combining partition trees and cutting trees the right way, one can get data structures that have storage in between that of partition trees and cutting trees. In particular, for any $n \leqslant m \leqslant n^d$, a data structure of size $O(m^{1+\varepsilon})$ has $O(n^{1+\varepsilon}/m^{1/d})$ query time, close to the lower bound: Exercise 16.13 shows how to do this.

In the discussion above we have concentrated on simplex range searching. Half-space range searching is, of course, a special case of this. It turns out that for this special case better results can be achieved. For example, for half-plane range reporting (not for counting) in the plane, there is a data structure with $O(\log n + k)$ query time that uses $O(n)$ storage [86]. Here k is the number of reported points. Improved results are possible in higher dimensions as well: the points in a query half-space can be reported in $O(n^{1-1/\lfloor d/2 \rfloor}(\log n)^{O(1)} + k)$ time with a data structure that uses $O(n \log \log n)$ storage [227].

Finally, Agarwal and Matoušek [3] generalized the results on range searching to query ranges that are semi-algebraic sets.

16.5 Exercises

16.1 Prove that the selected nodes in a partition tree are exactly the nodes v with the following property: $t(v) \subset h$ (or, in case v is a leaf, the point stored at v lies in h) and there is no ancestor μ of v such that $t(\mu) \subset h$. Use this to prove that $S \cap h$ is the disjoint union of the canonical subsets of the selected nodes.

16.2 Prove that the recurrence for $M(n)$ given in the proof of Lemma 16.2 solves to $M(n) = O(n)$.

16.3 Prove that the recurrence for $Q(n)$ given in the proof of Lemma 16.3 solves to $Q(n) = O(n^{1/2+\varepsilon})$.

16.4 Lemma 16.3 shows that for any $\varepsilon > 0$ we can build a partition tree with $O(n^{1/2+\varepsilon})$ query time, by choosing the parameter r that determines the branching degree of the tree to be a large enough constant. We can do even better if we choose r depending on n. Show that the query time reduces to $O(\sqrt{n} \log n)$ if we choose $r = \sqrt{n}$. (Note that the value of r is not the same any more at different nodes in the tree. However, this is not a problem.)

16.5 Prove that the recurrence for $M(n)$ given in the proof of Lemma 16.5 solves to $M(n) = O(n \log n)$.

16.6 Prove that the recurrence for $Q(n)$ given in the proof of Lemma 16.6 solves to $Q(n) = O(n^{1/2+\varepsilon})$.

16.7 Let T be a set of n triangles in the plane. An *inverse range counting query* asks to count the number of triangles from T containing a query point q.

    a.   Design a data structure for inverse range counting queries that uses roughly linear storage (for example, $O(n \log^c n)$ for some constant c). Analyze the amount of storage and the query time of your data structure.

    b.   Can you do better if you know that all triangles are disjoint?

16.8 Prove that the recurrences for $Q(n)$ and $M(n)$ given in the proof of Lemma 16.8 solve to $Q(n) = O(\log n)$ and $M(n) = O(n^{2+\varepsilon})$.

16.9 Prove that the recurrences for $Q(n)$ and $M(n)$ given in the proof of Lemma 16.9 solve to $Q(n) = O(\log^2 n)$ and $M(n) = O(n^{2+\varepsilon})$.

16.10 A query in a two-level cutting tree visits the associated structures of the nodes on one path in the tree. The query time in an associated structure storing m lines is $O(\log m)$ by Lemma 16.8. Because the depth of the main tree is $O(\log n)$, the total query time is $O(\log^2 n)$. If we could choose the value of the parameter r of the main cutting tree larger than constant, for example n^δ for some small $\delta > 0$, then the depth of the main tree would become smaller, resulting in a reduction of the query time. Unfortunately, there is an $O(r^2)$ term in the recurrence for $Q(n)$ in the proof of Lemma 16.9.

    a.   Describe a way to get around this problem, so that you can choose $r := n^\delta$.

    b.   Prove that the query time of your two-level data structure is $O(\log n)$.

    c.   Prove that the amount of storage of the data structure is still $O(n^{2+\varepsilon})$.

16.11 Design a data structure for triangular range searching that has $O(\log^3 n)$ query time. Describe the data structure as well as the query algorithm precisely, and analyze both storage and query time.

16.12 Let S be a set of n points in the plane, each having a positive real weight associated with them. Describe two data structures for the following query problem: find the point in a query half-plane with the largest weight. One data structure should use linear storage, and the other data structure should have logarithmic query time. Analyze the amount of storage and the query time of both data structures.

16.13 In this chapter we have seen a data structure for half-plane range searching with linear storage but a rather high query time (the partition tree) and a data structure with logarithmic query time but a rather high use of storage (the cutting tree). Sometimes one would like to have something in between: a data structure that has a better query time than partition trees, but uses less storage than a cutting tree. In this exercise we show how to design such a structure.

Suppose that we have $O(m^{1+\varepsilon})$ storage available, for some m between n and n^2. We want to build a structure for selecting points in half-planes that uses $O(m^{1+\varepsilon})$ storage and has as fast a query time as possible. The idea is to start with the fastest structure we have (the cutting tree) and switch to the slow structure (the partition tree) when we run out of storage. That is, we continue the construction of the cutting tree recursively until the number of lines we have to store drops below some threshold $\hat{n}$.

a. Describe the data structure and the query algorithm in detail.
b. Compute the value of the threshold $\hat{n}$ such that the amount of storage is $O(m^{1+\varepsilon})$.
c. Analyze the query time of the resulting data structure.

Bibliography

[1] P. K. Agarwal, M. de Berg, J. Matoušek, and O. Schwarzkopf. Constructing levels in arrangements and higher order Voronoi diagrams. In *Proc. 10th Annu. ACM Sympos. Comput. Geom.*, pages 67–75, 1994.

[2] P. K. Agarwal and M. van Kreveld. Implicit point location in arrangements of line segments, with an application to motion planning. *Internat. J. Comput. Geom. Appl.*, 4:369–383, 1994.

[3] P. K. Agarwal and J. Matoušek. On range searching with semialgebraic sets. *Discrete Comput. Geom.*, 11:393–418, 1994.

[4] A. Aggarwal. *The art gallery problem: Its variations, applications, and algorithmic aspects*. Ph.D. thesis, Johns Hopkins Univ., Baltimore, MD, 1984.

[5] O. Aichholzer, F. Aurenhammer, M. Taschwer, and G. Rote. Triangulations intersect nicely. In *Proc. 11th Annu. ACM Sympos. Comput. Geom.*, pages 220–229, 1995.

[6] V. Akman. *Unobstructed Shortest Paths in Polyhedral Environments. Lecture Notes in Computer Science*, vol. 251. Springer-Verlag, 1987.

[7] N. Amenta. Helly-type theorems and generalized linear programming. *Discrete Comput. Geom.*, 12:241–261, 1994.

[8] A. M. Andrew. Another efficient algorithm for convex hulls in two dimensions. *Inform. Process. Lett.*, 9:216–219, 1979.

[9] E. M. Arkin, J. S. B. Mitchell, and S. Suri. Logarithmic-time link path queries in a simple polygon. *Internat. J. Comput. Geom. Appl.*, 5:369–395, 1995.

[10] B. Aronov and M. Sharir. On translational motion planning in 3-space. In *Proc. 10th Annu. ACM Sympos. Comput. Geom.*, pages 21–30, 1994.

[11] T. Asano, T. Asano, L. J. Guibas, J. Hershberger, and H. Imai. Visibility of disjoint polygons. *Algorithmica*, 1:49–63, 1986.

[12] T. Asano, D. Kirkpatrick, and C. K. Yap. d_1-optimal motion for a rod. In *Proc. 12th Annu. ACM Sympos. Comput. Geom.*, pages 252–263, 1996.

[13] F. Aurenhammer. A criterion for the affine equality of cell complexes in R^d and convex polyhedra in R^{d+1}. *Discrete Comput. Geom.*, 2:49–64, 1987.

[14] F. Aurenhammer. Power diagrams: properties, algorithms and applications. *SIAM J. Comput.*, 16:78–96, 1987.

[15] F. Aurenhammer. Linear combinations from power domains. *Geom. Dedicata*, 28:45–52, 1988.

[16] F. Aurenhammer. Voronoi diagrams: A survey of a fundamental geometric data structure. *ACM Comput. Surv.*, 23:345–405, 1991.

[17] F. Aurenhammer and H. Edelsbrunner. An optimal algorithm for constructing the weighted Voronoi diagram in the plane. *Pattern Recogn.*, 17:251–257, 1984.

[18] F. Aurenhammer, F. Hoffmann, and B. Aronov. Minkowski-type theorems and least-squares partitioning. In *Proc. 8th Annu. ACM Sympos. Comput. Geom.*, pages 350–357, 1992.

[19] F. Aurenhammer and O. Schwarzkopf. A simple on-line randomized incremental algorithm for computing higher order Voronoi diagrams. *Internat. J. Comput. Geom. Appl.*, 2:363–381, 1992.

[20] D. Avis and G. T. Toussaint. An efficient algorithm for decomposing a polygon into star-shaped polygons. *Pattern Recogn.*, 13:395–398, 1981.

[21] F. Avnaim, J.-D. Boissonnat, and B. Faverjon. A practical exact motion planning algorithm for polygonal objects amidst polygonal obstacles. In *Proc. 5th IEEE Internat. Conf. Robot. Autom.*, pages 1656–1661, 1988.

[22] C. Bajaj and T. K. Dey. Convex decomposition of polyhedra and robustness. *SIAM J. Comput.*, 21:339–364, 1992.

[23] I. J. Balaban. An optimal algorithm for finding segment intersections. In *Proc. 11th Annu. ACM Sympos. Comput. Geom.*, pages 211–219, 1995.

[24] C. Ballieux. Motion planning using binary space partitions. Technical Report Inf/src/93-25, Utrecht University, 1993.

[25] B. Barber and M. Hirsch. A robust algorithm for point in polyhedron. In *Proc. 5th Canad. Conf. Comput. Geom.*, pages 479–484, Waterloo, Canada, 1993.

[26] R. E. Barnhill. Representation and approximation of surfaces. In J. R. Rice, editor, *Math. Software III*, pages 69–120. Academic Press, New York, 1977.

[27] J. Barraquand and J.-C. Latombe. Robot motion planning: A distributed representation approach. *Internat. J. Robot. Res.*, 10:628–649, 1991.

[28] B. G. Baumgart. A polyhedron representation for computer vision. In *Proc. AFIPS Natl. Comput. Conf.*, vol. 44, pages 589–596, 1975.

[29] H. Baumgarten, H. Jung, and K. Mehlhorn. Dynamic point location in general subdivisions. *J. Algorithms*, 17:342–380, 1994.

[30] P. Belleville, M. Keil, M. McAllister, and J. Snoeyink. On computing edges that are in all minimum-weight triangulations. In *Proc. 12th Annu. ACM Sympos. Comput. Geom.*, pages V7–V8, 1996.

[31] R. V. Benson. *Euclidean Geometry and Convexity*. McGraw-Hill, New York, 1966.

[32] J. L. Bentley. Multidimensional binary search trees used for associative searching. *Commun. ACM*, 18:509–517, 1975.

[33] J. L. Bentley. Solutions to Klee's rectangle problems. Technical report, Carnegie-Mellon Univ., Pittsburgh, PA, 1977.

[34] J. L. Bentley. Decomposable searching problems. *Inform. Process. Lett.*, 8:244–251, 1979.

[35] J. L. Bentley and T. A. Ottmann. Algorithms for reporting and counting geometric intersections. *IEEE Trans. Comput.*, C-28:643–647, 1979.

[36] J. L. Bentley and J. B. Saxe. Decomposable searching problems I: Static-to-dynamic transformation. *J. Algorithms*, 1:301–358, 1980.

[37] M. de Berg. Linear size binary space partitions for fat objects. In *Proc. 3rd Annu. European Sympos. Algorithms (ESA '95)*, pages 252–263, 1995.

[38] M. de Berg. Computing half-plane and strip discrepancy of planar point sets. *Comput. Geom. Theory Appl.*, 6:69–83, 1996.

[39] M. de Berg, M. de Groot, and M. Overmars. New results on binary space partitions in the plane. In *Proc. 4th Scand. Workshop Algorithm Theory. Lecture Notes in Computer Science*, vol. 824, pages 61–72. Springer-Verlag, 1994.

[40] M. de Berg, M. van Kreveld, B. J. Nilsson, and M. H. Overmars. Shortest path queries in rectilinear worlds. *Internat. J. Comput. Geom. Appl.*, 2:287–309, 1992.

[41] M. de Berg, M. van Kreveld, and J. Snoeyink. Two- and three-dimensional point location in rectangular subdivisions. *J. Algorithms*, 18:256–277, 1995.

[42] M. de Berg, J. Matoušek, and O. Schwarzkopf. Piecewise linear paths among convex obstacles. *Discrete Comput. Geom.*, 14:9–29, 1995.

[43] M. de Berg and O. Schwarzkopf. Cuttings and applications. *Internat. J. Comput. Geom. Appl.*, 5:343–355, 1995.

[44] M. Bern and D. Eppstein. Mesh generation and optimal triangulation. In D.-Z. Du and F. K. Hwang, editors, *Computing in Euclidean Geometry. Lecture Notes Series on Computing*, vol. 1, pages 23–90. World Scientific, Singapore, 1992.

[45] M. Bern and D. Eppstein. Polynomial-size nonobtuse triangulation of polygons. *Internat. J. Comput. Geom. Appl.*, 2:241–255, 449–450, 1992.

[46] M. Bern, D. Eppstein, and J. Gilbert. Provably good mesh generation. In *Proc. 31st Annu. IEEE Sympos. Found. Comput. Sci.*, pages 231–241, 1990.

[47] M. Bern, S. Mitchell, and J. Ruppert. Linear-size nonobtuse triangulation of polygons. In *Proc. 10th Annu. ACM Sympos. Comput. Geom.*, pages 221–230, 1994.

[48] B. K. Bhattacharya and J. Zorbas. Solving the two-dimensional findpath problem using a line-triangle representation of the robot. *J. Algorithms*, 9:449–469, 1988.

[49] J.-D. Boissonnat, O. Devillers, R. Schott, M. Teillaud, and M. Yvinec. Applications of random sampling to on-line algorithms in computational geometry. *Discrete Comput. Geom.*, 8:51–71, 1992.

[50] J.-D. Boissonnat, O. Devillers, and M. Teillaud. A semidynamic construction of higher-order Voronoi diagrams and its randomized analysis. *Algorithmica*, 9:329–356, 1993.

[51] J.-D. Boissonnat and M. Teillaud. On the randomized construction of the Delaunay tree. *Theoret. Comput. Sci.*, 112:339–354, 1993.

[52] P. Bose and G. Toussaint. Geometric and computational aspects of manufacturing processes. *Comput. & Graphics*, 18:487–497, 1994.

[53] R. A. Brooks and T. Lozano-Pérez. A subdivision algorithm in configuration space for findpath with rotation. *IEEE Trans. Syst. Man Cybern.*, 15:224–233, 1985.

[54] G. Brown. Point density in stems per acre. *New Zealand Forestry Service Research Notes*, 38:1–11, 1965.

[55] J. L. Brown. Vertex based data dependent triangulations. *Comput. Aided Geom. Design*, 8:239–251, 1991.

[56] K. Q. Brown. Comments on "Algorithms for reporting and counting geometric intersections". *IEEE Trans. Comput.*, C-30:147–148, 1981.

[57] C. Burnikel, K. Mehlhorn, and S. Schirra. On degeneracy in geometric computations. In *Proc. 5th ACM-SIAM Sympos. Discrete Algorithms*, pages 16–23, 1994.

[58] P. A. Burrough. *Principles of Geographical Information Systems for Land Resources Assessment*. Oxford University Press, New York, 1986.

[59] A. Bykat. Convex hull of a finite set of points in two dimensions. *Inform. Process. Lett.*, 7:296–298, 1978.

[60] J. Canny. *The Complexity of Robot Motion Planning*. MIT Press, Cambridge, MA, 1987.

[61] J. Canny, B. R. Donald, and E. K. Ressler. A rational rotation method for robust geometric algorithms. In *Proc. 8th Annu. ACM Sympos. Comput. Geom.*, pages 251–260, 1992.

[62] T. M. Y. Chan. Output-sensitive results on convex hulls, extreme points, and related problems. In *Proc. 11th Annu. ACM Sympos. Comput. Geom.*, pages 10–19, 1995.

[63] D. R. Chand and S. S. Kapur. An algorithm for convex polytopes. *J. ACM*, 17:78–86, 1970.

[64] B. Chazelle. A theorem on polygon cutting with applications. In *Proc. 23rd Annu. IEEE Sympos. Found. Comput. Sci.*, pages 339–349, 1982.

[65] B. Chazelle. Convex partitions of polyhedra: a lower bound and worst-case optimal algorithm. *SIAM J. Comput.*, 13:488–507, 1984.

[66] B. Chazelle. Filtering search: a new approach to query-answering. *SIAM J. Comput.*, 15:703–724, 1986.

[67] B. Chazelle. Reporting and counting segment intersections. *J. Comput. Syst. Sci.*, 32:156–182, 1986.

[68] B. Chazelle. Lower bounds on the complexity of polytope range searching. *J. Amer. Math. Soc.*, 2:637–666, 1989.

[69] B. Chazelle. Lower bounds for orthogonal range searching, I: the reporting case. *J. ACM*, 37:200–212, 1990.

[70] B. Chazelle. Lower bounds for orthogonal range searching, II: the arithmetic model. *J. ACM*, 37:439–463, 1990.

[71] B. Chazelle. Triangulating a simple polygon in linear time. In *Proc. 31st Annu. IEEE Sympos. Found. Comput. Sci.*, pages 220–230, 1990.

[72] B. Chazelle. An optimal convex hull algorithm and new results on cuttings. In *Proc. 32nd Annu. IEEE Sympos. Found. Comput. Sci.*, pages 29–38, 1991.

[73] B. Chazelle. Triangulating a simple polygon in linear time. *Discrete Comput. Geom.*, 6:485–524, 1991.

[74] B. Chazelle. Cutting hyperplanes for divide-and-conquer. *Discrete Comput. Geom.*, 9:145–158, 1993.

[75] B. Chazelle. Geometric discrepancy revisited. In *Proc. 34th Annu. IEEE Sympos. Found. Comput. Sci.*, pages 392–399, 1993.

[76] B. Chazelle. An optimal convex hull algorithm in any fixed dimension. *Discrete Comput. Geom.*, 10:377–409, 1993.

[77] B. Chazelle and H. Edelsbrunner. An improved algorithm for constructing kth-order Voronoi diagrams. *IEEE Trans. Comput.*, C-36:1349–1354, 1987.

[78] B. Chazelle and H. Edelsbrunner. An optimal algorithm for intersecting line segments in the plane. In *Proc. 29th Annu. IEEE Sympos. Found. Comput. Sci.*, pages 590–600, 1988.

[79] B. Chazelle and H. Edelsbrunner. An optimal algorithm for intersecting line segments in the plane. *J. ACM*, 39:1–54, 1992.

[80] B. Chazelle, H. Edelsbrunner, L. Guibas, and M. Sharir. Algorithms for bichromatic line segment problems and polyhedral terrains. Report UIUCDCS-R-90-1578, Dept. Comput. Sci., Univ. Illinois, Urbana, IL, 1989.

[81] B. Chazelle, H. Edelsbrunner, L. Guibas, and M. Sharir. A singly-exponential stratification scheme for real semi-algebraic varieties and its applications. *Theoret. Comput. Sci.*, 84:77–105, 1991.

[82] B. Chazelle, H. Edelsbrunner, L. Guibas, and M. Sharir. Algorithms for bichromatic line segment problems and polyhedral terrains. *Algorithmica*, 11:116–132, 1994.

[83] B. Chazelle and J. Friedman. Point location among hyperplanes and unidirectional ray-shooting. *Comput. Geom. Theory Appl.*, 4:53–62, 1994.

[84] B. Chazelle and L. J. Guibas. Fractional cascading: I. A data structuring technique. *Algorithmica*, 1:133–162, 1986.

[85] B. Chazelle and L. J. Guibas. Fractional cascading: II. Applications. *Algorithmica*, 1:163–191, 1986.

[86] B. Chazelle, L. J. Guibas, and D. T. Lee. The power of geometric duality. *BIT*, 25:76–90, 1985.

[87] B. Chazelle and J. Incerpi. Triangulating a polygon by divide and conquer. In *Proc. 21st Allerton Conf. Commun. Control Comput.*, pages 447–456, 1983.

[88] B. Chazelle and J. Incerpi. Triangulation and shape-complexity. *ACM Trans. Graph.*, 3:135–152, 1984.

[89] B. Chazelle and L. Palios. Triangulating a non-convex polytope. *Discrete Comput. Geom.*, 5:505–526, 1990.

[90] B. Chazelle, M. Sharir, and E. Welzl. Quasi-optimal upper bounds for simplex range searching and new zone theorems. *Algorithmica*, 8:407–429, 1992.

[91] B. Chazelle and E. Welzl. Quasi-optimal range searching in spaces of finite VC-dimension. *Discrete Comput. Geom.*, 4:467–489, 1989.

[92] D. Z. Chen, K. S. Klenk, and H.-Y. T. Tu. Shortest path queries among weighted obstacles in the rectilinear plane. In *Proc. 11th Annu. ACM Sympos. Comput. Geom.*, pages 370–379, 1995.

[93] Y.-B. Chen and D. Ierardi. Optimal motion planning for a rod in the plane subject to velocity constraints. In *Proc. 9th Annu. ACM Sympos. Comput. Geom.*, pages 143–152, 1993.

[94] S. W. Cheng and R. Janardan. New results on dynamic planar point location. *SIAM J. Comput.*, 21:972–999, 1992.

[95] L. P. Chew. Guaranteed-quality mesh generation for curved surfaces. In *Proc. 9th Annu. ACM Sympos. Comput. Geom.*, pages 274–280, 1993.

[96] L. P. Chew and R. L. Drysdale, III. Voronoi diagrams based on convex distance functions. In *Proc. 1st Annu. ACM Sympos. Comput. Geom.*, pages 235–244, 1985.

[97] L. P. Chew and K. Kedem. A convex polygon among polygonal obstacles: Placement and high-clearance motion. *Comput. Geom. Theory Appl.*, 3:59–89, 1993.

[98] Y.-J. Chiang, F. P. Preparata, and R. Tamassia. A unified approach to dynamic point location, ray shooting, and shortest paths in planar maps. *SIAM J. Comput.*, 25:207–233, 1996.

[99] Y.-J. Chiang and R. Tamassia. Dynamic algorithms in computational geometry. *Proc. IEEE*, 80:1412–1434, September 1992.

[100] Y.-J. Chiang and R. Tamassia. Optimal shortest path and minimum-link path queries between two convex polygons inside a simple polygonal obstacle. *Internat. J. Comput. Geom. Appl.*, 1995.

[101] F. Chin, J. Snoeyink, and C.-A. Wang. Finding the medial axis of a simple polygon in linear time. In *Proc. 6th Annu. Internat. Sympos. Algorithms Comput. (ISAAC 95). Lecture Notes in Computer Science*, vol. 1004, pages 382–391. Springer-Verlag, 1995.

[102] N. Chin and S. Feiner. Near real time shadow generation using bsp trees. In *Proc. SIGGRAPH '89*, pages 99–106, 1989.

[103] J. Choi, J. Sellen, and C. K. Yap. Approximate Euclidean shortest path in 3-space. In *Proc. 10th Annu. ACM Sympos. Comput. Geom.*, pages 41–48, 1994.

[104] J. Choi, J. Sellen, and C.-K. Yap. Precision-sensitive Euclidean shortest path in 3-space. In *Proc. 11th Annu. ACM Sympos. Comput. Geom.*, pages 350–359, 1995.

[105] V. Chvátal. A combinatorial theorem in plane geometry. *J. Combin. Theory Ser. B*, 18:39–41, 1975.

[106] V. Chvátal. *Linear Programming*. W. H. Freeman, New York, 1983.

[107] K. C. Clarke. *Analytical and Computer Cartography*. Prentice Hall, Englewood Cliffs, NJ, 2nd edition, 1995.

[108] K. L. Clarkson. Linear programming in $O(n3^{d^2})$ time. *Inform. Process. Lett.*, 22:21–24, 1986.

[109] K. L. Clarkson. New applications of random sampling in computational geometry. *Discrete Comput. Geom.*, 2:195–222, 1987.

[110] K. L. Clarkson. A Las Vegas algorithm for linear programming when the dimension is small. In *Proc. 29th Annu. IEEE Sympos. Found. Comput. Sci.*, pages 452–456, 1988.

[111] K. L. Clarkson and P. W. Shor. Applications of random sampling in computational geometry, II. *Discrete Comput. Geom.*, 4:387–421, 1989.

[112] K. L. Clarkson, R. E. Tarjan, and C. J. Van Wyk. A fast Las Vegas algorithm for triangulating a simple polygon. *Discrete Comput. Geom.*, 4:423–432, 1989.

[113] R. Cole. Searching and storing similar lists. *J. Algorithms*, 7:202–220, 1986.

[114] G. E. Collins. Quantifier elimination for real closed fields by cylindrical algebraic decomposition. In *Proc. 2nd GI Conference on Automata Theory and Formal Languages. Lecture Notes in Computer Science*, vol. 33, pages 134–183. Springer-Verlag, 1975.

[115] T. H. Cormen, C. E. Leiserson, and R. L. Rivest. *Introduction to Algorithms*. The MIT Press, Cambridge, MA, 1990.

[116] F. d'Amore and P. G. Franciosa. On the optimal binary plane partition for sets of isothetic rectangles. *Inform. Process. Lett.*, 44:255–259, 1992.

[117] G. B. Dantzig. *Linear Programming and Extensions*. Princeton University Press, Princeton, NJ, 1963.

[118] O. Devillers. Randomization yields simple $O(n \log^* n)$ algorithms for difficult $\Omega(n)$ problems. *Internat. J. Comput. Geom. Appl.*, 2:97–111, 1992.

[119] T. K. Dey, K. Sugihara, and C. L. Bajaj. Delaunay triangulations in three dimensions with finite precision arithmetic. *Comput. Aided Geom. Design*, 9:457–470, 1992.

[120] M. T. Dickerson, S. A. McElfresh, and M. H. Montague. New algorithms and empirical findings on minimum weight triangulation heuristics. In *Proc. 11th Annu. ACM Sympos. Comput. Geom.*, pages 238–247, 1995.

[121] M. T. Dickerson and M. H. Montague. A (usually?) connected subgraph of the minimum weight triangulation. In *Proc. 12th Annu. ACM Sympos. Comput. Geom.*, pages 204–213, 1996.

[122] G. L. Dirichlet. Über die reduktion der positiven quadratischen formen mit drei unbestimmten ganzen zahlen. *J. Reine Angew. Math.*, 40:209–227, 1850.

[123] D. Dobkin and D. Eppstein. Computing the discrepancy. In *Proc. 9th Annu. ACM Sympos. Comput. Geom.*, pages 47–52, 1993.

[124] D. Dobkin and D. Mitchell. Random-edge discrepancy of supersampling patterns. In *Graphics Interface '93*, 1993.

[125] M. E. Dyer. On a multidimensional search technique and its application to the Euclidean one-centre problem. *SIAM J. Comput.*, 15:725–738, 1986.

[126] N. Dyn, D. Levin, and S. Rippa. Data dependent triangulations for piecewise linear interpolation. *IMA Journal of Numerical Analysis*, 10:137–154, 1990.

[127] W. F. Eddy. A new convex hull algorithm for planar sets. *ACM Trans. Math. Softw.*, 3:398–403 and 411–412, 1977.

[128] H. Edelsbrunner. Dynamic data structures for orthogonal intersection queries. Report F59, Inst. Informationsverarb., Tech. Univ. Graz, Graz, Austria, 1980.

[129] H. Edelsbrunner. *Algorithms in Combinatorial Geometry. EATCS Monographs on Theoretical Computer Science*, vol. 10. Springer-Verlag, 1987.

[130] H. Edelsbrunner and L. J. Guibas. Topologically sweeping an arrangement. *J. Comput. Syst. Sci.*, 38:165–194, 1989. Corrigendum in 42 (1991), 249–251.

[131] H. Edelsbrunner, L. J. Guibas, J. Hershberger, R. Seidel, M. Sharir, J. Snoeyink, and E. Welzl. Implicitly representing arrangements of lines or segments. *Discrete Comput. Geom.*, 4:433–466, 1989.

[132] H. Edelsbrunner, L. J. Guibas, and J. Stolfi. Optimal point location in a monotone subdivision. *SIAM J. Comput.*, 15:317–340, 1986.

[133] H. Edelsbrunner, G. Haring, and D. Hilbert. Rectangular point location in d dimensions with applications. *Comput. J.*, 29:76–82, 1986.

[134] H. Edelsbrunner and H. A. Maurer. On the intersection of orthogonal objects. *Inform. Process. Lett.*, 13:177–181, 1981.

[135] H. Edelsbrunner and E. P. Mücke. Simulation of simplicity: a technique to cope with degenerate cases in geometric algorithms. *ACM Trans. Graph.*, 9:66–104, 1990.

[136] H. Edelsbrunner, J. O'Rourke, and R. Seidel. Constructing arrangements of lines and hyperplanes with applications. *SIAM J. Comput.*, 15:341–363, 1986.

[137] H. Edelsbrunner and M. H. Overmars. Batched dynamic solutions to decomposable searching problems. *J. Algorithms*, 6:515–542, 1985.

[138] H. Edelsbrunner and R. Seidel. Voronoi diagrams and arrangements. *Discrete Comput. Geom.*, 1:25–44, 1986.

[139] H. Edelsbrunner, R. Seidel, and M. Sharir. On the zone theorem for hyperplane arrangements. *SIAM J. Comput.*, 22:418–429, 1993.

[140] H. Edelsbrunner and E. Welzl. Halfplanar range search in linear space and $O(n^{0.695})$ query time. *Inform. Process. Lett.*, 23:289–293, 1986.

[141] H. ElGindy and G. T. Toussaint. On triangulating palm polygons in linear time. In N. Magnenat-Thalmann and D. Thalmann, editors, *New Trends in Computer Graphics*, pages 308–317. Springer-Verlag, 1988.

[142] I. Emiris and J. Canny. A general approach to removing degeneracies. In *Proc. 32nd Annu. IEEE Sympos. Found. Comput. Sci.*, pages 405–413, 1991.

[143] I. Emiris and J. Canny. An efficient approach to removing geometric degeneracies. In *Proc. 8th Annu. ACM Sympos. Comput. Geom.*, pages 74–82, 1992.

[144] I. D. Faux and M. J. Pratt. *Computational Geometry for Design and Manufacture*. Ellis Horwood, Chichester, U.K., 1979.

[145] U. Finke and K. Hinrichs. Overlaying simply connected planar subdivisions in linear time. In *Proc. 11th Annu. ACM Sympos. Comput. Geom.*, pages 119–126, 1995.

[146] R. A. Finkel and J. L. Bentley. Quad trees: a data structure for retrieval on composite keys. *Acta Inform.*, 4:1–9, 1974.

[147] S. Fisk. A short proof of Chvàtal's watchman theorem. *J. Combin. Theory Ser. B*, 24:374, 1978.

[148] J. D. Foley, A. van Dam, S. K. Feiner, and J. F. Hughes. *Computer Graphics: Principles and Practice*. Addison-Wesley, Reading, MA, 1990.

[149] S. Fortune. Numerical stability of algorithms for 2-d Delaunay triangulations and Voronoi diagrams. In *Proc. 8th Annu. ACM Sympos. Comput. Geom.*, pages 83–92, 1992.

[150] S. Fortune and V. Milenkovic. Numerical stability of algorithms for line arrangements. In *Proc. 7th Annu. ACM Sympos. Comput. Geom.*, pages 334–341, 1991.

[151] S. Fortune and C. J. Van Wyk. Efficient exact arithmetic for computational geometry. In *Proc. 9th Annu. ACM Sympos. Comput. Geom.*, pages 163–172, 1993.

[152] S. J. Fortune. A sweepline algorithm for Voronoi diagrams. *Algorithmica*, 2:153–174, 1987.

[153] A. Fournier and D. Y. Montuno. Triangulating simple polygons and equivalent problems. *ACM Trans. Graph.*, 3:153–174, 1984.

[154] H. Fuchs, Z. M. Kedem, and B. Naylor. On visible surface generation by a priori tree structures. *Comput. Graph.*, 14:124–133, 1980. Proc. SIGGRAPH '80.

[155] K. R. Gabriel and R. R. Sokal. A new statistical approach to geographic variation analysis. *Systematic Zoology*, 18:259–278, 1969.

[156] M. R. Garey, D. S. Johnson, F. P. Preparata, and R. E. Tarjan. Triangulating a simple polygon. *Inform. Process. Lett.*, 7:175–179, 1978.

[157] S. K. Ghosh and D. M. Mount. An output-sensitive algorithm for computing visibility graphs. *SIAM J. Comput.*, 20:888–910, 1991.

[158] K. Goldberg, D. Halperin, J.-C. Latombe, and R. Wilson, editors. *Algorithmic Foundations of Robotics*. A. K. Peters, Wellesley, MA, 1995.

[159] J. E. Goodman and J. O'Rourke. *Handbook of Discrete and Computational Geometry*. CRC Press, Boca Raton, Florida, 1997.

[160] R. L. Graham. An efficient algorithm for determining the convex hull of a finite planar set. *Inform. Process. Lett.*, 1:132–133, 1972.

[161] P. J. Green and B. W. Silverman. Constructing the convex hull of a set of points in the plane. *Comput. J.*, 22:262–266, 1979.

[162] B. Grünbaum. *Convex Polytopes*. Wiley, New York, 1967.

[163] L. J. Guibas, J. Hershberger, D. Leven, M. Sharir, and R. E. Tarjan. Linear-time algorithms for visibility and shortest path problems inside triangulated simple polygons. *Algorithmica*, 2:209–233, 1987.

[164] L. J. Guibas, D. E. Knuth, and M. Sharir. Randomized incremental construction of Delaunay and Voronoi diagrams. *Algorithmica*, 7:381–413, 1992.

[165] L. J. Guibas, L. Ramshaw, and J. Stolfi. A kinetic framework for computational geometry. In *Proc. 24th Annu. IEEE Sympos. Found. Comput. Sci.*, pages 100–111, 1983.

[166] L. J. Guibas, D. Salesin, and J. Stolfi. Epsilon geometry: building robust algorithms from imprecise computations. In *Proc. 5th Annu. ACM Sympos. Comput. Geom.*, pages 208–217, 1989.

[167] L. J. Guibas and R. Sedgewick. A dichromatic framework for balanced trees. In *Proc. 19th Annu. IEEE Sympos. Found. Comput. Sci.*, pages 8–21, 1978.

[168] L. J. Guibas and R. Seidel. Computing convolutions by reciprocal search. *Discrete Comput. Geom.*, 2:175–193, 1987.

[169] L. J. Guibas and M. Sharir. Combinatorics and algorithms of arrangements. In J. Pach, editor, *New Trends in Discrete and Computational Geometry. Algorithms and Combinatorics*, vol. 10, pages 9–36. Springer-Verlag, 1993.

[170] L. J. Guibas, M. Sharir, and S. Sifrony. On the general motion planning problem with two degrees of freedom. *Discrete Comput. Geom.*, 4:491–521, 1989.

[171] L. J. Guibas and J. Stolfi. Primitives for the manipulation of general subdivisions and the computation of Voronoi diagrams. *ACM Trans. Graph.*, 4:74–123, 1985.

[172] L. J. Guibas and J. Stolfi. Ruler, compass and computer: the design and analysis of geometric algorithms. In R. A. Earnshaw, editor, *Theoretical Foundations of Computer Graphics and CAD. NATO ASI Series F*, vol. 40, pages 111–165. Springer-Verlag, 1988.

[173] D. Halperin. *Algorithmic Motion Planning via Arrangements of Curves and of Surfaces*. Ph.D. thesis, Computer Science Department, Tel-Aviv University, Tel Aviv, 1992.

[174] D. Halperin and M. Sharir. Almost tight upper bounds for the single cell and zone problems in three dimensions. In *Proc. 10th Annu. ACM Sympos. Comput. Geom.*, pages 11–20, 1994.

[175] D. Halperin and M. Sharir. Arrangements and their applications in robotics: Recent developments. In K. Goldbergs, D. Halperin, J.-C. Latombe, and R. Wilson, editors, *Proc. Workshop on Algorithmic Foundations of Robotics*. A. K. Peters, Boston, MA, 1995.

[176] D. Haussler and E. Welzl. Epsilon-nets and simplex range queries. *Discrete Comput. Geom.*, 2:127–151, 1987.

[177] J. Hershberger and S. Suri. Offline maintenance of planar configurations. In *Proc. 2nd ACM-SIAM Sympos. Discrete Algorithms*, pages 32–41, 1991.

[178] J. Hershberger and S. Suri. Efficient computation of Euclidean shortest paths in the plane. In *Proc. 34th Annu. IEEE Sympos. Found. Comput. Sci.*, pages 508–517, 1993.

[179] J. Hershberger and S. Suri. An optimal algorithm for Euclidean shortest paths in the plane. Manuscript, Washington University, 1995.

[180] S. Hertel, M. Mäntylä, K. Mehlhorn, and J. Nievergelt. Space sweep solves intersection of convex polyhedra. *Acta Inform.*, 21:501–519, 1984.

[181] S. Hertel and K. Mehlhorn. Fast triangulation of simple polygons. In *Proc. 4th Internat. Conf. Found. Comput. Theory. Lecture Notes in Computer Science*, vol. 158, pages 207–218. Springer-Verlag, 1983.

[182] K. Ho-Le. Finite element mesh generation methods: A review and classification. *Comput. Aided Design*, 20:27–38, 1988.

[183] C. Hoffmann. *Geometric and Solid Modeling*. Morgan Kaufmann, San Mateo, CA, 1989.

[184] J. E. Hopcroft, J. T. Schwartz, and M. Sharir. *Planning, Geometry, and Complexity of Robot Motion*. Ablex Publishing, Norwood, NJ, 1987.

[185] C. Icking, G. Rote, E. Welzl, and C. Yap. Shortest paths for line segments. *Algorithmica*, 10:182–200, 1993.

[186] H. Inagaki and K. Sugihara. Numerically robust algorithm for constructing constrained Delaunay triangulation. In *Proc. 6th Canad. Conf. Comput. Geom.*, pages 171–176, 1994.

[187] R. A. Jarvis. On the identification of the convex hull of a finite set of points in the plane. *Inform. Process. Lett.*, 2:18–21, 1973.

[188] G. Kalai. A subexponential randomized simplex algorithm. In *Proc. 24th Annu. ACM Sympos. Theory Comput.*, pages 475–482, 1992.

[189] M. Kallay. The complexity of incremental convex hull algorithms in R^d. *Inform. Process. Lett.*, 19:197, 1984.

[190] R. G. Karlsson. Algorithms in a restricted universe. Report CS-84-50, Univ. Waterloo, Waterloo, ON, 1984.

[191] R. G. Karlsson and J. I. Munro. Proximity on a grid. In *Proceedings 2nd Symp. on Theoretical Aspects of Computer Science. Lecture Notes in Computer Science*, vol. 182, pages 187–196. Springer-Verlag, 1985.

[192] R. G. Karlsson and M. H. Overmars. Scanline algorithms on a grid. *BIT*, 28:227–241, 1988.

[193] N. Karmarkar. A new polynomial-time algorithm for linear programming. *Combinatorica*, 4:373–395, 1984.

[194] L. Kavraki, P. Švestka, J.-C. Latombe, and M. H. Overmars. Probabilistic roadmaps for path planning in high dimensional configuration spaces. *IEEE Trans. on Robotics and Automation*, 12:566–580, 1996.

[195] K. Kedem, R. Livne, J. Pach, and M. Sharir. On the union of Jordan regions and collision-free translational motion amidst polygonal obstacles. *Discrete Comput. Geom.*, 1:59–71, 1986.

[196] K. Kedem and M. Sharir. An efficient algorithm for planning collision-free translational motion of a convex polygonal object in 2-dimensional space amidst polygonal obstacles. In *Proc. 1st Annu. ACM Sympos. Comput. Geom.*, pages 75–80, 1985.

[197] K. Kedem and M. Sharir. An efficient motion planning algorithm for a convex rigid polygonal object in 2-dimensional polygonal space. *Discrete Comput. Geom.*, 5:43–75, 1990.

[198] L. G. Khachiyan. Polynomial algorithm in linear programming. *U.S.S.R. Comput. Math. and Math. Phys.*, 20:53–72, 1980.

[199] O. Khatib. Real-time obstacle avoidance for manipulators and mobile robots. *Internat. J. Robot. Res.*, 5:90–98, 1985.

[200] D. G. Kirkpatrick. Optimal search in planar subdivisions. *SIAM J. Comput.*, 12:28–35, 1983.

[201] D. G. Kirkpatrick, M. M. Klawe, and R. E. Tarjan. Polygon triangulation in $O(n \log \log n)$ time with simple data structures. In *Proc. 6th Annu. ACM Sympos. Comput. Geom.*, pages 34–43, 1990.

[202] D. G. Kirkpatrick and R. Seidel. The ultimate planar convex hull algorithm? *SIAM J. Comput.*, 15:287–299, 1986.

[203] V. Klee. On the complexity of d-dimensional Voronoi diagrams. *Archiv der Mathematik*, 34:75–80, 1980.

[204] R. Klein. Abstract Voronoi diagrams and their applications. In *Computational Geometry and its Applications. Lecture Notes in Computer Science*, vol. 333, pages 148–157. Springer-Verlag, 1988.

[205] R. Klein. *Concrete and Abstract Voronoi Diagrams. Lecture Notes in Computer Science*, vol. 400. Springer-Verlag, 1989.

[206] R. Klein, K. Mehlhorn, and S. Meiser. Randomized incremental construction of abstract Voronoi diagrams. *Comput. Geom. Theory Appl.*, 3:157–184, 1993.

[207] J.-C. Latombe. *Robot Motion Planning*. Kluwer Academic Publishers, Boston, 1991.

[208] J.-P. Laumond and M. H. Overmars, editors. *Algorithms for Robotic Motion and Manipulation*. A. K. Peters, Wellesley, MA, 1996.

[209] C. L. Lawson. Transforming triangulations. *Discrete Math.*, 3:365–372, 1972.

[210] C. L. Lawson. Software for C^1 surface interpolation. In J. R. Rice, editor, *Math. Software III*, pages 161–194. Academic Press, New York, 1977.

[211] D. Lee and A. Lin. Computational complexity of art gallery problems. *IEEE Trans. Inform. Theory*, 32:276–282, 1986.

[212] D. T. Lee. Proximity and reachability in the plane. Report R-831, Dept. Elect. Engrg., Univ. Illinois, Urbana, IL, 1978.

[213] D. T. Lee. Two-dimensional Voronoi diagrams in the L_p-metric. *J. ACM*, 27:604–618, 1980.

[214] D. T. Lee. On k-nearest neighbor Voronoi diagrams in the plane. *IEEE Trans. Comput.*, C-31:478–487, 1982.

[215] D. T. Lee and F. P. Preparata. Location of a point in a planar subdivision and its applications. *SIAM J. Comput.*, 6:594–606, 1977.

[216] D. T. Lee and C. K. Wong. Quintary trees: a file structure for multidimensional database systems. *ACM Trans. Database Syst.*, 5:339–353, 1980.

[217] D. T. Lee and C. K. Wong. Voronoi diagrams in L_1 (L_∞) metrics with 2-dimensional storage applications. *SIAM J. Comput.*, 9:200–211, 1980.

[218] D. T. Lee, C. D. Yang, and C. K. Wong. Rectilinear paths among rectilinear obstacles. *Discrete Appl. Math.*, 70:185–215, 1996.

[219] J. van Leeuwen and D. Wood. Dynamization of decomposable searching problems. *Inform. Process. Lett.*, 10:51–56, 1980.

[220] D. Leven and M. Sharir. Planning a purely translational motion for a convex object in two-dimensional space using generalized Voronoi diagrams. *Discrete Comput. Geom.*, 2:9–31, 1987.

[221] T. Lozano-Pérez. Automatic planning of manipulator transfer movements. *IEEE Trans. Syst. Man Cybern.*, SMC-11:681–698, 1981.

[222] T. Lozano-Pérez. Spatial planning: A configuration space approach. *IEEE Trans. Comput.*, C-32:108–120, 1983.

[223] T. Lozano-Pérez and M. A. Wesley. An algorithm for planning collision-free paths among polyhedral obstacles. *Commun. ACM*, 22:560–570, 1979.

[224] G. S. Lueker. A data structure for orthogonal range queries. In *Proc. 19th Annu. IEEE Sympos. Found. Comput. Sci.*, pages 28–34, 1978.

[225] D. J. Maguire, M. F. Goodchild, and D. W. Rhind, editors. *Geographical Information Systems: Principles and Applications*. Longman, London, 1991.

[226] H. G. Mairson and J. Stolfi. Reporting and counting intersections between two sets of line segments. In R. A. Earnshaw, editor, *Theoretical Foundations of Computer Graphics and CAD. NATO ASI Series F*, vol. 40, pages 307–325. Springer-Verlag, 1988.

[227] J. Matoušek. Reporting points in halfspaces. In *Proc. 32nd Annu. IEEE Sympos. Found. Comput. Sci.*, pages 207–215, 1991.

[228] J. Matoušek. Efficient partition trees. *Discrete Comput. Geom.*, 8:315–334, 1992.

[229] J. Matoušek. Reporting points in halfspaces. *Comput. Geom. Theory Appl.*, 2:169–186, 1992.

[230] J. Matoušek. Range searching with efficient hierarchical cuttings. *Discrete Comput. Geom.*, 10:157–182, 1993.

[231] J. Matoušek and O. Schwarzkopf. On ray shooting in convex polytopes. *Discrete Comput. Geom.*, 10:215–232, 1993.

[232] J. Matoušek, O. Schwarzkopf, and J. Snoeyink. Non-canonical randomized incremental construction. Manuscript, 1997.

[233] J. Matoušek, M. Sharir, and E. Welzl. A subexponential bound for linear programming. In *Proc. 8th Annu. ACM Sympos. Comput. Geom.*, pages 1–8, 1992.

[234] H. A. Maurer and T. A. Ottmann. Dynamic solutions of decomposable searching problems. In U. Pape, editor, *Discrete Structures and Algorithms*, pages 17–24. Carl Hanser Verlag, München, Germany, 1979.

[235] E. M. McCreight. Efficient algorithms for enumerating intersecting intervals and rectangles. Report CSL-80-9, Xerox Palo Alto Res. Center, Palo Alto, CA, 1980.

[236] E. M. McCreight. Priority search trees. *SIAM J. Comput.*, 14:257–276, 1985.

[237] P. J. McKerrow. *Introduction to Robotics*. Addison-Wesley, Reading, MA, 1991.

[238] R. Mead. A relation between the individual plant-spacing and yield. *Annals of Botany, N.S.*, 30:301–309, 1966.

[239] N. Megiddo. Linear programming in linear time when the dimension is fixed. *J. ACM*, 31:114–127, 1984.

[240] K. Mehlhorn, S. Meiser, and C. Ó'Dúnlaing. On the construction of abstract Voronoi diagrams. *Discrete Comput. Geom.*, 6:211–224, 1991.

[241] K. Mehlhorn and S. Näher. Dynamic fractional cascading. *Algorithmica*, 5:215–241, 1990.

[242] K. Mehlhorn and M. H. Overmars. Optimal dynamization of decomposable searching problems. *Inform. Process. Lett.*, 12:93–98, 1981.

[243] G. Meisters. Polygons have ears. *Amer. Math. Monthly*, 82:648–651, 1975.

[244] E. A. Melissaratos and D. L. Souvaine. Coping with inconsistencies: a new approach to produce quality triangulations of polygonal domains with holes. In *Proc. 8th Annu. ACM Sympos. Comput. Geom.*, pages 202–211, 1992.

[245] V. Milenkovic. Robust construction of the Voronoi diagram of a polyhedron. In *Proc. 5th Canad. Conf. Comput. Geom.*, pages 473–478, Waterloo, Canada, 1993.

[246] N. Miller and M. Sharir. Efficient randomized algorithms for constructing the union of fat triangles and pseudodiscs. Unpublished manuscript.

[247] J. S. B. Mitchell. Shortest paths among obstacles in the plane. In *Proc. 9th Annu. ACM Sympos. Comput. Geom.*, pages 308–317, 1993.

[248] J. S. B. Mitchell. Shortest paths among obstacles in the plane. *Internat. J. Comput. Geom. Appl.*, 6:309–332, 1996.

[249] J. S. B. Mitchell and C. H. Papadimitriou. The weighted region problem: finding shortest paths through a weighted planar subdivision. *J. ACM*, 38:18–73, 1991.

[250] J. S. B. Mitchell, G. Rote, and G. Woeginger. Minimum-link paths among obstacles in the plane. *Algorithmica*, 8:431–459, 1992.

[251] M. E. Mortenson. *Geometric Modeling*. Wiley, New York, 1985.

[252] D. E. Muller and F. P. Preparata. Finding the intersection of two convex polyhedra. *Theoret. Comput. Sci.*, 7:217–236, 1978.

[253] H. Müller. Rasterized point location. In *Proceedings Workshop on Graph-theoretic Concepts in Computer Science*, pages 281–293. Trauner Verlag, Linz, Austria, 1985.

[254] K. Mulmuley. A fast planar partition algorithm, I. In *Proc. 29th Annu. IEEE Sympos. Found. Comput. Sci.*, pages 580–589, 1988.

[255] K. Mulmuley. A fast planar partition algorithm, I. *Journal of Symbolic Computation*, 10:253–280, 1990.

[256] K. Mulmuley. *Computational Geometry: An Introduction Through Randomized Algorithms*. Prentice Hall, Englewood Cliffs, NJ, 1994.

[257] B. Naylor, J. A. Amanatides, and W. Thibault. Merging BSP trees yields polyhedral set operations. *Comput. Graph.*, 24:115–124, August 1990. Proc. SIG-GRAPH '90.

[258] J. Nievergelt and F. P. Preparata. Plane-sweep algorithms for intersecting geometric figures. *Commun. ACM*, 25:739–747, 1982.

[259] N. Nilsson. A mobile automaton: An application of artificial intelligence techniques. In *Proc. IJCAI*, pages 509–520, 1969.

[260] C. Ó'Dúnlaing and C. K. Yap. A "retraction" method for planning the motion of a disk. *J. Algorithms*, 6:104–111, 1985.

[261] A. Okabe, B. Boots, and K. Sugihara. *Spatial Tessellations: Concepts and Applications of Voronoi Diagrams*. John Wiley & Sons, Chichester, U.K., 1992.

[262] J. O'Rourke. *Art Gallery Theorems and Algorithms*. Oxford University Press, New York, 1987.

[263] M. H. Overmars. *The Design of Dynamic Data Structures. Lecture Notes in Computer Science*, vol. 156. Springer-Verlag, 1983.

[264] M. H. Overmars. Efficient data structures for range searching on a grid. *J. Algorithms*, 9:254–275, 1988.

[265] M. H. Overmars. Geometric data structures for computer graphics: an overview. In R. A. Earnshaw, editor, *Theoretical Foundations of Computer Graphics and CAD. NATO ASI Series F*, vol. 40, pages 21–49. Springer-Verlag, 1988.

[266] M. H. Overmars. Point location in fat subdivisions. *Inform. Process. Lett.*, 44:261–265, 1992.

[267] M. H. Overmars and J. van Leeuwen. Further comments on Bykat's convex hull algorithm. *Inform. Process. Lett.*, 10:209–212, 1980.

[268] M. H. Overmars and J. van Leeuwen. Dynamization of decomposable searching problems yielding good worst-case bounds. In *Proc. 5th GI Conf. Theoret. Comput. Sci. Lecture Notes in Computer Science*, vol. 104, pages 224–233. Springer-Verlag, 1981.

[269] M. H. Overmars and J. van Leeuwen. Maintenance of configurations in the plane. *J. Comput. Syst. Sci.*, 23:166–204, 1981.

[270] M. H. Overmars and J. van Leeuwen. Some principles for dynamizing decomposable searching problems. *Inform. Process. Lett.*, 12:49–54, 1981.

[271] M. H. Overmars and J. van Leeuwen. Two general methods for dynamizing decomposable searching problems. *Computing*, 26:155–166, 1981.

[272] M. H. Overmars and J. van Leeuwen. Worst-case optimal insertion and deletion methods for decomposable searching problems. *Inform. Process. Lett.*, 12:168–173, 1981.

[273] M. H. Overmars and A. F. van der Stappen. Range searching and point location among fat objects. In J. van Leeuwen, editor, *Algorithms – ESA'94. Lecture Notes in Computer Science*, vol. 855, pages 240–253. Springer-Verlag, 1994.

[274] M. H. Overmars and P. Švestka. A probabilistic learning approach to motion planning. In *Algorithmic Foundations of Robotics*, pages 19–38. A. K. Peters, Boston, MA, 1995.

[275] M. H. Overmars and C.-K. Yap. New upper bounds in Klee's measure problem. *SIAM J. Comput.*, 20:1034–1045, 1991.

[276] J. Pach and M. Sharir. On vertical visibility in arrangements of segments and the queue size in the Bentley-Ottman line sweeping algorithm. *SIAM J. Comput.*, 20:460–470, 1991.

[277] J. Pach, W. Steiger, and E. Szemerédi. An upper bound on the number of planar k-sets. *Discrete Comput. Geom.*, 7:109–123, 1992.

[278] L. Palazzi and J. Snoeyink. Counting and reporting red/blue segment intersections. In *Proc. 3rd Workshop Algorithms Data Struct. Lecture Notes in Computer Science*, vol. 709, pages 530–540. Springer-Verlag, 1993.

[279] C. H. Papadimitriou. An algorithm for shortest-path motion in three dimensions. *Inform. Process. Lett.*, 20:259–263, 1985.

[280] M. S. Paterson and F. F. Yao. Efficient binary space partitions for hidden-surface removal and solid modeling. *Discrete Comput. Geom.*, 5:485–503, 1990.

[281] M. S. Paterson and F. F. Yao. Optimal binary space partitions for orthogonal objects. *J. Algorithms*, 13:99–113, 1992.

[282] M. Pocchiola and G. Vegter. Computing the visibility graph via pseudo-triangulations. In *Proc. 11th Annu. ACM Sympos. Comput. Geom.*, pages 248–257, 1995.

[283] M. Pocchiola and G. Vegter. The visibility complex. *Internat. J. Comput. Geom. Appl.*, 6:279–308, 1996.

[284] F. P. Preparata. An optimal real-time algorithm for planar convex hulls. *Commun. ACM*, 22:402–405, 1979.

[285] F. P. Preparata and S. J. Hong. Convex hulls of finite sets of points in two and three dimensions. *Commun. ACM*, 20:87–93, 1977.

[286] F. P. Preparata and M. I. Shamos. *Computational Geometry: An Introduction*. Springer-Verlag, 1985.

[287] F. P. Preparata and R. Tamassia. Efficient point location in a convex spatial cell-complex. *SIAM J. Comput.*, 21:267–280, 1992.

[288] E. Quak and L. Schumaker. Cubic spline fitting using data dependent triangulations. *Comput. Aided Geom. Design*, 7:293–302, 1990.

[289] J. H. Reif and J. A. Storer. A single-exponential upper bound for finding shortest paths in three dimensions. *J. ACM*, 41:1013–1019, 1994.

[290] S. Rippa. Minimal roughness property of the Delaunay triangulation. *Comput. Aided Geom. Design*, 7:489–497, 1990.

[291] H. Rohnert. Shortest paths in the plane with convex polygonal obstacles. *Inform. Process. Lett.*, 23:71–76, 1986.

[292] J. Ruppert and R. Seidel. On the difficulty of triangulating three-dimensional non-convex polyhedra. *Discrete Comput. Geom.*, 7:227–253, 1992.

[293] J.-R. Sack and J. Urrutia. *Handbook of Computational Geometry*. Elsevier Science Publishers, Amsterdam, 1997.

[294] H. Samet. The quadtree and related hierarchical data structures. *ACM Comput. Surv.*, 16, June 1984.

[295] H. Samet. An overview of quadtrees, octrees, and related hierarchical data structures. In R. A. Earnshaw, editor, *Theoretical Foundations of Computer Graphics and CAD. NATO ASI Series F*, vol. 40, pages 51–68. Springer-Verlag, 1988.

[296] H. Samet. *Applications of Spatial Data Structures: Computer Graphics, Image Processing, and GIS*. Addison-Wesley, Reading, MA, 1990.

[297] H. Samet. *The Design and Analysis of Spatial Data Structures*. Addison-Wesley, Reading, MA, 1990.

[298] N. Sarnak and R. E. Tarjan. Planar point location using persistent search trees. *Commun. ACM*, 29:669–679, 1986.

[299] J. B. Saxe and J. L. Bentley. Transforming static data structures to dynamic structures. In *Proc. 20th Annu. IEEE Sympos. Found. Comput. Sci.*, pages 148–168, 1979.

[300] R. J. Schilling. *Fundamentals of Robotics, Analysis and Control*. Prentice Hall, Englewood Cliffs, NJ, 1990.

[301] H. W. Scholten and M. H. Overmars. General methods for adding range restrictions to decomposable searching problems. *J. Symbolic Comput.*, 7:1–10, 1989.

[302] A. Schrijver. *Theory of Linear and Integer Programming*. Wiley-Interscience, New York, 1986.

[303] J. T. Schwartz and M. Sharir. On the "piano movers" problem I: the case of a two-dimensional rigid polygonal body moving amidst polygonal barriers. *Commun. Pure Appl. Math.*, 36:345–398, 1983.

[304] J. T. Schwartz and M. Sharir. On the "piano movers" problem II: general techniques for computing topological properties of real algebraic manifolds. *Adv. Appl. Math.*, 4:298–351, 1983.

[305] J. T. Schwartz and M. Sharir. A survey of motion planning and related geometric algorithms. In D. Kapur and J. Mundy, editors, *Geometric Reasoning*, pages 157–169. MIT Press, Cambridge, MA, 1989.

[306] J. T. Schwartz and M. Sharir. Algorithmic motion planning in robotics. In J. van Leeuwen, editor, *Algorithms and Complexity. Handbook of Theoretical Computer Science*, vol. A, pages 391–430. Elsevier, Amsterdam, 1990.

[307] R. Seidel. *Output-size sensitive algorithms for constructive problems in computational geometry*. Ph.D. thesis, Dept. Comput. Sci., Cornell Univ., Ithaca, NY, 1986. Technical Report TR 86-784.

[308] R. Seidel. A simple and fast incremental randomized algorithm for computing trapezoidal decompositions and for triangulating polygons. *Comput. Geom. Theory Appl.*, 1:51–64, 1991.

[309] R. Seidel. Small-dimensional linear programming and convex hulls made easy. *Discrete Comput. Geom.*, 6:423–434, 1991.

[310] M. I. Shamos. *Computational Geometry*. Ph.D. thesis, Dept. Comput. Sci., Yale Univ., New Haven, CT, 1978.

[311] M. I. Shamos and D. Hoey. Closest-point problems. In *Proc. 16th Annu. IEEE Sympos. Found. Comput. Sci.*, pages 151–162, 1975.

[312] M. I. Shamos and D. Hoey. Geometric intersection problems. In *Proc. 17th Annu. IEEE Sympos. Found. Comput. Sci.*, pages 208–215, 1976.

[313] M. Sharir and P. K. Agarwal. *Davenport-Schinzel Sequences and Their Geometric Applications*. Cambridge University Press, New York, 1995.

[314] M. Sharir and E. Welzl. A combinatorial bound for linear programming and related problems. In *Proc. 9th Sympos. Theoret. Aspects Comput. Sci. Lecture Notes in Computer Science*, vol. 577, pages 569–579. Springer-Verlag, 1992.

[315] T. C. Shermer. Recent results in art galleries. *Proc. IEEE*, 80:1384–1399, September 1992.

[316] P. Shirley. Discrepancy as a quality measure for sample distributions. In F. H. Post and W. Barth, editors, *Proc. Eurographics'91*, pages 183–194. Elsevier Science, Vienna, Austria, September 1991.

[317] R. Sibson. Locally equiangular triangulations. *Comput. J.*, 21:243–245, 1978.

[318] A. F. van der Stappen, D. Halperin, and M. H. Overmars. The complexity of the free space for a robot moving amidst fat obstacles. *Comput. Geom. Theory Appl.*, 3:353–373, 1993.

[319] A. F. van der Stappen and M. H. Overmars. Motion planning amidst fat obstacles. In *Proc. 10th Annu. ACM Sympos. Comput. Geom.*, pages 31–40, 1994.

[320] S. Suri. *Minimum link paths in polygons and related problems*. Ph.D. thesis, Dept. Comput. Sci., Johns Hopkins Univ., Baltimore, MD, 1987.

[321] R. E. Tarjan and C. J. Van Wyk. An $O(n \log \log n)$-time algorithm for triangulating a simple polygon. *SIAM J. Comput.*, 17:143–178, 1988. Erratum in 17 (1988), 106.

[322] S. J. Teller and C. H. Séquin. Visibility preprocessing for interactive walkthroughs. *Comput. Graph.*, 25:61–69, July 1991. Proc. SIGGRAPH '91.

[323] W. C. Thibault and B. F. Naylor. Set operations on polyhedra using binary space partitioning trees. *Comput. Graph.*, 21:153–162, 1987. Proc. SIGGRAPH '87.

[324] G. T. Toussaint. The relative neighbourhood graph of a finite planar set. *Pattern Recogn.*, 12:261–268, 1980.

[325] V. K. Vaishnavi and D. Wood. Rectilinear line segment intersection, layered segment trees and dynamization. *J. Algorithms*, 3:160–176, 1982.

[326] G. Vegter. The visibility diagram: A data structure for visibility problems and motion planning. In *Proc. 2nd Scand. Workshop Algorithm Theory. Lecture Notes in Computer Science*, vol. 447, pages 97–110. Springer-Verlag, 1990.

[327] R. Volpe and P. Khosla. Artificial potential with elliptical isopotential contours for obstacle avoidance. In *Proc. 26th IEEE Conf. on Decision and Control*, pages 180–185, 1987.

[328] G. M. Voronoi. Nouvelles applications des paramètres continus à la théorie des formes quadratiques. premier Mémoire: Sur quelques propriétés des formes quadratiques positives parfaites. *J. Reine Angew. Math.*, 133:97–178, 1907.

[329] G. M. Voronoi. Nouvelles applications des paramètres continus à la théorie des formes quadratiques. deuxième Mémoire: Recherches sur les parallélloèdres primitifs. *J. Reine Angew. Math.*, 134:198–287, 1908.

[330] A. Watt. *3D Computer Graphics*. Addison-Wesley, Reading, MA, 1993.

[331] E. Welzl. Constructing the visibility graph for n line segments in $O(n^2)$ time. *Inform. Process. Lett.*, 20:167–171, 1985.

[332] E. Welzl. Partition trees for triangle counting and other range searching problems. In *Proc. 4th Annu. ACM Sympos. Comput. Geom.*, pages 23–33, 1988.

[333] E. Welzl. Smallest enclosing disks (balls and ellipsoids). In H. Maurer, editor, *New Results and New Trends in Computer Science. Lecture Notes in Computer Science*, vol. 555, pages 359–370. Springer-Verlag, 1991.

[334] D. E. Willard. *Predicate-oriented database search algorithms*. Ph.D. thesis, Aiken Comput. Lab., Harvard Univ., Cambridge, MA, 1978. Report TR-20-78.

[335] D. E. Willard. The super-*b*-tree algorithm. Report TR-03-79, Aiken Comput. Lab., Harvard Univ., Cambridge, MA, 1979.

[336] D. E. Willard. Polygon retrieval. *SIAM J. Comput.*, 11:149–165, 1982.

[337] D. E. Willard. Log-logarithmic worst case range queries are possible in space $O(n)$. *Inform. Process. Lett.*, 17:81–89, 1983.

[338] D. E. Willard. New trie data structures which support very fast search operations. *J. Comput. Syst. Sci.*, 28:379–394, 1984.

[339] D. E. Willard and G. S. Lueker. Adding range restriction capability to dynamic data structures. *J. ACM*, 32:597–617, 1985.

[340] A. C. Yao. A lower bound to finding convex hulls. *J. ACM*, 28:780–787, 1981.

[341] A. C. Yao and F. F. Yao. A general approach to *D*-dimensional geometric queries. In *Proc. 17th Annu. ACM Sympos. Theory Comput.*, pages 163–168, 1985.

[342] C. K. Yap. A geometric consistency theorem for a symbolic perturbation scheme. *J. Comput. Syst. Sci.*, 40:2–18, 1990.

[343] C. K. Yap. Towards exact geometric computation. In *Proc. 5th Canad. Conf. Comput. Geom.*, pages 405–419, 1993.

[344] D. Zhu and J.-C. Latombe. New heuristic algorithms for efficient hierarchical path planning. *IEEE Trans. on Robotics and Automation*, 7:9–20, 1991.

[345] G. M. Ziegler. *Lectures on Polytopes. Graduate Texts in Mathematics*, vol. 152. Springer-Verlag, 1994.

Index

Springer
and the
environment

At Springer we firmly believe that an international science publisher has a special obligation to the environment, and our corporate policies consistently reflect this conviction.
We also expect our business partners – paper mills, printers, packaging manufacturers, etc. – to commit themselves to using materials and production processes that do not harm the environment. The paper in this book is made from low- or no-chlorine pulp and is acid free, in conformance with international standards for paper permanency.

 Springer

Druck: Strauss Offsetdruck, Mörlenbach
Verarbeitung: Schäffer, Grünstadt